Horizontal Wells
Focus on the
Reservoir

Edited by

Timothy R. Carr

Erik P. Mason

Charles T. Feazel

AAPG Methods in Exploration Series No. 14

Published by

The American Association of Petroleum Geologists

Tulsa, Oklahoma, U.S.A.

Printed in the U.S.A.

AAPG Editor: John C. Lorenz
Geoscience Director: J. B. "Jack" Thomas
Special Publications Editor: Hazel Rowena Mills
Copy Editor: Sharon Mason
Editorial Assistant: Gary F. Stewart
Cover Design: Rusty Johnson
Production: ProType Inc., Tulsa, Oklahoma
Printing: The Covington Group, Kansas City, Missouri

On the cover: Upper image — 3-D display of Fault Block II, Wilmington oil field, Los Angeles County, California, showing locations of wells drilled for a steamflood project. The figure has a 2x vertical exaggeration, and the wellbores are greatly exaggerated to enhance the visual impact of the well pattern. Image courtesy of Christopher C. Phillips, Tidelands Oil Production Company, and Donald D. Clarke, city of Long Beach, Department of Oil Properties, Long Beach, California. Lower image — Cross section of well A-3ST1, Ram Powell field, eastern deep-water Gulf of Mexico. The well design and relative placement in the oil rim are typical of the three J sandstone horizontal producers. Image courtesy of Peter A. Craig, Shell Exploration and Production Company, New Orleans, Louisiana.

This and other AAPG publications are available from:

The AAPG Bookstore
P.O. Box 979
Tulsa, OK 74101-0979
U.S.A.
Telephone: 1-918-584-2555 or 1-800-364-AAPG (U.S.A.)
Fax: 1-918-560-2652 or 1-800-898-2274 (U.S.A.)
www.aapg.org

Canadian Society of Petroleum Geologists
No. 160, 540 Fifth Avenue S.W.
Calgary, Alberta T2P 0M2
Canada
Telephone: 1-403-264-5610
Fax: 1-403-264-5898
www.cspg.org

Geological Society Publishing House
Unit 7, Brassmill Enterprise Centre
Brassmill Lane, Bath BA13JN
U.K.
Telephone: +44-1225-445046
Fax: +44-1225-442836
www.geolsoc.org.uk

Affiliated East-West Press Private Ltd.
G-1/16 Ansari Road, Darya Ganj
New Delhi 110 002
India
Telephone: +91-11-3279113
Fax: +91-11-3260538
E-mail: affiliat@nda.vsnl.in

Acknowledgments

This volume is an outgrowth of the AAPG Reservoir Development Committee's efforts to encourage and present work in reservoir geology, reservoir characterization, and development/production geology. In late 1999, one result of the committee's efforts was an AAPG/SWPLA Hedberg Research Symposium, International Horizontal and Extended Reach Well Symposium: Focus on the Reservoir. More than 50 examples of horizontal and extended-reach wells and drilling programs in different geologic settings from all over the world were presented at the symposium, held in Houston, Texas.

The papers collected in this volume represent only a small fraction of those presented at the symposium. We have tried to include a good sampling of summary papers and case studies of single- and multiwell drilling programs during exploration appraisal, primary field development, and field redevelopment, organized by geologic setting.

The editors wish to acknowledge the contribution made by all participating authors. Their time and effort are truly appreciated. We wish to thank the AAPG Publications Department, specifically Rowena Mills for editorial guidance. Sharon Mason, copy editor, crossed every *t* without one author crossing the Rubicon. Ray Sorenson deserves special recognition for reviewing all abstracts submitted for the symposium and every paper in this volume.

The initial symposium never would have occurred without the organizational skills of Debbi Boonstra of the AAPG Education Department and the efforts of Rami Kamal (Saudi Aramco), Fernando Chacartegui (Petroleos de Venezuela), John Sneider (Sneider Exploration), and Ray Sorenson (Anadarko Petroleum). Appreciation also is extended to the sponsors of the symposium and of this volume.

Timothy R. Carr

Erik P. Mason

Charles T. Feazel

AAPG wishes to thank the following

for their generous contributions

to

Horizontal Wells
Focus on the Reservoir

HM Research Associates Ltd.

IHS Energy Group

Kansas Geological Survey

Phillips Petroleum Company

Shell E&P Company

Contributions are applied toward the production costs of
publication, thus directly reducing the book's purchase price and
making the volume available to a larger readership.

About the Editors

Timothy R. Carr is chief of the Petroleum Research Section and a senior scientist for Kansas Geological Survey at the University of Kansas, in Lawrence. He is also codirector of the Energy Research Center and courtesy professor in the Department of Geology. Carr holds a B.A. in economics from the University of Wisconsin, an M.S. in geology from Texas Tech University, and a Ph.D. in geology from the University of Wisconsin.

In 1980, Carr joined Atlantic Richfield (ARCO), where he worked for 12 years in research, operations, and management positions. At ARCO, he was involved in exploration and development projects in several geographic locations, including Alaska, the North Sea, East Greenland, California, and Kansas. In 1992, Carr joined Kansas Geological Survey. He works closely with operators and public agencies interested in Kansas. Current projects include research and public-service efforts in subsurface petroleum geology and geophysics, environmental geology, and developing on-line information systems to improve public access to petroleum information and technology.

Erik P. Mason is a geologist for Shell International E&P, presently exploring for oil in the Perdido fold belt in the Gulf of Mexico. He holds a B.A. in geology from Principia College in Elsah, Illinois, and an M.S. in geology from Oklahoma State University.

In 1982, Mason joined Phillips Petroleum Company, where he worked for six years in production and international exploration. He joined Shell in 1988 in New Orleans, working in various production and exploration assignments. In 2002, Mason moved to Houston to assume his present position as team leader of Shell International's Gulf of Mexico New Ventures team.

Charles T. Feazel is manager of reservoir characterizations for ConocoPhillips Petroleum Company. He has served Phillips for more than 27 years in many capacities in research and management, with assignments in Oklahoma, Texas, and Norway. He earned a B.A. in geology from Ohio Wesleyan University and M.A. and Ph.D. degrees in geology from Johns Hopkins University. His specialties include carbonate sedimentology, reservoir description, field development, and the spectrum of reservoir characterization from depositional facies to flow units. Feazel has worked in numerous geologic settings, including the Nevada desert, various Caribbean islands, Greenland, the Beaufort Sea, the North Sea, Alaska, the Gulf of Mexico, the U.S. Midcontinent, the Caspian Sea, and the Middle East.

Feazel has a strong interest in science education and for many years has conducted workshops and field trips for science teachers. In addition, he is the author of a book and more than 20 magazine articles explaining science and exploration to the public.

Table of Contents

Fluvial and Eolian Reservoirs

Chalk Reservoirs

Carbonate Reservoirs

Techniques

Introduction

Horizontal Wells: Focus on the Reservoir

INTRODUCTION

During the last quarter of the 20th century, the rapid increase in progressively deeper-water offshore exploration presented extremely strong economic incentives and technical challenges that encouraged efficient exploitation of an entire field from a single central drilling and production facility. These economic and technical challenges were the drivers that increased the mechanical precision and control needed for directional drilling. The petroleum industry quickly developed the technology to drill horizontal and extended-reach wells efficiently. However, the understanding of reservoir geology and reservoir engineering required to best use horizontal and extended-reach technology in different reservoir situations was slower to mature. Our understanding of petroleum reservoirs has been enhanced significantly by moving beyond a perspective constrained by vertical penetrations to a horizontal viewpoint. This volume highlights some of these changes in our understanding of petroleum reservoirs. Another common theme running through this volume is the evolution of technical information required from geoscientists, engineers, and managers to develop a detailed and integrated approach to the design and execution of horizontal wells. Static prespud drill plans originating from the home office require near-real-time integrated decisions that incorporate uncertainty and variability in the reservoir to modify well trajectories while drilling.

Today, horizontal and extended-reach wells are used to overcome various challenges imposed by reservoir geometry, fluid characteristics, economic conditions, and environmental constraints. These challenges include:

1) coning (gas or water)
2) waterflood conformance
3) improved recovery from thin beds
4) economic and technical limitations (e.g., offshore limitations on number of slots)
5) environmental restrictions
6) heterogeneous reservoirs (e.g., fractured or restricted-flow units)
7) recovery from tar sands

As with any technical and scientific area, new terminology is created and older terminology is modified to the specific perspectives required for horizontal and extended-reach wells. A glossary at the end of this chapter includes some of the most common terminology.

Horizontal drilling has provided a new tool to create new economic opportunities and has helped to change the orientation of the geosciences.

After the success in the 1980s of horizontal wells in areas such as the Rospo Mare Field and the Austin Chalk trend of Texas, global interest in application and improvement of horizontal-well technology blossomed. Global statistics show that horizontal drilling has become a key technology to improve cost-effective recovery of additional reserves (**Philip H. Stark**, Chapter 1). Through 2001, 34,777 horizontal wells had been drilled in 72 countries spanning the globe. Approximately 85% of those wells were concentrated in a small number of plays in the United States and Canada. In the onshore area of the U.S. Gulf Coast, the Austin Chalk play accounted for 4320 wells and 6887 completions. The Austin Chalk play also illustrates the tremendous increase of horizontal wells with multiple laterals.

TURBIDITE AND DELTAIC RESERVOIRS

Turbidite-fan sandstone reservoirs are prolific producers in many areas of the world. However, economic recovery of reserves from relatively small fields and stratigraphically and structurally complex reservoirs, coupled with high-cost environments such as deep water offshore, can present significant challenges. Turbidite reservoir quality, net pay, and flow-unit continuity can vary significantly across a field, especially along fan margins. In many mature fields, bypassed oil and oil trapped by waterflooding along stratigraphic discontinuities and faults, as well as relatively thin oil rims in low-permeability, shaley-sandstone intervals along fan margins, can provide targets with significant remaining reserves. Coupled with a detailed understanding of the reservoir architecture, horizontal wells provide a powerful tool to cost-effectively exploit marginally economic fields and bypassed reserves.

Michael S. Clark et al. (Chapter 2) use log-derived petrophysical data constrained by core analyses to develop a detailed 3-D reservoir model for the western depositional margin of the Stevens turbidite complex in Yowlumne Field of the southern San Joaquin Basin, California. An improved understanding of the reservoirs was critical to targeting a horizontal well. The external and internal geometries of the reservoir at Yowlumne field were modeled by using digital data to move through a well-understood but seldom undertaken analysis of reservoir characterization and simulation that includes:

1) description of the reservoir architecture (i.e., stratigraphic geometries, flow units, and facies distributions) using well-log correlation, pressure data, and existing 3-D seismic data
2) determination of reservoir properties using core, well logs, borehole breakout, and microseismic data
3) construction of a digital petrophysical database (e.g., reservoir thickness, porosity, permeability, net/gross sandstone ratio, and water saturation)
4) contour mapping of petrophysical properties by individual flow units

5) location and quantification of bypassed reserves using a computer model to simulate fluid flow in the reservoir

Based on the understanding of reservoir architecture, a horizontal well was designed and drilled along depositional strike of the distal margin of the Yowlumne fan.

Donald D. Clarke and Christopher C. Phillips (Chapter 3) describe a horizontal-well drilling program to tap bypassed heavy oil and add new reserves in California's mature supergiant, the Wilmington oil field. More than 5000 wells have been drilled in the Wilmington field since its discovery in 1932. They have yielded more than 2.5 billion barrels of oil of the original 9 billion barrels of oil in place. The field produces from stacked semi- and unconsolidated Pliocene and Miocene clastic slope and basin turbidite sandstones. The highly heterogeneous reservoir consists of numerous rhythmically deposited sequences of sandstone interlayered with siltstones and shales deposited as lenticular lobate shapes that are complicated by basal scour, amalgamation, onlapping, and channeling. The entire sequence of stacked reservoir units has been folded and faulted into 10 large fault blocks with numerous smaller faults. Despite more than 60 years of primary and secondary recovery efforts, a substantial amount of bypassed oil remains in this complex reservoir, which consists of more than 50 individual flow units.

The first horizontal well in the Wilmington field was completed in November 1993 as part of an optimized waterflood project. Since then, 39 additional horizontal wells have been drilled. Horizontal wells have been used as part of a steam-assisted gravity-drainage project in the shallower, heavier oil zones. The laterals for the steamflood wells were placed at the bottom of the sandstone to maximize capture of oil through steam-assisted gravity drainage. Each horizontal steamflood well replaces three or four vertical wells. Other horizontal wells targeted in waterflood patterns were placed at the top of the sandstone to recover attic reserves. Horizontal wells also have been drilled in plays dominated by shaley-sandstone intervals.

Three-dimensional modeling and visualization were critical, from planning through completion. To be effective in a highly mature field, horizontal wells require precision placement. A combination of computerized mapping and digitized injection surveys was used in the Wilmington horizontal-drilling program to identify the dominant unswept flow units prior to drilling and to design the most advantageous well paths. Analysis of measurement-while-drilling (MWD) data, supported by logging-while-drilling (LWD) data, was used to update computerized cross sections extracted from the mapping and to make real-time geosteering decisions that would ensure accurate wellbore placement. The drilling process was complemented by rapid postdrilling analysis for completion-interval selection based on integration of LWD analysis and the computer model. The accuracy and flexibility generated by the 3-D geologic model, combined with the efficiency of the computerized tools to extract and modify the modeled information, significantly enhanced the success of development projects in the Wilmington oil field. The program described has the potential to increase ultimate reserves by hundreds of millions of barrels.

E. P. Mason et al. (Chapter 4) describe application of horizontal wells in the offshore Gulf of Mexico around the flanks of the mature South Pass 62 field. South Pass 62, discovered in 1965, is a giant field 48 km (30 mi) east of the Mississippi River delta in 91 m (300 ft) of water. The field was developed with 61 directionally drilled wells from three platforms and redeveloped from 1986 to 1988 with 31 wells from a fourth platform. A 3-D seismic-based field study completed in 1994 identified additional reservoir targets for a third drilling program that used horizontal and directionally drilled slim-hole sidetracks. The horizontal-well redevelopment program resulted in significant successes and some costly failures.

The South Pass 62 field is located on the north flank of a mushroom-shaped, south-leaning salt dome, which rises from below 7620 m (25,000 ft) to near the seafloor. Typical formation structural dips decrease from 70° adjacent to the salt to 10° off structure. Several generations of faults exist, with throws ranging from inches to several hundred feet. Nearly 60 stacked reservoir sandstones combine with steep formation dips and nearly 200 seismically identified faults to create a complex field of hundreds of reservoir compartments with widely variable reservoir characteristics. Approximately 60 Pliocene and Miocene age deltaic and turbidite sandstone reservoirs onlap the salt from a depth of 1158 to 5791 m (3800 to 19,000 ft). Horizontal wells were designed to drain underpressured reservoirs, increase production rates in laminated/thin-bedded intervals, connect multiple fault blocks with a single lateral, and eliminate multiple recompletions. The majority of the successful wells exploited thin oil-filled shoreface sands, partially depleted zones, and massive sandstone-filled channels. Reservoirs in upper shoreface sandstones—which are typically 3 to 6 m thick, continuous, upward coarsening, well connected, and easy to correlate—were the program's most productive and profitable targets. Failures, measured by production that was less than expected, resulted from mechanical failure, large cost overruns in drilling the well, or geologic "surprises" (e.g., undetected faults, perched water, unexpectedly low pressure). Failures were much more common when horizontal wells were targeted to connect multiple fault blocks and drain low-resistivity/laminated sandstone reservoirs.

In contrast to the application of horizontal wells in mature fields, the approach described by **Liz Jolley et al.** (Chapter 5) for the development of Andrew field—a small oil and gas field on the U.K. continental shelf in the North Sea, discovered in 1974—exclusively used horizontal wells. The relatively small size and remote location of the field made development using vertical wells uneconomic. Integrated reservoir characterization, combined with a development program of only horizontal wells, challenged traditional North Sea development scenarios. The program resulted in small, simple facility design with reduced manpower requirements and shortened construction times. Platform commissioning times were accelerated. The development team, which was integrated across companies, worked to realize cost benefits. Project uncertainties were ranked and managed by incorporating design contingencies and acquiring critical data. Reservoir-characterization uncertainties were identified and expressed in terms of development risks, possible solutions, and their impact on project value (in reserves and development costs). Sensitivity analysis was done by running multiple scenarios on a full-field reservoir model. The model, based on an integrated reservoir and fluids description of the Andrew field, strongly influenced the location, standoff, perforation strategy, and drawdown management of the resulting horizontal oil producers.

As a result of an integrated program, from reservoir characterization to well design, the Andrew field development was successful; plateau oil production extended 18 months beyond the predicted onset of field decline. The focus of design activity on critical reservoir uncertainties and astute business decisions contributed to improved development success. Key components of the development program were to drill all horizontal producers, optimize low gas-oil ratio (GOR) oil recovery, closely manage the reservoir under production to delay gas coning and water breakthrough, and collect sufficient surveillance data to allow regular updating of the reservoir-management plan. Project economics were improved by reducing well numbers from 24 (in a conventional-well case) to 10 horizontal producers. Careful positioning of wells relative to the gas-oil contacts (GOC) and oil-water contacts (OWC) maximized oil production. Other strategies employed included (1) drilling long wells that entered the reservoir on the crest and exited through the flank of the field, (2) completion design, (3) perforation strategy, (4) careful well management, and (5) rapid turnaround of knowledge gained during the development program to target two additional infill wells. As a result, the recovery factor for the Andrew field rose from 45% original oil in place

(OOIP) at the initiation of development in 1994 to 49% OOIP at the end of 1999. The final field recovery factor for the Andrew field has reached 53% by employing horizontal-well design that facilitates low-cost sidetracking and multilaterals, and by continued active management of the reservoir drawdown.

In the Gulf of Mexico, Ram Powell, a major tension-leg platform on the modern slope in 980 m (3214 ft) of water, was developed in a manner similar to that of Andrew field. As described by **Peter A. Craig et al.** (Chapter 6), three primary reservoirs are under development from five horizontal, open-hole, gravel-packed wells. A program of five horizontal wells was included in an initial development plan for the field because of the necessity of increasing the field's profitability. The relatively thin oil rim, reservoirs with low permeability and thickness (kh), and multiple reservoir compartments presented challenges to providing oil rates and recovery efficiency sufficient to justify development. These challenges were met by combining horizontal wells with five conventional, high-angle wells to produce multiple stacked pay zones and reservoirs.

Middle Miocene (approximately 12.0 Ma) slope or "bypass" deposition dominates reservoirs at Ram Powell. Updip, the pinch-out of the multiple sandstone units form a stratigraphy that dips south-southeast at 2° to 4°. To the south, the reservoirs truncate against a salt diapir, forming a syncline.

Because of the complex reservoir architecture and relatively thin oil rim, several existing conventional exploratory and appraisal wells were used as pilots to design potential horizontal-well paths, which reduced the number and cost of new pilot wells by half. Knowledge of reservoir architecture developed from the pilot wells allowed optimal horizontal-well path design and placement in the reservoir; survey accuracy was a critical component of success. In reservoir facies characterized by well-understood lateral variations (such as levees or sheet sandstones), horizontal wells were drilled oblique to strike to transect the entire feature. In more laterally variable facies systems, such as channels, the need for accurate directional control and active geosteering increased. In these variable systems, the horizontal-well path was designed to cut through multiple reservoir compartments.

At Ram Powell, petrophysical tools unique to horizontal wells were used to geosteer wells and to provide near-real-time reservoir analysis. Density imaging provided a basis for structural and stratigraphic evaluation in thin-bedded reservoir units. Quantitative fluorescence technique (QFT) was used to quantify fluorescence in cuttings, and as a method for pay evaluation in long-reach laterals without LWD log confirmation.

The demands imposed by real-time evaluation and geosteering decisions while drilling a horizontal well require a well-designed but flexible plan and tight integration of engineering and geoscience disciplines in an efficient decision-making process. The incorporation of horizontal wells from the beginning of the development program at Ram Powell significantly improved the field's profitability by reducing the number of wells drilled (reducing development costs by 40%) and increasing production rates (performance improvement factors [PIFs] of two to five).

FLUVIAL AND EOLIAN RESERVOIRS

Development of fluvial reservoirs can be one of the most challenging and rewarding applications of horizontal technologies. By the very nature of fluvial reservoirs, one is faced with relatively thin and discontinuous flow units. A detailed and predictive understanding of the lateral and vertical distribution of rapidly changing reservoir facies is critical to successful application of horizontal-drilling technology.

Horizontal wells and other enhanced-oil-recovery (EOR) processes were used to recover significant additional untapped or bypassed reserves in Prudhoe Bay field, Alaska (**R. S. Tye et al.**, Chapter 7). An integrated geologic and engineering approach was used in the Triassic Ivishak sandstone to characterize the depositional environments and to identify opportunities to maintain production. The results of this approach contributed to improvements in horizontal-well planning and completion practices and to increased oil recovery. At Prudhoe Bay, the basal Ivishak is characterized by progradation and abandonment of fluvial-dominated deltaic depositional environments. The reservoir units consist of en echelon, relatively thin distributary-mouth bar and distributary-channel sandstone. Laterally extensive flow barriers that impede the gravity-drainage process are formed by onlapping, retrogradational mudstone. In addition, Ivishak reservoirs are structurally isolated in small fault blocks covering 100 acres or less. As a result of small drainage areas and rapid gas coning, conventional vertical wells sacrifice significant reserves. Detailed mapping was used to design horizontal-well trajectories that maximize lateral wellbore exposure to productive sandstone facies. As the understanding of individual reservoir flow units in the Ivishak increased, a unique EOR program was developed. Miscible injectant stimulation treatment (MIST) used a horizontal well in a watered-out pattern. Based on the reservoir parameters, a quantity of miscible injectant was introduced at the toe of the horizontal lateral, forming a gas bubble that pushed oil to the producers. After achieving designed injection quantities, the perforations were squeezed and the well was reperforated nearer to the heel of the lateral. Sequentially injecting gas, squeezing perforations, and reperforating along the horizontal lateral in more proximal position stripped significant additional mobile oil that is bypassed by conventional waterflood.

The fluvial facies of the Faja region of the Eastern Venezuela Basin form numerous reservoir compartments that contain widely varying fluid compositions. The challenges of identifying preferred targets in a highly heterogeneous reservoir containing heavy oil that can be recovered economically only with horizontal-well technologies is well illustrated by **Douglas S. Hamilton et al.** (Chapter 8). A detailed reservoir characterization prior to drilling the first horizontal well is critical. In terms of producibility, success in highly heterogeneous and discontinuous fluvial reservoirs is controlled strongly by the optimal location of the kickoff point (KOP) and the relative position of the horizontal lateral of the wellbore in flow units. Small changes in facies and structural position strongly affect relative permeability, and water resistivity affects the producibility and determines economics of individual wells.

N. F. Hurley et al. (Chapter 9) describe application of a medium-radius horizontal well in the eolian facies of the Tensleep Sandstone at Byron field in the Bighorn Basin, Wyoming. The Byron field, a mature field, has been on production for more than 60 years and has been waterflooded aggressively. Knowledge gleaned from older vertical wellbores about the paleowind direction and trend of fractures that contribute to a rapid increase in water cut was used to design and drill the horizontal well. The lateral wellbore was oriented approximately perpendicular to the trend of eolian architectural elements and parallel to northeast-trending fractures. The approximately 150-m (500-ft) open-hole lateral also was designed and drilled toe up in a 6-m- (20-ft-) thick stratigraphic interval of the uppermost Tensleep Sandstone.

Cores, borehole images, and outcrop descriptions were used to characterize the reservoir. Borehole-image log interpretation in the horizontal well showed two sets of roughly orthogonal fractures, and at least five dune-related architectural elements that define reservoir compartments. Outcrop analogs and core studies indicate that flow barriers or baffles between reservoir compartments commonly exist at bounding surfaces between dune-related architectural elements.

Only the last 46 m (150 ft) was saturated with oil, and oil-water contact was recognized in the lateral. The present-day contact is tens of meters above the original oil-water contact, and is interpreted as the height of the oil-water contact in the fractures. The well, when put on production, had a very high water cut, suggesting that it may not be possible to control water influx in the Byron field by using a horizontal well to minimize fracture intersections. Hurley et al. use the results of their study to suggest a novel but untested way to complete a horizontal well. They would run tubing to total depth and use the borehole as a downhole oil-water separator. This completion technique might reduce water cuts and help prolong the life of Byron field. It also may have potential application in other fractured reservoirs with strong water drives or active waterfloods.

CHALK RESERVOIRS

Three papers in this volume describe chalk reservoirs (**Charles T. Feazel and Hardy H. Nielsen**, Chapter 10; **O Jørgensen and N. W. Petersen**, Chapter 11; and **N. W. Petersen et al.**, Chapter 12). Chalk is a reservoir rock ideally suited to horizontal drilling; despite very high porosity, it commonly has low permeability. As a result, high hydrocarbon deliverability requires long-exposed reservoir intervals. Chalk drills easily, leaving reasonably smooth and stable wellbore walls. In addition, in fields that produce prolifically, chalk reservoirs commonly are fractured intensely, making horizontal wellbore orientation a critical determinant to strategies that optimize production.

Since 1971, Ekofisk, a giant oil and gas field in the Norwegian sector of the North Sea, has produced oil and gas from chalk of Danian and Maastrichtian age. The field has been under waterflood since 1987. A major redevelopment of the field involved planning and drilling as many as 50 new wells—many of them horizontal. With approximately 225 existing wellbores, it was a challenge to design well paths that avoided collisions in the middle of an offshore field undergoing dynamic subsidence, while simultaneously targeting areas of the reservoir that had not been swept adequately by injected seawater. A detailed and integrated visualization of the Ekofisk reservoir was used to design wells that would optimize economic recovery of the remaining reserve potential (**Feazel and Nielsen**, Chapter 10). Targets and paths for new infill horizontal wells were based on a reservoir flow-simulation model upscaled from a 3-D geocellular model. The project, known as Ekofisk II, included new surface facilities and additional wells with total measured lengths exceeding 7772 m (25,500 ft) (including horizontal reservoir sections of more than 2377 m [7800 ft]). Collision-avoidance issues complicated the designing of well paths in a gas-obscured crestal area of the seismic volume, but once the drill bit exited the gas chimney, a well was geosteered using a combination of 3-D seismic data and biostratigraphic control. With approximately half of the redevelopment program complete, the reservoir-characterization effort and the new horizontal wells have added significantly to the field's original hydrocarbons in place, enhanced daily production rates, and increased estimated ultimate recovery during the field's next 30 years of anticipated production. As a result of using reservoir flow simulation to integrate 3-D reservoir characterization with dynamic production and pressure data, initial production rates of more than 12,000 BOPD were observed in some of the infill horizontal wells.

Jørgensen and Petersen (Chapter 11) present another view of the North Sea Danian and Maastrichtian chalks in their paper on the Dan field in the Danish sector. In the Dan field, natural fractures are less abundant than in other Central Graben chalk fields. Discriminating natural open fractures from more abundant drilling-induced fractures on wellbore image logs is critical to designing and drilling horizontal wells. Drilling-induced fractures are predominantly shear failures oriented 45° relative to the wellbore, but natural fractures are confined to a limited strike range determined by the regional stress field. The strike of natural fractures derived from image logs agreed with core analyses and with the strike of hydraulic fractures generated during injection and stimulation. The Dan waterflood uses horizontal production wells with multiple sand-propped fractures. Injection at high rates has produced large vertical fractures as much as several hundred meters (several thousand feet) long that improve injectivity and reduce the number of injection wells required. Knowledge of fracture direction is a key component for optimizing well patterns to affect maximum areal sweep efficiency.

A second paper on the Dan field, by **Petersen et al.** (Chapter 12), focuses on the practical application of depositional cycles in the chalk. For horizontal wells drilled on the field's flanks, the main reservoir target is a thin (<3-m) high-porosity zone, part of a sequence of cyclic beds in the low-permeability Maastrichtian chalk. Cyclic sequences can be correlated stratigraphically across the field. So as not to jeopardize well reach and thus appraised area, severe doglegs and extensive open-hole sidetracking were avoided. This resulted in a gently undulating well trajectory crossing the target zone several times. Geosteering using biostratigraphy and LWD proved more difficult than anticipated, because biostratigraphic zones in the target unit varied along the well path. It proved difficult to determine whether the well was above or below the highly porous target zone. The LWD log ultimately was used to determine the position of the well relative to the highly porous zone by means of cyclostratigraphic correlation.

CARBONATE RESERVOIRS

Operations in a mature province require continued and creative application of new technologies to maintain production. Advances in drilling technology have made horizontal-well applications economic for the independent producer in mature provinces. Proper application of horizontal drilling can contribute to the revival of the productive potential of mature areas, such as the Midcontinent of the United States. Independent producers with limited technical and economic resources operate a significant portion of the oil and gas fields in these mature areas. Horizontal wells can be used to recover unswept mobile hydrocarbons that cannot be drained economically by vertical wells. One of the principal causes of failure for horizontal wells in mature fields has been inadequate evaluation and selection of optimum targets. The increased cost of a horizontal well makes identification of viable reservoir candidates for horizontal drilling of crucial importance, especially for an independent producer with limited resources.

Reevaluation of a well that previously had been deemed unsuccessful resulted in undertaking a 60-well horizontal-drilling program in the Silurian (Niagaran) Northern Pinnacle Reef Trend (NRT) in the Michigan Basin. **Lester A. Pearce et al.** (Chapter 13) discuss the geologic, drilling, and operational challenges of the various horizontal-drilling opportunity types in the NRT. Applications for horizontal wells include avoiding gas and water coning, contacting additional pay in heterogeneous reservoirs, gravity drainage in low-pressure reservoirs, and increasing rate in low-porosity reservoirs. In pinnacle reefs, horizontal wells require an integrated approach from technical, operational, and management personnel to identify, drill, complete, produce, and even undertake *a posteriori* evaluation.

Saibal Bhattacharya et al. (Chapter 14) discuss and demonstrate cost-effective approaches and technologies that integrate the limited, older geoscience and engineering data commonly available to the operator in mature fields to evaluate candidate reservoirs for horizontal-well applications. The small, independent producer can apply these screening techniques at the field scale, lease level, and well level to identify candidate reservoirs and to predict horizontal-well performance. Field examples from several mature carbonate reservoirs in Kansas are used to demonstrate the application of each technique. The demonstrated tools use readily available, low-cost spreadsheet and mapping packages to analyze production data, map geologic data, integrate and compare geologic and production data, conduct detailed petrophysical analyses, carry out field- and lease-level volumetric analyses, and conduct material-balance calculations. The use of PC-based freeware simulators is described to history-match well and field production, map residual reserves on a field scale, and predict performance of targeted horizontal in-fill wells in candidate reservoirs and leases. This process of identifying candidate reservoirs or leases and evaluating their productive potential for horizontal infill drilling enables independent producers to evaluate the technical and economic challenges of potential horizontal applications prior to drilling and helps them to optimize target selection.

TECHNIQUES

All the studies in this volume focus on the difficulty of understanding exactly where in the reservoir a horizontal wellbore is (and where it is headed next). A common theme is the need for a detailed and integrated approach to the design and execution of a horizontal well. At the same time, planning, execution, and evaluation are fraught with surprises for even the most meticulous team. Uncertainty and variability are expected in any geologic setting, but understanding and dealing with unexpected situations plays a pivotal role in positioning horizontal wells to efficiently access reserves. To maximize net present value, a horizontal well requires a plan for well-path design and placement of the wellbore in the reservoir that is rigorously defined but that maximizes flexibility. This is in contrast to the traditional static plan passed from the home office to well floor. In horizontal wells, well-path adjustments while drilling are to be expected, to handle geologic uncertainty, borehole-position uncertainty, and unanticipated variations in ability to control build rate and direction of the curve (**E. J. Stockhausen et al.**, Chapter 15). Geosteering—the near-real-time action of developing adjustments in the horizontal-well profile using field observations to place the wellbore in the designated target—is a new and demanding role for the geoscientist. Geosteering is not restricted to the employment of downhole geophysical tools. It uses a variety of observations in a geologic and engineering framework to design and modify a wellbore path in three dimensions, from the kickoff point to a landing point in the geologic target. Data such as drilling parameters, biostratigraphy, and cuttings must be integrated by the geoscientist with the mechanical limitations of changes in the well trajectory and desired completion.

Stockhausen et al. discuss the new and challenging real-time role for the geologist as a member of the horizontal-well construction team. The geologist should have an adequate understanding of all aspects of drilling, completion, and operations, to reasonably contribute to the well-plan design and to probable well-path adjustments while drilling. The geoscientist defines and evaluates geologic uncertainties and contributes to the design of a site-specific plan, using a dynamic set of adjustable marker-bed tangents. During drilling, the geologist must be capable of analyzing and reacting to real-time geologic and directional data, with an appreciation of both drilling and completion constraints, to land the well in the reservoir target while maintaining completion and production options. A rigorous but flexible well-path planning method in drilling horizontal and extended-reach wells can increase the precision of well placements, thus enhancing recoveries and sweep efficiencies. A well-designed horizontal project can significantly increase value and decrease costs. The geologist, as a member of the well-construction team, must recognize and respond to the uncertainties and inaccuracies that exist in geologic models and directional surveys, and must precisely control buildup rates and doglegs. The geologist, as part of a larger technical team of scientists, engineers, and managers, must be prepared to develop and communicate trajectory adjustments while drilling a horizontal well.

In a detailed example of active geosteering, **Andrew C. Morton et al.** (Chapter 16) discuss the use of heavy-mineral analysis for geosteering horizontal wells targeted to land in an eolian-dominated unit that forms the best reservoir interval. The reservoir of the Clair field is located in a faulted sequence of Devonian and Carboniferous fluvial, eolian, and lacustrine cycles. Initial drilling was promising; however, appraisal drilling using vertical wells did not provide the hydrocarbon rates necessary to make the project commercially viable. Three-dimensional seismic data were used to increase understanding of the geometry of the reservoir and could have been used to design the horizontal-well trajectories. In the Clair field, reservoir quality depends on the presence of both good-quality matrix and open fractures; optimal flow was restricted to a single eolian-dominated reservoir unit. A suboptimal horizontal well that landed in either the overlying or underlying units was normally uneconomic. This optimal target unit, at 45 m thick, was near the vertical-resolution limits of seismic depth-conversion uncertainty. In addition, because the interval surrounding the target was barren of fossils, biostratigraphic geosteering was not viable. To successfully land the horizontal well in the reservoir, timely decision making was required to review and modify changes in the well profile. Using core and cutting samples, a series of heavy-mineral units was recognized and related to depositional environments. Hydraulic and diagenetic processes subsequently modified heavy-mineral assemblages inherited from the sediment source area (provenance). Variations in heavy-mineral

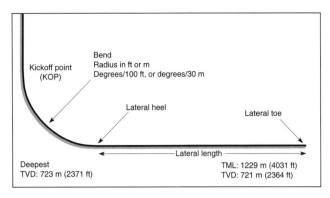

FIGURE 1. Simple vertical profile of a horizontal well, illustrating basic terms.

suites were described in terms of both ratio and varietal data. Techniques were developed to provide, within a short time frame, high-resolution stratigraphic results based on heavy-mineral analysis. Heavy-mineral analysis was a significant factor in the geosteering process that allowed the team to modify the build rate as the wellbore approached the top of the reservoir unit, and to make adjustments as the lateral crossed numerous fault zones in the reservoir.

GLOSSARY OF COMMON HORIZONTAL-DRILLING TERMS

Bend (curve)—the length of the horizontal well from the kickoff point to the heel, in which the well path is designed to change orientation from vertical or high-angle to horizontal (Figure 1).

Build rate—the rate, measured in degrees per unit length, at which the well path of a horizontal well moves through the bend (curve) from vertical or high-angle to the desired low-angle or horizontal attitude.

Extended-reach wells—wells that have relatively long lateral offsets compared to true vertical depth. The term primarily refers to wells restricted to a platform or pad (because of environmental or economic constraints) that are attempting to access reserves at a significant distance from the surface access point.

Geosteering—the action of developing adjustments in the well profile by using field observations to place the wellbore in the designated target. Geosteering can use a large variety of tools and techniques, including logging while drilling (LWD), drilling parameters, biostratigraphy, heavy minerals, and cuttings.

Heel—the point at which the wellbore achieves its design orientation in the targeted productive interval (Figure 1). The heel is normally at the end of the curve and the beginning of the lateral section(s) of a horizontal well.

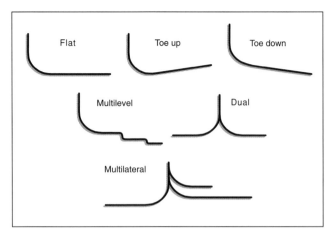

FIGURE 2. Basic types of vertical profiles for horizontal wells.

Horizontal well—a directional well, usually exceeding 80° of departure from vertical, which is designed and steered to maximize the productive interval of exposure in the wellbore. The horizontal lateral(s) can go beyond true 90° horizontal and actually drill upward.

Kickoff point (KOP)—the location in a vertical wellbore or inclined section of a slant well where directional drilling operations commence in order to build the wellbore to the design orientation (Figure 1). The KOP defines the beginning of the build curve.

Lateral—a term applied to either a single branch or individual multiple (multilateral) branches off the vertical wellbore in a targeted, potentially productive interval. In a single-branch case, a lateral runs from the end of the curve (heel) to the toe.

Logging while drilling (LWD)—sensors integrated into the drilling assembly that transmit data while drilling other than guidance data (MWD). Data can include standard measurements of the formation (electric logs such as gamma ray, resistivity, or porosity), drilling parameters (e.g., downhole torque, weight on bit, temperature, or pressure), or geophysical information (look-ahead seismic profiles). Data can be transmitted to the surface or stored for later retrieval.

Measurement while drilling (MWD)—tools that transmit guidance data while drilling from the drilling assembly to the surface. Data transmitted by telemetry include azimuth, inclination, and measured depth.

Multilevel well—a horizontal well constructed with multiple levels in a single lateral separated by higher-angle bends (Figure 2). A multilevel horizontal well typically is used to encounter multiple reservoir compartments.

Multilateral well—a horizontal well constructed with more than one productive lateral branching off a main wellbore (Figure 2). A well with two productive laterals is referred to as a dual lateral well (Figure 2).

Productivity improvement factor (PIF)—a multiplier of the production increase over offset vertical wells.

Sidetrack—in the context of horizontal wells, an additional wellbore drilled away from an original vertical or horizontal wellbore. A sidetracking operation may be done intentionally or may occur accidentally. A multilateral or multilevel well will have multiple intentional sidetracks drilled to provide multiple productive laterals.

Toe—the end point in a wellbore of the lateral section or sections.

Total measured length (TML)—the total length of a horizontal well from the surface datum to the toe. In general, it is referred to as measured depth (MD) and is determined by the sum of lengths of individual joints of drill pipe, drill collars, and other drill-string components.

Stark, P. H., 2003, Horizontal drilling—A global perspective, *in* T. R. Carr, E. P. Mason, and C. T. Feazel, eds., Horizontal wells: Focus on the reservoir: AAPG Methods in Exploration No. 14, p. 1–7.

Horizontal Drilling—A Global Perspective

Philip H. Stark

IHS Energy Group, Denver, Colorado, U.S.A.

ABSTRACT

Horizontal drilling has become a key technology used to reduce costs and enhance recoveries from producing reservoirs. Through 2001, commercial databases contained records on 34,777 horizontal wells from 72 countries. Canada (18,005 wells) and the United States (11,344 wells) were the leading countries for horizontal drilling. More than 5400 horizontal wells were recorded outside of North America. Russia, Venezuela, Oman, United Arab Emirates, Nigeria, Saudi Arabia, and Indonesia were the leading countries in terms of numbers of wells.

Although the concept of horizontal drilling emerged in the 1920s, economic viability was not demonstrated until the 1980s, when pilot projects at Rospo Mare field in Italy (1982) and Prudhoe Bay field (1984) and in the Austin Chalk of Texas (1985–1987) achieved three- to fourfold productivity increases with less than twofold cost increases. From a base of 51 wells in 1987, horizontal drilling increased rapidly; it expanded to the world's active producing provinces and peaked during 1997 with 4990 wells.

Horizontal drilling, which increases wellbore exposure to the reservoir, has delivered multiple benefits. Operators have used horizontal wells to revive economic production, to increase and speed recoveries, to reduce costs, and to increase rate of return. These benefits are critical for operators that must cope with increasing competition and volatile oil prices. The objective of this paper is to characterize the global geographic and geologic distribution of horizontal wells and to illustrate some of the benefits of horizontal drilling with examples from key fields and trends.

DATA SOURCES AND METHODOLOGY

Because of their global coverage, IHS Energy Group databases and reports were used as the primary sources of statistical and technical data in preparing this paper. A joint 1990 study (Horizontal Wells: Global Survey, Case Histories and Strategies) prepared by Erico Petroleum Information Ltd. and MRI, Ltd., is the source for reservoir and geologic aspects cited. Commercial database sources are not complete. Limited records have been reported from former Communist-bloc countries and from heavy-oil development projects in Venezuela. Other countries, such as Saudi Arabia, do not release individual well records, and the completeness and accuracy of reported horizontal-well records varies from region to region. Many horizontal-well records from the U.S. Gulf of Mexico, for instance, do not report reservoir formation names. Most of the charts and graphs in this paper were custom-generated from IHS Energy Group databases and therefore are subject to these limitations. Nevertheless, the magnitude and scope of coverage of these sources provide a representative framework that allows characterization of the global geographic and geologic distribution of horizontal wells and their benefits.

HORIZONTAL DRILLING WORLDWIDE

HORIZONTAL-WELL DISTRIBUTION

After a modern rebirth during the early 1980s, horizontal drilling has been used to reduce costs and enhance recoveries in producing reservoirs worldwide. As shown in Table 1, through 2001, IHS Energy Group databases contained records for 34,777 horizontal wells in 72 countries. Canada recorded the most horizontal-drilling activity, with 18,005 (including original and redrilled boreholes). The United States ranked second in drilling activity, with 11,344 horizontal wells. More than 5400 horizontal wells were reported outside of North America. Through May 1999, horizontal wells had been drilled in 2322 producing fields; more than half (1235) were in the United States, where operators tried horizontal completions in a variety of fields and reservoirs. Canadian operators appear to have been more selective in choosing their targets, as shown by the relatively high percentage of fields in which 10 or more horizontal wells were drilled. Operators were not summed in Table 1 because of the large number that drilled horizontal wells in more than one of the regions.

Horizontal wells have been drilled throughout the world's leading producing provinces. The map in Figure 1 shows the global distribution of 20,430 original horizontal wells (8998, United States; 8221, Canada; 3211, rest of the world) completed through May 1999. (The map does not include horizontal legs that were drilled from prior vertical boreholes.)

HISTORY

Although horizontal drain holes emerged in the 1920s, technological development was slow, and the practice was seldom used. In the 1950s, use of horizontal drilling was still limited to very short-radius (~50 ft.) drain holes in shallow, unconsolidated reservoirs. However, during the late 1970s and early 1980s, several factors—including access to enhanced horizontal technologies from the former Soviet Union; posi-

tive results in drilling long-radius, deviated wells from offshore platforms; and higher oil prices—helped trigger renewed interest in horizontal drilling. Pilot projects by Elf Aquitaine during 1982 at Rospo Mare field in Italy (Bosio, 1990); by British Petroleum and ARCO in 1984 at Prudhoe Bay field in Alaska (Montgomery, 1990a); by ARCO in 1996–1997 at Bima and Arjuna fields in Indonesia; and by Oryx, Mobil, Amoco, and Union Pacific Resources in 1985–1987 in the Austin Chalk trend of Texas (Montgomery, 1990b) finally demonstrated the economic viability of horizontal drilling. From a base of 51 wells in 1987, horizontal drilling increased rapidly in the world's active producing provinces and peaked during 1997, when more than 4000 wells were completed.

BENEFITS

Horizontal drilling, which fundamentally increases wellbore exposure to the reservoir, has delivered multiple benefits. Operators have used horizontal wells to revive economic production and to increase recovery from old producing fields. Horizontal wells also have been used to speed recovery and increase rate of return—a primary driver for North American projects. Other operators have used horizontal wells to reduce capital and operating costs significantly or to minimize the drilling/production operations impact and possible damage to fragile environments. Outside North America, maximizing total return (through enhanced reservoir management, maximizing recoveries, and minimizing long-term costs) has been a primary driver for using horizontal-drilling technologies. These benefits are critical to operators who must compete in an era of chronic, excess oil supplies and low or volatile oil prices.

Horizontal- and extended-reach wells have been drilled to address various reservoir, economic, and geologic aspects. The characteristics of eight important reservoir and/or economic aspects are listed in Table 2. Seven important geologic and reservoir types that are prime horizontal-drilling targets are listed in Table 3. These characteristics have yielded positive results in horizontal-well completions.

TABLE 1. GLOBAL HORIZONTAL-WELL COUNTS. WELL COUNTS ARE AS OF DECEMBER 31, 2001. FIELD AND OPERATOR COUNTS ARE THROUGH MAY 1999. (Source: IHS Energy Group databases.)

Country	Wells	Fields	Fields >10 wells	Operators
Canada	18,005	1235	94	994
United States	11,344	294	69	300
Rest of world	5428	693	64	306
(72 countries)	**34,777**	**2322**	**227**	

HORIZONTAL DRILLING IN THE UNITED STATES

The success of horizontal wells in the fractured Upper Cretaceous Austin Chalk trend in Texas and Louisiana sustained high levels of horizontal-drilling activity in the United States through 1998. As shown in Figure 2, 4230 horizontal wells were completed in the Austin Chalk trend through May 1999. Collectively, these wells generated 6887 completions, which indicates a proliferation of wells with multiple laterals.

Other leading historic reservoir targets include heterogeneous and/or fractured-carbonate reservoirs in the Ordovician Red River in North Dakota, the Permian San Andres and Devonian in the Permian Basin in west Texas, and the Lower Cretaceous Buda limestone in central Texas. Horizontal wells have been used to increase productivity while controlling water and gas coning in the Triassic Sadlerochit sandstone reservoir at the Prudhoe Bay Field in Alaska. The early hope that horizontal wells would yield improved recoveries at Prudhoe Bay, however, has not materialized. Standing (2000) showed that

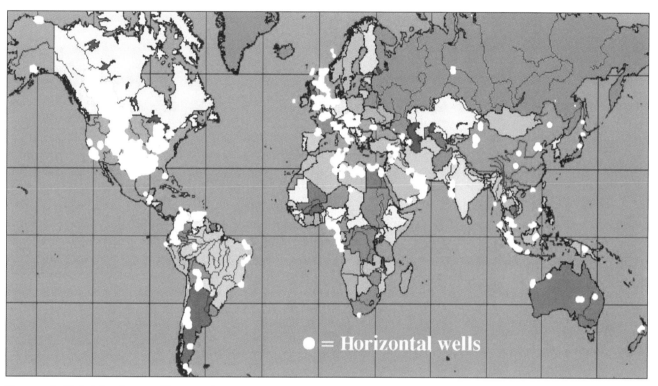

FIGURE 1. Global distribution of horizontal wells through May 1999. (Source: IHS Energy Group databases, May 1999.) This map shows only new wells drilled from the surface with horizontal borehole legs for which latitude-longitude coordinates were available. Only two new countries, Jordan and Vietnam, have reported horizontal wells since this map was generated.

.

TABLE 2. HORIZONTAL DRILLING PLAYS: RESERVOIR/ECONOMIC ASPECTS.

Characteristic	Field/area example
Coning	Prudhoe Bay (Alaska)
Thermal recovery	Belridge (California)
Waterflood	Weyburn-Estevan (Saskatchewan)
Thin beds	Cedar Hills (South Dakota)
Slot limitations (cost)	Dan (Denmark offshore)
Environment	Prudhoe Bay (Alaska)
Limited compartment	Uracoa (Venezuela)
Tar sands	Hamaca (Venezuela)

Sources: Fjeldgaard, 1990; Montgomery, 1990a, b, c, 1995, 1996; MRI Ltd. et al., 1990.

TABLE 3. HORIZONTAL DRILLING PLAYS: GEOLOGIC ASPECTS

Reservoir type	North American examples
Source rocks	Bakken Shale (North Dakota)
Fractured chalk	Austin Chalk (Texas)
Stratigraphic traps	Niagara Reef (Michigan)
Paleokarst reservoirs	San Andres/Yates (Texas)
Thin beds	Red River (North Dakota)
Tight-gas formations	Appalachian Basin
Heterogeneous sands	Tulare Formation (California)

Sources: Montgomery, 1989, 1990a, b, 1995, 1996; MRI Ltd. et al., 1990.

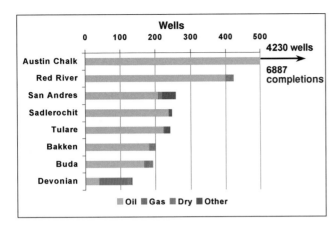

FIGURE 2. Leading horizontal-well formation/reservoir targets in the United States through May 1999. (Source: IHS Energy Group databases, May 1999.)

Prudhoe Bay horizontal wells produced at higher rates in their early years, but may not yield more ultimate recovery than neighboring vertical wells. Horizontal wells also are used to produce oil from heterogeneous upper Miocene Stevens turbidite sandstone (Kuespert et al., 1992) and Pleistocene Tulare reservoirs in California. Late Devonian Bakken shale source rocks in North Dakota's Williston Basin were targeted from 1988 through 1994 in a significant horizontal-drilling play. The Niagran reef trend in the Michigan Basin also has been a prime horizontal-well target. All but the Sadlerochit sandstone and Bakken shale continued to be active horizontal-well targets through 1999 and 2000.

The abundance and importance of horizontal drilling in offshore Gulf of Mexico should not be overlooked. Approximately 210 horizontal wells were drilled in this province through May 1999 but were not included in the chart in Figure 2 because the wells were completed in multiple Pleistocene and Tertiary Pliocene and Miocene reservoirs. Christensen et al. (1999) pointed out that horizontal wells were successful in exploiting oil rims, bioturbated sands, and stratigraphic compartments and in accelerating production to reduce life-cycle costs. Highly layered, low net-to-gross-pay reservoirs, however, were not found to be good candidates for horizontal wells.

Although fractured- and heterogeneous-carbonate reservoirs have been the primary horizontal-drilling targets in the United States, horizontal drilling has been applied to most of the reservoir, eco-

nomic, and geologic aspects identified in Tables 2 and 3. Moreover, horizontal wells have been drilled in virtually all U.S. producing regions, as shown on the map in Figure 3. Key reservoir trends that have been the focus of horizontal-drilling programs are marked on the map, including Canada's leading field for number of horizontal wells, the Weyburn-Estevan field in southeastern Saskatchewan. The main fairway for Devonian horizontal wells is in west Texas. In central Texas, early Cretaceous Buda carbonates have been drilled horizontally in and along the Austin Chalk fairway. The Permian San Andres fairway runs north to south along the Central Basin Platform and west of the Devonian trend.

AUSTIN CHALK TREND

The Austin Chalk trend has undergone three development cycles during the last 60 years, and horizontal drilling has played an integral role in the latest development cycle. The benefit of using horizontal-drilling technology in this area is apparent from a comparison of recoveries from vertical wells in the second cycle, which was triggered by increasing oil prices during the late 1970s, and by recoveries in the most recent cycle, which was triggered by horizontal-well success. At the Pearsall field in the southwest part of the Austin Chalk trend (P on the map in Figure 3), the horizontal-well productivity index ranged from 2.7 to 4.7 (mean = 3.75) times greater than the production from vertical wells drilled during the second cycle. Giddings field, located near the northeast end of the Austin Chalk fairway in east central Texas (G on

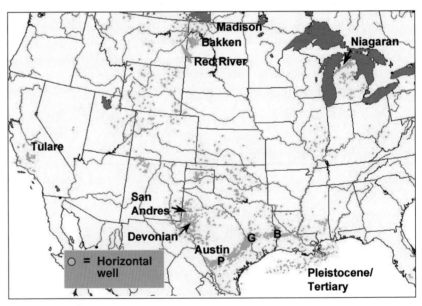

FIGURE 3. Map showing horizontal wells in the United States and southern Canada. Data extracted from IHS Energy Group databases as of May 1999. Leading reservoirs and fields for horizontal wells are annotated. P = Pearsall field; G = Giddings field; B = Brookeland field.

the map in Figure 3), produced 381 MMBOE from 2749 vertical wells during the second cycle (1979–1989). The latest cycle yielded an additional 605 MMBOE through 1998 from 2317 horizontal wells that have been completed since 1990. Figure 4 shows annual oil and gas production for Giddings field and illustrates the relevance of the recent horizontal-drilling cycle.

Since 1994, gas production at Giddings field has increased from downdip extension of the field, where horizontal wells yielded large production rates (20 MMCFD to 30 MMCFD) from deep, overpressured Austin Chalk. Horizontal wells also are being used to extend Austin Chalk production into Louisiana, where parts of the reservoir must be tapped below 15,000 ft. Prevailing high oil and gas prices likely will be required to stimulate continued drilling in this part of the trend.

Stark (1991, 1992) and Fritz (1991) provide additional information about the performance of the Austin Chalk and other United States horizontal-drilling plays and potential reservoir targets.

HORIZONTAL DRILLING IN CANADA GAINS MOMENTUM

As shown in Table 1, Canada ranks first in total number of horizontal wells reported as of December 2001. Canadian operators have led the world in annual horizontal-drilling activity since 1993, and they recorded a peak of 2814 horizontal wells during 1997. This compares with the peak U.S. count of 1383 horizontal wells, also recorded during 1997.

Horizontal completions have been used to exploit key reservoirs throughout the western Canadian sedimentary basin. Principal targets include the widespread Early Cretaceous Manneville Group sandstones, the Mississippian Madison Group carbonates, and Middle and Late Devonian carbonates, several of which produce from reefs. Eight of the top 10 fields (ranked by number of horizontal wells), including the well-known Athabaska and Cold Lake heavy-oil deposits, produce from Early Cretaceous sandstones. Farquharson et al. (1992) summarized early horizontal-drilling developments in many of these reservoirs.

The top-ranked field, however, the Weyburn-Estevan field in southeastern Saskatchewan, produces from Mississippian carbonates. This field provides an excellent example of combining horizontal-well technology and waterflooding to increase productivity. For purposes of comparison, production histories for 250 vertical wells and 250 horizontal wells were extracted from the same sector of the Weyburn-Estevan field. These production histories were normalized to the same start date and projected. Results indicated that the average vertical well would yield 395.7 MBO and 59 MMCFG, compared with 1027.6 MBO and 196.6 MMCFG from the average horizontal well. A normalized production-decline plot that includes 250 horizontal wells in the Weyburn-Estevan Field is shown in Figure 5.

Martin (2000) analyzed production in 12 western Canadian pools and found that horizontal wells achieved a first-year average productivity improvement factor (PIF) of 6.08 compared with vertical wells in the same reservoir. He also showed how peak and total first-year production could be used to provide insight concerning ultimate horizontal-well recoveries. Such positive results continue to stimulate Canadian horizontal drilling.

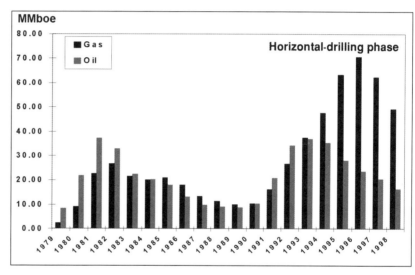

FIGURE 4. Austin Chalk oil and gas production, Giddings field, 1979–1998. Oil is in black, gas in gray.

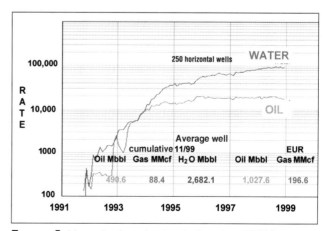

FIGURE 5. Normalized production-decline plot of 250 horizontal wells, Mississippian Madison Group, Weyburn-Estevan field, Saskatchewan, Canada. EUR = estimated ultimate recovery.

HORIZONTAL DRILLING IN THE REST OF THE WORLD

Horizontal-drilling and completion techniques also have been applied widely and successfully in most of the important producing basins and fields throughout the rest of the world. Outside North America, IHS Energy Group records indicate that approximately 5400 horizontal wells have been drilled in more than 700 fields that are credited with 450,000 MMBO of remaining reserves. If Ghawar and other Saudi Arabian fields with active horizontal-well development programs are added, the total oil reserves being produced at least in part by horizontal wells could exceed 600 billion bbl.

The leading countries (excluding the United States and Canada) for horizontal-drilling activity through June 2000 are shown in Figure 6. Venezuela, Oman, the United Arab Emirates, and Nigeria were the leading countries for which individual well records are reported. R. A. Kamal (personal communication, 2000) noted that Saudi Aramco has drilled more than 200 horizontal-development wells, in which case Saudi Arabia would rank fifth in horizontal-drilling activity, just ahead of Indonesia, which reported 212 horizontal wells. New information (Matlashov and Ustinov, 2001) indicates that 1177 horizontal wells were drilled in Russia through 2001. Peak activity was reported during 2000, when 198 horizontal wells were drilled. Based on this information, Russia vies with Venezuela as the leading country for horizontal drilling outside North America.

Four of the leading fields for horizontal-well drilling outside North America, as reported through June 2000,

are in Venezuela. Three of these fields, Hamaca, Uracoa, and Melones, produce from the prolific early and middle Miocene Oficina sandstone reservoirs. Lagunillas, the fourth Venezuelan field, is part of the giant Bolivar Coastal fields complex in Lake Maracaibo. Other leading fields, including Yibal in Oman and Zakum in the United Arab Emirates, use horizontal wells to exploit prolific carbonate-platform reservoirs. The Nimr field in Oman, the Rabi-Kounga field in Gabon, the Bab field in the U.A.E., and the Attaka field in Indonesia's Kutei Basin round out the list of leading fields for numbers of horizontal wells.

EXAMPLES FROM KEY FIELDS AND TRENDS

The following examples illustrate the diversity of reservoir conditions in which horizontal wells have made positive contributions:

- Elf Aquitaine achieved early and pivotal economic success with its horizontal-well experiments at the Rospo Mare field in Italy during the early 1980s. The Early Cretaceous karsted-carbonate reservoir at Rospo Mare contains heavy, 11° API-gravity oil. The horizontal-well program (21 wells) helped to boost oil production from approximately 2000 BOPD in 1986, to a peak of nearly 29,000 BOPD in 1990. Cumulative oil production hit 68.1 MMBO in 1997.
- The Dan field in the Danish offshore sector provides an interesting comparison and contrast to Austin Chalk reservoirs in the United States. The Early Cretaceous Dan Chalk reservoir records 28% porosity and 0.5 md permeability. Reported reserves are 645 MMBOE and 860 bcfg. Most horizontal boreholes radiate from the center of the nearly circular, faulted-domal structure. The 46-well horizontal program helped to boost production from approximately 60,000 BOPD in 1991 to more than 100,000 BOPD in 1998.
- The Hamaca area in Venezuela illustrates the use of horizontal wells and thermal (steam) processes to boost recoveries of heavy oil. Hamaca field is part of the well-known "Orinoco Tar Belt" and is estimated to contain 40 billion bbl OIP, with 5.6 billion bbl in recoverable reserves. The heavy, 7° API-gravity crude oil is trapped in a faulted and folded monocline near the southern limit of the fluviodeltaic early and middle Miocene Oficina sandstone reservoir. Recovering the heavy oil with vertical

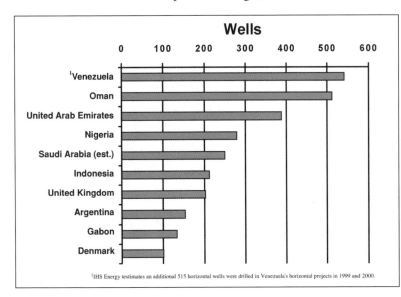

FIGURE 6. Horizontal wells—leading countries outside North America through June 1999. (Source: IHS Energy Group databases, June 2000.) Saudi Arabia well count is estimated.

wells is difficult, although the Oficina records 32% porosity and as much as 10 d permeability. Approximately 780 wells, including at least 205 horizontal boreholes, have been drilled in the Hamaca area. After the introduction of horizontal drilling at Hamaca field in 1993, production more than doubled, from 40,000 BOPD to approximately 90,000 BOPD in 1996. Horizontal wells at Hamaca field reportedly produced as much as 2000 BOPD when combined with steam injection.

- At Widuri field, Indonesia's largest offshore field in the West Java Sea, Maxus used 3-D seismic data to target fluvial channels in the upper Gita Formation (lower Miocene). The B-14 horizontal well established initial production at 7000 BOPD, approximately three to four times that of vertical wells in the same reservoir facies.

SUMMARY

Horizontal-drilling and completion technologies have been deployed successfully in a variety of reservoir and operating conditions throughout most of the world's producing provinces. Case histories indicate that horizontal completions generally enhance oil and gas productivity but do not necessarily increase total recovery. Benefits such as increased cash flow, lower development and producing costs, and reduced environmental impact should continue to drive the use of global horizontal drilling. Sharing of new horizontal technologies and completion practices should allow operators to continue to lower costs, to enhance recoveries, and to successfully deploy horizontal completions into additional conventional and unconventional reservoirs.

REFERENCES CITED

Bosio, J., 1990, Horizontal wells—On the up and up, *in* Exploration & Production Technology International 1990: London, Sedgwick Energy Ltd., p. 23–26.

Christiansen, D., S. Whitney, T. Bernardi, and L. Young, 1999, Shelf horizontal lookback 1993–1996 results and findings (abs.): Proceedings, AAPG/Society of Professional Well Log Analysts Hedberg Research Symposium, Horizontal wells—Focus on the reservoir, The Woodlands, Texas, October 10–13, 1999.

Farquharson, R., l. Spratt, and P. Wang, 1992, Perspectives on horizontal drilling in Canada, *in* J. W. Schmoker, E. B. Coalson, and C. A. Brown, eds., Geological studies relevant to horizontal drilling: Examples from western North America: Rocky Mountain Association of Geologists, p. 15–24.

Fjeldgaard, K., 1990, Steep learning curve for horizontal wells, *in* Exploration & Production Technology International 1990: London, Sedgwick Energy Ltd., p. 29–36.

Fritz, R., 1991, Analyses show potential of horizontal drilling: American Oil and Gas Reporter, v. 34, no. 6, p. 18–27.

Kuespert, J., S. Reid, and G. McJannet, 1992, Horizontal well development in the Upper Miocene 26R Stevens Pool, Elk Hills Field, Naval Petroleum Reserve No. 1, San Joaquin Basin, California, *in* J. W. Schmoker, E. B. Coalson, and C. A. Brown, eds., Geological studies relevant to horizontal drilling: Examples from western North America: Rocky Mountain Association of Geologists, p. 267–284.

Martin, I., 2000, Study evaluates horizontal well ultimate recovery: Oil & Gas Journal, v. 98, no. 29, p. 4–56.

Matlashov, I. A., and S. K. Ustinov, 1991, About development of horizontal drilling—Problems and prospects, *in* Collection of Reports of VI International Conference on Horizontal Drilling, Izhevsk, Russia, October 23–25, 2001.

Montgomery, S., 1989, Williston Basin Bakken shale: Learning curve on the horizontal: Petroleum Frontiers, v. 6, no. 4, 52 p.

Montgomery, S., 1990a, Horizontal drilling: A survey of domestic case histories: Petroleum Frontiers, v. 7, no. 2, p. 1–13.

Montgomery, S., 1990b, Horizontal drilling in the Austin Chalk, part 1: Geology, drilling history and field rules: Petroleum Frontiers, v. 7, no. 3, 44 p.

Montgomery, S. 1990c, Horizontal drilling in the Austin Chalk, part 2: Patterns of production: Petroleum Frontiers, v. 7, no. 4, 58 p.

Montgomery, S., 1995, Louisiana Austin Chalk: Petroleum Frontiers, v. 12, no. 3, 68 p.

Montgomery, S., 1996, Williston Basin horizontal Red River play, Petroleum Frontiers, v. 13, no. 3, 61 p.

MRI Ltd. and Erico Petroleum Information Ltd., 1990, Horizontal wells: Global survey, case histories and strategies: Tulsa, Oklahoma, MRI Ltd., 4 vols., 1219 p.

Standing, T., 2000, Data shows steep Prudhoe Bay production decline: Oil & Gas Journal, v. 98, no. 40, p. 86–96.

Stark, P., 1992, Perspectives on horizontal drilling in western North America, *in* J. W. Schmoker, E. B. Coalson, and C.A.Brown, eds., Geological studies relevant to horizontal drilling: Examples from western North America: Rocky Mountain Association of Geologists, p. 3–14.

Turbidite
and
Deltaic
Reservoirs

Clark, M. S., R. K. Prather, and J. D. Melvin, 2003, Characterization and exploitation of the distal margin of a fan-shaped turbidite reservoir—The ARCO-DOE 91X-3 horizontal well project, Yowlumne field, San Joaquin Basin, California, *in* T. R. Carr, E. P. Mason, and C. T. Feazel, eds., Horizontal wells: Focus on the reservoir: AAPG Methods in Exploration No. 14, p. 11–26.

2

Characterization and Exploitation of the Distal Margin of a Fan-shaped Turbidite Reservoir—The ARCO-DOE 91X-3 Horizontal Well Project, Yowlumne Field, San Joaquin Basin, California

Michael S. Clark

Chevron USA, Bakersfield, California, U.S.A.

Rick K. Prather

Aera Energy, Bakersfield, California, U.S.A.

John D. Melvin

Netherlands and Sewell and Associates Dallas, Texas, U.S.A.

ABSTRACT

The deepest onshore horizontal well in California is the ARCO-DOE 91X-3, which was drilled at Yowlumne field in the San Joaquin Basin, California, to exploit the thinning, distal margin of a fan-shaped, layered turbidite complex. Yowlumne is a giant oil field that has produced more than 17.2 million m^3 (108 million bbl) of oil from the Stevens Sandstone, a clastic facies of the Miocene Monterey Shale source rock. Most Yowlumne production is from the Yowlumne Sandstone, a layered, fan-shaped, prograding Stevens turbidite complex deposited in a slope-basin setting. Well-log, seismic, and pressure data indicate seven depositional lobes with both left-stepping and basinward-stepping geometries.

To facilitate cost-effective exploitation of remaining field reserves, a 3-D model of the reservoir architecture was constructed from log-derived petrophysical data, constrained by core analyses. This model indicates concentration of channel and lobe facies along the axis and west (left) margin of the Yowlumne fan to result in average net/gross sandstone ratios of 80%, porosity (Ø) of 16%, and liquid permeability (K_{liquid}) of 10–20 md. By contrast, more abundant levee and distal margin facies along the east margin result in shale-bounded reservoir layers with higher clay contents and lower net/gross sandstone ratio (65%), porosity (12%), and permeability (2 md). Thus, the distal fan margin is not an attractive place to drill for oil.

However, modeling indicates that most remaining field reserves exist along the east fan margin. Although a waterflood during the last 20 years will enable recovery of 45% of original oil in place along the fan axis, about 480,000 m^3 (3 million bbl) of oil trapped at the thinning fan margins will be abandoned with the current well distribution. The ARCO 91X-3 was a Department of Energy–funded well to test economic recovery of this bypassed oil by utilizing a high-angle deviated well with three hydraulic fracture stimulations to provide connectivity between reservoir layers, thereby providing the same productive capacity as three vertical wells. The well was drilled along strike to a measured depth of 4360 m (14,300 ft) to tangentially penetrate at angles up to 85° as much as 600 m (2000 ft) of the distal margin of the Yowlumne fan. Because of drilling and completion difficulties, the proposed multiple fracture stimulations were not attempted. Nonetheless, use of highly deviated to horizontal wells with multiple fracture stimulations remains an economically viable option for maximizing productivity from the thinning, distal margins of layered, low-permeability turbidite reservoirs.

INTRODUCTION

Although fan-shaped reservoirs of low-permeability (10–100 md) turbidite sandstones are prolific producers in many oil fields, many of these reservoirs are characterized by declining production. Typically, reservoir quality, measured as increasing porosity and permeability, improves with increasing sandstone thickness. Because fans thicken in the middle, production tends to increase toward the depositional axes and decrease toward the fan margins where net pay thickness and reservoir quality decrease. Consequently, it is more difficult to produce oil economically from fan margins, and oil reservoired there is more likely to be bypassed and abandoned than oil reservoired elsewhere within the fan. Despite this observation, many waterflood patterns tend to sweep oil toward the fan margins, where it is banked against the zero edge of the reservoir. Thus, bypassed oil stored along thinning fan margins and oil trapped there by waterflooding represent significant remaining reserves in many mature fields.

Economic recovery of oil trapped along fan margins requires a detailed understanding of the reservoir architecture to cost-effectively exploit these bypassed reserves. This paper summarizes an integrated study of the distal margin of a layered, low-permeability, fan-shaped turbidite complex that led to drilling of the deepest horizontal well in onshore California. Geologic modeling, characterization of reservoir properties, and flow simulation were used to locate the well. Modeling also facilitated the design of multiple "frac jobs" (hydraulic fracture stimulations) in the wellbore to maximize production rates by improving vertical connectivity in a layered, low-permeability turbidite reservoir.

YOWLUMNE FIELD

Yowlumne is a giant oil field in the San Joaquin Basin, California (Figure 1), that has produced, through December 2000, more than 17.2 million m³ (108 million bbl) of oil and 2.7 billion m³ (94 billion ft³) of gas from

FIGURE 1. Structure map of Yowlumne field drawn on the N-point marker, a regional correlation horizon that marks the approximate top of the Stevens Sandstone. Note the fanlike shape of the Yowlumne Sandstone and the relationship of Yowlumne Units A and B to the anticlinal closure.

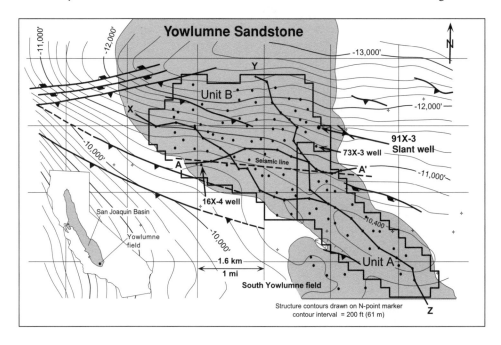

upper Miocene, deep-water sandstones known as the "Stevens" (Figure 1). These sandstones, which represent clastic facies of organic-rich shales in the Monterey Formation (Figure 2), are some of the most prolific reservoirs in the basin and have contributed about 15% of more than 1.9 billion m³ (12.3 billion bbl) of oil produced here since 1864. Because Stevens oils derive from Monterey Shale source rocks, Yowlumne is part of a Monterey-Stevens petroleum system (Graham and Williams, 1985).

Most production at Yowlumne is from the Yowlumne Sandstone, one of several discontinuous sandstone bodies collectively referred to as the Stevens (Figure 3). A better understanding of this reservoir is needed to reduce a 35% annual decline rate in the field and to develop remaining reserves before aging facilities require replacement or abandonment.

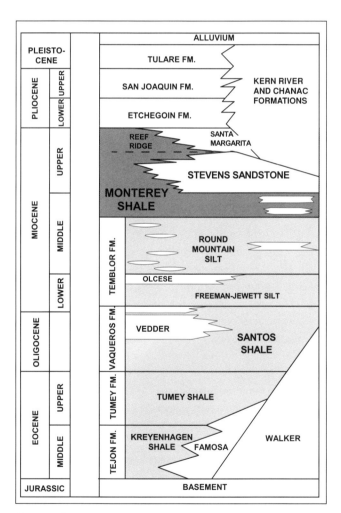

FIGURE 2. Stratigraphic section showing nomenclature for the Yowlumne field area. The Stevens is an informal unit that represents a deep-marine, clastic facies of the Monterey Formation, an organic-rich shale that is considered to be the main source rock for most of the oil produced from the basin.

METHODOLOGY

Field exploitation is facilitated with an accurate understanding of the reservoirs. A five-step analysis was performed to model the geometries, facies distributions, and reservoir properties of the main producing reservoir at Yowlumne field.

1) description of the reservoir architecture (i.e., stratigraphic geometries, flow units, and facies distributions) using well-log correlations, pressure data, and existing 3-D seismic data
2) determination of reservoir properties using core, well-log, borehole-breakout, and microseismic data
3) construction of a digital petrophysical database (e.g., reservoir thickness, porosity, permeability, net/gross sandstone ratio, and water saturation)
4) contour mapping of petrophysical properties by individual flow units
5) location and quantification of bypassed reserves using a computer model to simulate fluid flow in the reservoir

A digital petrophysical database was constructed from well-log data constrained by core analyses. Total porosity ($\emptyset_{total}$) is derived from sonic data, and shale volumes (Vsh) are derived from gamma-ray logs by normalizing the log data against statistically determined shale and sand baselines. Effective porosity ($\emptyset_{eff}$) is obtained from $\emptyset_{eff} = \emptyset_{total} - (\emptyset_{sh} \times Vsh)$ where $\emptyset_{sh}$ is the average porosity of the shale fraction ($Vsh > 90\%$). Liquid permeability (K_{liquid}) is calculated from $\emptyset_{eff}$ using an algorithm derived from core data, and validity of log-derived K_{liquid} and $\emptyset_{eff}$ distributions (of net sandstone) is confirmed by comparison to core data corrected for overburden pressure. Net sandstone is defined as sandstone with $Vsh = 0$ to 30% and $\emptyset_{eff} = 8$ to 30%. Water saturation (Sw) is calculated from $\emptyset_{eff}$ using the Archie equation with water resistivity values corrected for formation temperatures.

RESERVOIR ARCHITECTURE

Reservoir architecture is a 3-D description of the reservoirs in a field and the nonproductive rocks that surround them. The reservoir architecture contains three principal components:

- structural configurations of time-significant stratigraphic surfaces (e.g., unconformities, downlap surfaces, and maximum flooding surfaces)
- geometries of reservoir flow units and compartments
- distributions of facies between surfaces and distributions of reservoir properties within facies

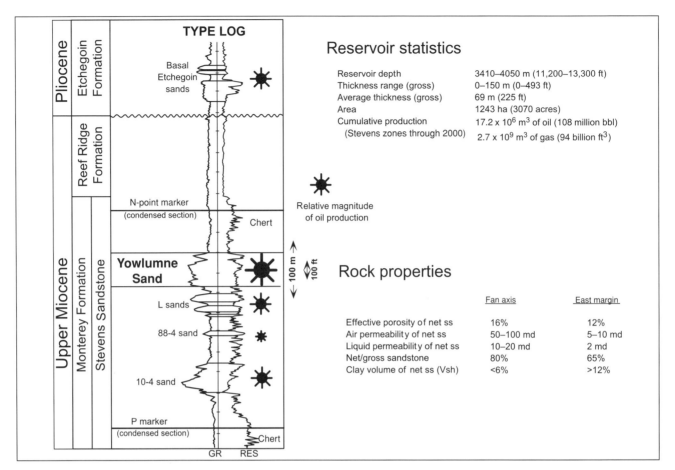

FIGURE 3. Type log for Yowlumne field with reservoir statistics and rock properties of the Yowlumne Sandstone. Reservoir quality decreases from the axis of the fan eastward toward the fan margin.

Understanding the reservoir architecture of a field is fundamental before reservoir models can be constructed to facilitate well placement, design completion programs, and implement effective waterflood management.

STRUCTURE

The oil accumulation at Yowlumne is controlled in part by an anticlinal closure formed during Miocene-Pliocene deformation of the south basin margin. Yowlumne Unit A was created to waterflood the area of structural closure, but subsequent drilling established production in a stratigraphic accumulation on the north-dipping flank of the anticline also (Figure 1). Consequently, Yowlumne Unit B was created to flood the flank accumulation (Burzlaff, 1983; Metz and Whitworth, 1984). Thus, the field is a combination structural-stratigraphic trap.

FLOW UNITS AND RESERVOIR COMPARTMENTS

More than 95% of production at Yowlumne field is from the Yowlumne Sandstone, a fan-shaped Stevens

Sandstone body as much as 150 m (493 ft) thick (Figure 3). Bouma sequences evident in cores from the field indicate deposition of this body by turbidity currents. Also, the Yowlumne Sandstone body is lens shaped in cross section and does not significantly incise underlying strata (Figure 4). Because large-scale channeling is absent, deposition was primarily as sheet sands transported by sediment-gravity flows.

Thin shales divide the fan into lobe-shaped reservoir layers (Figure 4). Five of these—the A, B, C, D, and E sands—produce oil from Unit B. The W sand is a basal sixth layer that is wet and isolated by pressure from overlying sandstones. Layers A through E merge into homogenous, clean sandstone on the west margin of the fan, yet contain interbedded shale on the east. For example, the 16x-4 horizontal well on the west side (Figure 1) penetrates a thick interval of clean sandstone (Marino and Schultz, 1992). By contrast, the 73x-3 well on the east side (Figure 1) penetrates shale layers, some of which are two or more meters thick, interbedded with the reservoir sandstones.

Well-log correlations and 3-D seismic data indicate

downlap within the fan, with basinward progradation to the north and lateral progradation to the west (Figures 4 and 5). In other words, lobe-shaped, shale-bounded reservoir layers in Unit B step to the left when facing basinward, in the direction of sediment transport (Jessup and Kamerling, 1991; Clark et al., 1996b). Thus, the basal productive layer (sand E) is thickest on the right (east) side of the fan, whereas the top layer (sand A) is thickest on the left (west).

Variations in reservoir pressures and injection of radioactive tracers indicate weak compartmentalization of the reservoir, which results in separate permeability pathways along which fluids flow at different rates (Metz and Whitworth, 1984, Figure 9; Berg and Royo, 1990). Most likely, these pathways represent different flow units that, for the most part, are in pressure communication over geologic time. Consequently, these compartments, which correlate to the shale-bounded, lobe-shaped reservoir layers already discussed, develop the same oil-water contact over thousands of years yet acquire slightly different pressures as the field is rapidly produced over tens of years.

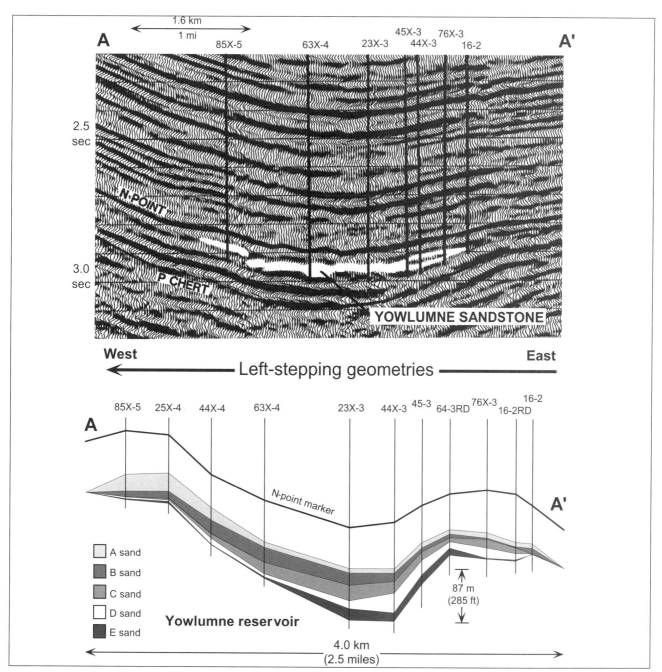

FIGURE 4. Seismic line and cross section A-A′ showing left-stepping geometries and cross-sectional lens shape of the Yowlumne fan. The line and section transect the fan from west to east, perpendicular to the direction of sediment transport.

Several observations demonstrate that compartmentalization in the field also exists on larger scales.

- The reservoir in Yowlumne Unit A has, over time, consistently exhibited reservoir pressures that differ from those in the same interval in Unit B (Figure 6).
- Gross sandstone isopach maps indicate separate northern (Unit B) and southern (Unit A) depocenters (loci of thickening), which appear to represent different depositional accumulations (Figure 7).
- Because sandstones in Unit B have more quartz, less clay, and higher original porosity than the equivalent sandstones in Unit A, different depositional histories are indicated (Whelan, 1984).

- The oil-water contact in Unit B is 660 m (2180 ft) structurally lower than the contact in Unit A (Figure 5).
- Although a few studies interpret a single oil-water contact steeply tilted to 5° (480 ft/mi, 90.9 m/km), a large density difference of 0.154 g/cc between the oils (32° API) and formation waters (TDS [total dissolved solids] of 22,000 ppm) is more consistent with a lower-gradient oil-water contact. More likely, the large density contrast results in buoyant oils unlikely to support a contact tilted more than 1° (100 ft/mi, 18.9 m/km), even in the presence of a strong water drive.
- Comparisons of oil analyses from different parts of the field indicate that Units A and B are not in fluid communication (Figure 8).

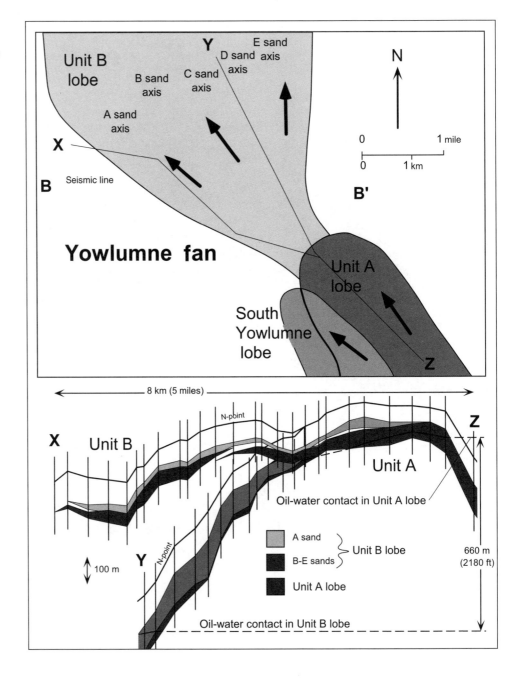

FIGURE 5. Map with cross sections X-Z and Y-Z showing basinward-stepping geometries exhibited by lobe-shaped sand bodies that make up the Yowlumne Sandstone. Note 660 m (2100 ft) of structural relief between oil-water contacts in the lobes of Units A and B.

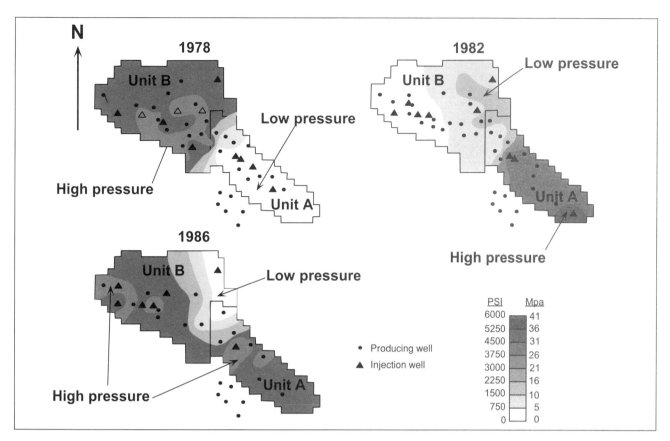

FIGURE 6. Maps showing fieldwide variations in the average reservoir pressure of the Yowlumne Sandstone. The variations indicate that Units A and B may represent separate compartments that are not in fluid communication with one another.

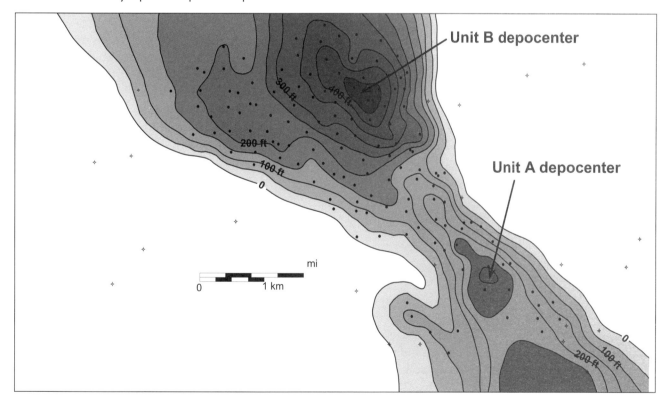

FIGURE 7. Gross-sandstone isopach map of the Yowlumne sandstone showing separate depocenter thicks in Units A and B. The different depocenters indicate that the units may represent separate depositional accumulations, an interpretation consistent with Units A and B representing separate compartments with different oil-water contacts.

FIGURE 8. Star plots of peak-height ratios from gas-liquid chromatograms of two representative oils in the field. These plots are consistent with plots of another six oils from the field that were also analyzed in the study, and they indicate that oils in Unit B are compositionally distinct from those in Unit A. Because variations at the D, K, M, N, and P axes are significant enough to indicate that oils in Units A and B are not in fluid communication with each other, Units A and B appear to represent separate reservoir compartments (Suhas Talukdar, Core Laboratories, Inc., written communication, 1998).

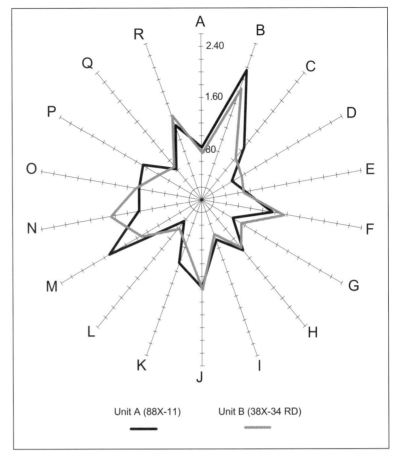

Thus, Units A and B are separate compartments, and basinward progradation to the north indicates that reservoir compartments in Unit B are younger than those in Unit A (Figure 5).

A common oil-water contact, tilted less than 1° (18.9 m/km, 100 ft/mi), exists between Unit A and South Yowlumne field, a lobate sandstone accumulation on the west side of Yowlumne field (Figure 1). This common contact indicates depositional continuity, with Unit A and South Yowlumne representing adjacent depositional lobes. If the system prograges laterally to the west, as comparison to Unit B indicates, then the South Yowlumne Sandstones are younger than those in Unit A (Figure 4).

RESERVOIR PROPERTIES

The Yowlumne Sandstone is a low-permeability, clay-bearing, arkosic wacke derived from crystalline source terranes (Tieh et al., 1986). Some mixed-layer, possibly detrital, smectite/illite is present. However, burial depths of 3000 to 4000 m (10,000 to 13,000 ft) and formation temperatures of 110° to 130°C (235° to 270°F) have transformed most of the clay to authigenic kaolinite (Jim Boles, written communication, 1997).

High-resolution, millimeter-scale permeability pro-files of cores cutting turbidite bedding in the reservoir indicate decreasing permeability with increasing clay content. For example, massive to graded turbidite sandstones of Bouma A intervals are a full order of magnitude more permeable than the laminated to rippled B and C intervals, presumably because of the presence of clay lamina in the latter (Figure 9). Slumped intervals also demonstrate low permeability. Because fluid flow is facilitated parallel to bedding, through permeable Bouma A intervals, and restricted across other facies, vertical permeabilities in the reservoir are much lower than horizontal permeabilities. Stochastic shales in the reservoir—discontinuous interwell shales below the resolution of logging tools—represent baffles that further reduce the effective vertical permeability (Begg et al., 1985).

Log-derived petrophysical data, constrained by core analyses, indicate decreasing reservoir quality toward the east margin of the fan (Figure 3). Values of net/gross sandstone = 80%, clay content of net sand (Vsh) < 6%, effective porosity ($\emptyset_{eff}$) = 16%, and liquid permeability (K_{liquid}) = 10–20 md along the fan axis decrease to net/gross = 65%, Vsh > 12%, $\emptyset_{eff}$ = 12%, and K_{liquid} = 2 md along the east fan margin (Figure 10). Most likely, this eastward degradation in reservoir quality results from an increase in the frequency and thickness of interbedded

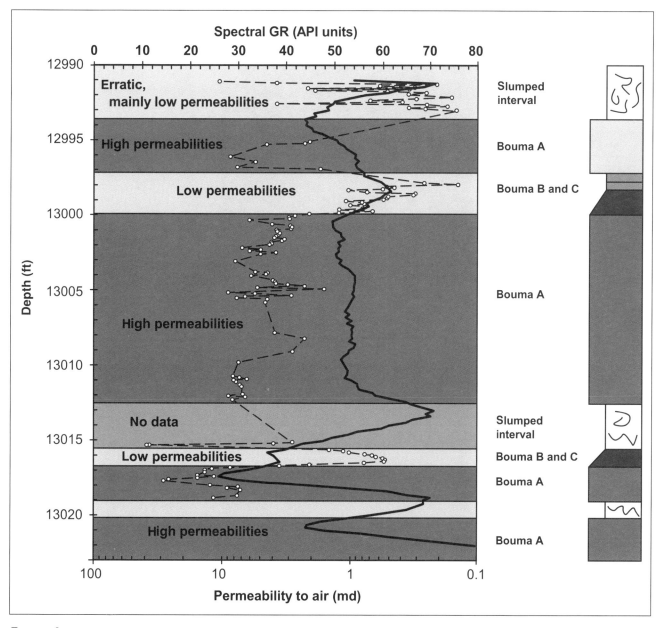

FIGURE 9. Permeability profile from a 30-ft cored interval in the Yowlumne 91X-3 horizontal well revealing greater permeabilities in Bouma A layers relative to Bouma B and C layers, and slumped intervals. These variations result in the creation of reservoir permeability pathways that parallel bedding.

shales (net/gross sandstone), increasing clay content (Vsh), and decreasing grain size. Because of high clay content and abundant interbedded shales, combined with thin reservoir, the east margin of the fan is not an attractive place to drill for oil.

DISCUSSION

DEPOSITIONAL CONTROLS

The Yowlumne Sandstone lobes are part of a linear trend of Stevens sandstones that includes Landslide field,

a northward-prograding complex located 6.4 km (4 mi) to the south of Yowlumne field. Seismic data indicate sandstone continuity between these fields (Stolle et al., 1988; Quinn, 1990). Therefore, depositional lobes should exist between Landslide and Yowlumne, and additional lobes may be present basinward (to the north) of Yowlumne Unit B.

Most likely, basinward-stepping geometries in the Yowlumne-Landslide system resulted in part from additional sediment input into the southern San Joaquin Basin during renewed uplift of the basin margins. Also, these geometries probably resulted in part from decreas-

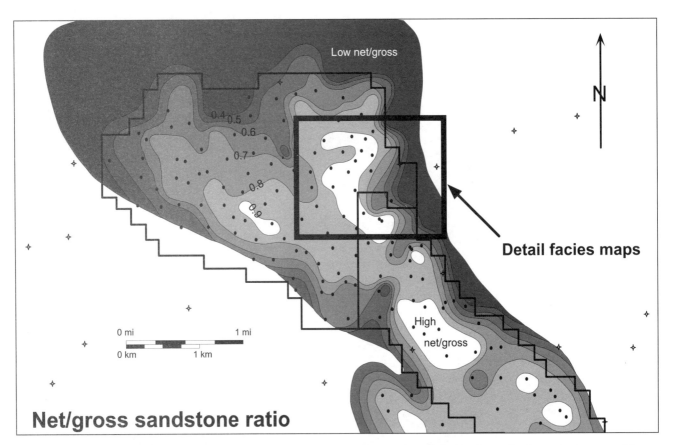

FIGURE 10a THROUGH 10d. Maps 10a through 10d, showing reservoir properties of the Yowlumne Sandstone in the area of the ARCO 91X-3 well.

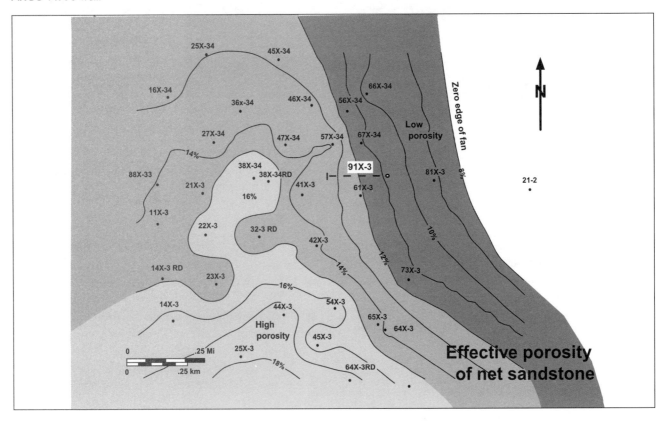

FIGURE 10b.

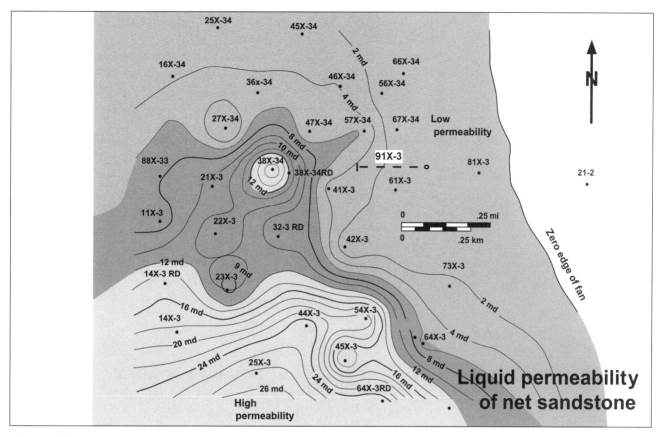

FIGURE 10c.

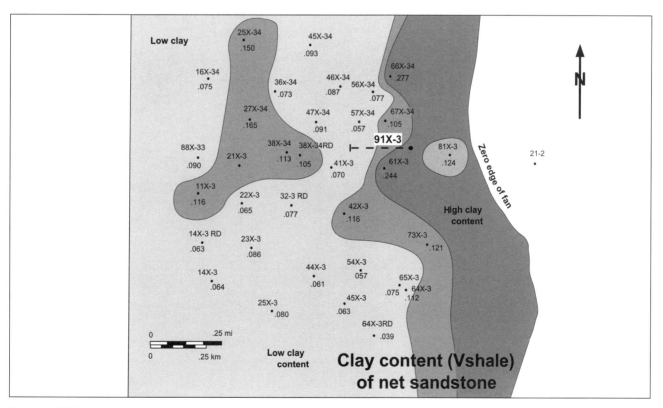

FIGURE 10d.

ing accommodation (i.e., tectonic subsidence plus eustatic sea-level change). Decreasing subsidence rates along the Bakersfield Arch, a regional high in the central part of the basin, resulted in decreased accommodation, which had a profound effect on aggradational and progradational geometries of Stevens sandstones deposited there (Clark et al., 1996a). Because the Miocene Yowlumne fan was located in a slope-basin setting close to a tectonically active transform margin represented by the San Andreas Fault, subsidence-driven accommodation was probably just as important at Yowlumne as along the Bakersfield Arch.

Left-stepping geometries probably resulted from Coriolis forces, which in the Northern Hemisphere favor clockwise-directed circulation patterns. Consequently, Northern Hemisphere fans preferentially deposit levees on the right sides of channels, forcing lobe deposition to the left (Menard, 1964). Abundant interbedded shales, characteristic of levee facies, on the east side of the Yowlumne fan (as in the 73x-3 well), and clean sandstones, characteristic of lobe facies, on the west (as in the 16x-4 well) are consistent with this interpretation. Alternatively, the clean sandstones represent channel facies instead of lobes (Metz and Whitworth, 1984; Berg and Royo, 1990). However, a channel complex is inconsistent with progradational geometries and lack of major incision beneath the fan. Also, left-stepping geometries in Unit A and South Yowlumne field indicate lateral shifting of sandstone lobes where the alternative model requires a channel axis. Furthermore, lobate sandstone distributions are more consistent with a lobe model.

Paleotopography may also explain depositional geometries in the Yowlumne Stevens. The reservoir on seismic data appears as an interval of thickening on the downthrown, east side of a parallel series of northwest-to-southeast-trending reverse faults (Figure 11). These faults also parallel the west margin of the fan (Figure 1). Although some thickening clearly represents less compaction of sandstone in the downthrown block, relative to shale in the upthrown block, thickening may also reflect fault movement during fan deposition, indicating growth of a paleobathymet-

ric high that constrained westward growth of the fan (Quinn, 1990).

In addition, a case can be made for another paleohigh on the east margin of the fan, resulting from differential compaction of lower Stevens sands (Dorman, 1980; Quinn, 1990). If so, confinement of fan growth to a depression between subtle paleohighs resulted in a linear trend of sandstone lobes that does not significantly incise underlying strata (Scott and Tillman, 1981). Although this model explains linearity of the Yowlumne-Landslide trend, it does not account for a greater concentration of interbedded shale on the east margin of the Yowlumne fan in Unit B.

Possibly, overbank deposition on the east margin of the fan, in response to Coriolis forces, built levees which forced lobe and channel deposition to the west, thereby confining sand-rich facies between levees on the east and a growing paleohigh on the west. Reid (1990) proposes a model for Stevens sandstones at Elk Hills, another field in

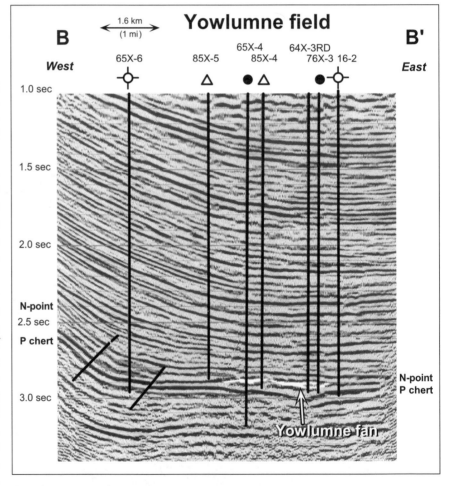

FIGURE 11. Seismic line showing two parallel northwest-to-southeast-trending reverse faults with deposition of the Yowlumne Sandstone on the downthrown, eastern side of the east fault. Failure of the east fault to offset the N-point marker indicates growth of a fault-bounded paleohigh on the west side of the Yowlumne fan contemporaneous with deposition. Location of the line is shown in Figure 5.

the basin, where channel confinement by low-relief anticlines resulted in deposition of overbank facies that grade laterally into basinal shales on the channel margins. This facies relation resembles the east margin of the Yowlumne fan. Interestingly, channels confined by growing subsea anticlines at Elk Hills resulted in sharp, well-defined channel margins with little or no overbank deposition (Reid, 1990), a facies relation analogous to the west margin of the Yowlumne fan.

EXPLOITATION

Remaining oil reserves at Yowlumne are identified using a computer model to simulate fluid flow in the reservoir. The geologic framework for this model was constructed by subdividing reservoir layers A, B, C, D, E, and W into 10 flow units. Digital contour maps (grids) for each flow unit were constructed of net/gross sandstone, $\varnothing_{eff}$, K_{liquid}, and Sw for net sandstone, and the digital data were exported to the model. To verify the model, it was run backward in time, and model output was matched to historical production. Running the model forward enables evaluation of future exploitation scenarios, such as changing producer and injector locations in the well grid or changing waterflood injection rates. Present reserves are evaluated by looking at model output for the present.

A waterflood initiated at Yowlumne 20 years ago will result in ultimate recovery of about 45% of the original oil in place along the main body of the fan (Burzlaff, 1983). Despite this high recovery, a full-field flow model indicates that as much as 480,000 m^3 of oil (3 million bbl) trapped along the thinning fan margins will be abandoned with the current well distribution. These bypassed reserves represent oil banked against the zero edge of the reservoir by the waterflood and reserves not exploited because of decreasing reservoir quality and thickness.

Although the Yowlumne Sandstone is a prolific reservoir, deep drilling depths and high formation temperature result in high drilling, completion, and production costs. When these costs are combined with decreasing reservoir quality and thickness on the east margin of the fan, economic exploitation of the bypassed reserves contained there requires additional challenges beyond those required to exploit the fan axis. Reservoir modeling of the fan margin is an effective tool for minimizing costs by guiding well placement and facilitating risk management.

Using a detailed (partial-field), 722-ac, 11,560-cell model[1] of a prospective area indicated by the full-field model, a high-angle, deviated well, the ARCO 91X-3, was located to exploit 74,000 m^3 (465,000 bbl) of bypassed oil banked against the east margin of the Yowlumne fan (Figure 1). This well was drilled roughly parallel to strike and deviated to the west as much as 85°, with a planned tangential penetration of 600 m (2000 ft) of the distal margin of the Yowlumne fan (Figure 12). The well successfully drilled 335 m (1100 ft) of reservoir with a true stratigraphic thickness of only 55 m (180 ft), but did not penetrate the lowermost, most prospective reservoir layers. With a measured depth of 4360 m (14,300 ft) and true vertical depth of 3815 m (12,515 ft), the 91X-3 is currently the deepest onshore horizontal well in California.

Borehole-breakout data indicate that maximum horizontal stresses in the northeast part of the field trend north-south (Castillo and Zoback, 1994). Also, microseismic data collected during fracture stimulation of the 84-32, a highly deviated well on the west margin of the field, indicate northwest-southeast fracture propagation in the subsurface. Most likely, close proximity of the field to the San Andreas Fault is the main influence controlling these stress orientations. Because the 91X-3 wellbore runs east-west, oblique to perpendicular to the most likely direction of fracture development, the well is favorably located to maximize drainage.

Originally, three "frac jobs" (hydraulic fracture stimulations), spaced 76 m (250 ft) apart, were planned in the 91X-3 to improve connectivity across shale-bounded reservoir layers by inducing fractures northwest-southeast, oblique to the well path. Thus, one deviated well would provide, at lower cost, the same productive capacity as several vertical wells, resulting in more cost-effective exploitation of the fan margin. Unfortunately, problems encountered during running of the production casing, low oil prices, and a corporate merger delayed the planned completion program. Although the well was put on production, no fracture stimulations were attempted. Nonetheless, the use of highly deviated to horizontal wells with multiple fracture stimulations remains an economically viable option for maximizing productivity from the thinning, distal margins of layered, low-permeability turbidite reservoirs.

---

[1]The model contains 10 layers with 34 × 34 cells per layer. Individual cells measure 50 m (165 ft) on a side.

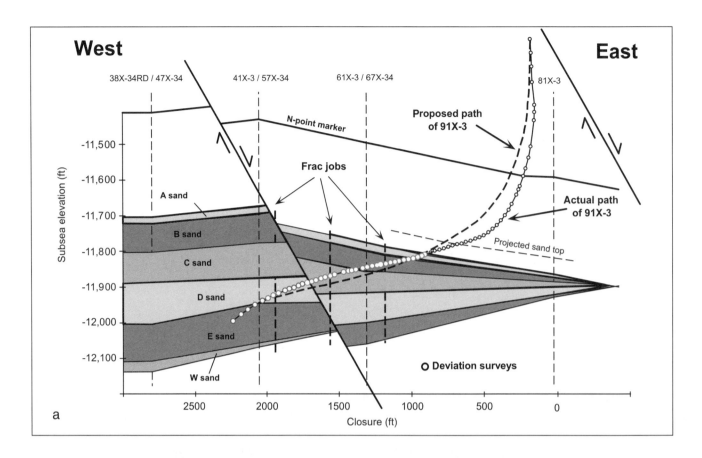

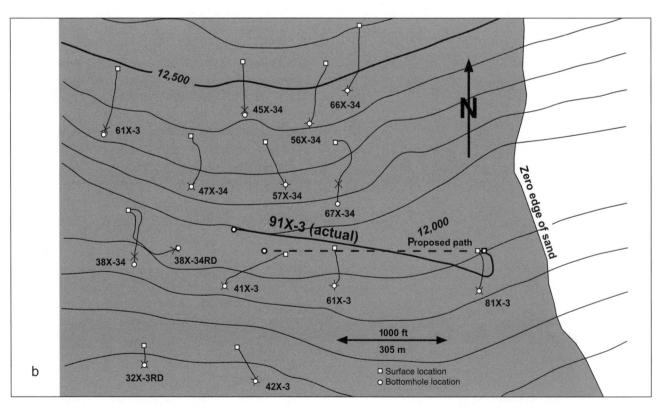

FIGURE 12. Cross section (a) and map (b) of the 91X-3 horizontal well showing proposed and actual drill paths and positions of proposed "frac jobs." The cross section is oriented parallel to the wellbore. The original well plan called for the well to penetrate the base of the Yowlumne Sandstone.

CONCLUSIONS

Interaction of climate with rapid uplift of source areas dramatically increased coarse sediment input into the southern San Joaquin Basin during late Miocene orogeny, resulting in northward progradation of a layered, fan-shaped, Stevens turbidite complex. Deposition of lobe-shaped reservoir compartments in this fan with basinward-stepping geometries is attributed to decreasing accommodation and high sediment flux. Deposition of compartments with left-stepping geometries is attributed to Coriolis forces.

More abundant shale-bearing levee facies characterize the east (right) margin of the fan, whereas sand-rich lobe facies characterize the west (left). This architecture results in decreasing reservoir quality from the fan axis eastward toward the fan margin. Because of high clay content, abundant interbedded shales, and thin reservoir, the east margin of the fan is not an attractive place to drill for oil.

Cost-effective exploitation of bypassed oil trapped along the margins of the Yowlumne fan is facilitated by 3-D computer modeling of rock properties and fluid flow to locate highly deviated to horizontal wells and to design completion programs. For example, flow simulation, constrained by the lobe model presented in this study, was used to locate the deepest horizontal well in onshore California, along the east margin of the fan. A planned completion program of three "frac jobs" would have provided this well, at lower cost, with the same productive capacity as three vertical wells. Unfortunately, mechanical problems and political complications delayed execution of the stimulation program. Nonetheless, exploitation programs using highly deviated to horizontal wells with multiple frac jobs remain a viable, cost-effective strategy for maximizing productivity from layered, low-permeability turbidite reservoirs that characterize thinning fan margins.

ACKNOWLEDGMENTS

This study was funded in part by a grant from the Class-III Reservoir Field Demonstration Program of the U. S. Department of Energy, Contract Number DE-FC22-95BC14940. We would like to thank Tom Berkman, Bill Fedewa, Mark Kamerling, Mike Laue, Tony Marino, Mike Simmons, and George Stewart for their contributions and many thoughtful suggestions. Comments by Roger Slatt and Tim Carr and an anonymous reviewer greatly improved the quality of this paper. Also, we extend our sincere thanks to ARCO Western Energy and the Department of Energy for permission to publish this paper.

REFERENCES CITED

Begg, S. H., D. M. Chang, and H. H. Haldorsen, 1985, A simple statistical method for calculating the effective vertical permeability of a reservoir containing discontinuous shales: Society of Petroleum Engineers, Paper 14271, presented at the 60th Annual Technical Conference and Exhibition in Las Vegas, Nevada, September 22–25, 1985, 15 p.

Berg, R. R., and G. R. Royo, 1990, Channel-fill turbidite reservoir, Yowlumne field, California, in J. H. Barwin, J. G. MacPherson, and J. R. Studlick, eds., Sandstone petroleum reservoirs: Casebooks in earth science: New York, Springer-Verlag, p. 467–487.

Burzlaff, A. A., 1983, Unitizing and waterflooding the California Yowlumne oil field: Society of Petroleum Engineers, SPE 11685, p. 187–194.

Castillo, D. A., and M. D. Zoback, 1994, Systematic variations in stress state in the southern San Joaquin valley: Inferences based on wellbore data and contemporary seismicity: AAPG Bulletin, v. 78, p. 1257–1275.

Clark, M. S., L. M. Beckley, T. J. Crebs, and M. T. Singleton, 1996a, Tectono-eustatic controls on reservoir compartmentalization: An example from the upper Miocene, California: Marine and Petroleum Geology, v. 13, p. 475–491.

Clark, M. S., J. Melvin, and M. Kamerling, 1996b, Growth patterns of a Miocene turbidite complex in an active-margin basin, Yowlumne field, San Joaquin Basin, California (abs.): AAPG Annual Convention Official Program, San Diego, 1996, v. 5, p. A27.

Dorman, J. H., 1980, Oil and gas in submarine channels in the Great Valley, California: San Joaquin Geological Society, Selected Papers, v. 5, p. 38–54.

Graham, S. A., and L. A. Williams, 1985, Tectonic, depositional, and diagenetic history of Monterey Formation (Miocene), central San Joaquin Basin, California: AAPG Bulletin, v. 69, p. 385–411.

Jessup, D. D., and M. Kamerling, 1991, Depositional style of the Yowlumne sands, Yowlumne oil field, southern San Joaquin Basin, California (abs.): AAPG Bulletin, v. 75, p. 368.

Marino, A. W., and S. M. Schultz, 1992, Case study of Stevens sand horizontal well: Society of Petroleum Engineers, SPE 24910, p. 549–563.

Menard, H. W., 1964, Marine geology of the Pacific: New York, McGraw-Hill, 271 p.

Metz, R. T., and J. L. Whitworth, 1984, Yowlumne oil field, in G. W. Kendall and S. C. Kiser, eds., Selected papers presented to San Joaquin Geological Society, v. 6, p. 3–23.

Quinn, M. J., 1990, Upper Miocene sands in the Maricopa depocenter, southern San Joaquin valley, California, in J. G. Kuespert and S. A. Reid, eds., Structure, stratigraphy and hydrocarbon occurrences of the San Joaquin Basin, California: Pacific Section, AAPG Guidebook 65, p. 97–113.

Reid, S. A., 1990, Trapping characteristics of upper Miocene turbidite deposits, Elk Hills field, Kern County, California, in J. G. Kuespert and S. A. Reid, eds., Structure, stratigraphy and hydrocarbon occurrences of the San Joaquin Basin, California: Pacific Section, AAPG Guidebook 65, p. 141–156.

Scott, R. M., and R. W. Tillman, 1981, Stevens Sandstone

(Miocene), San Joaquin Basin, California, *in* C. T. Siemers, R. W. Tillman, and C. R. Williamson, eds., Deep-water clastic sediments, a core workshop: Society of Sedimentary Geologists (SEPM), Core Workshop 2, p. 116–248.

Stolle, J., K. A. Nadolny, B. P. Collins, D. S. Greenfield, and K. A. March, 1988, Landslide oil field, Kern County, California, a success story—an exploration/development case history, *in* J. W. Randall and R. C. Countryman, eds., Selected papers presented to San Joaquin Geological Society, v. 7, p. 1–13.

Tieh, T. T., R. R. Berg, R. K. Popp, J. E. Brasher, and J. D. Pike, 1986, Deposition and diagenesis of upper Miocene arkoses, Yowlumne and Rio Viejo fields, Kern County, California: AAPG Bulletin, v. 70, p. 953–969.

Whelan, H. T. M., 1984, Geostatistical estimation of the spatial distributions of porosity and percent clay in a Miocene Stevens turbidite reservoir: Yowlumne field, California: Master's thesis, Stanford University, Stanford, California, 126 p.

3

Clarke, D. D., and C. C. Phillips, 2003, Three-dimensional geologic modeling and horizontal drilling
bring more oil out of the Wilmington oil field of southern California, *in* T. R. Carr, E. P. Mason,
and C. T. Feazel, eds., Horizontal wells: Focus on the reservoir: AAPG Methods in Exploration
No. 14, p. 27–47.

Three-dimensional Geologic Modeling and Horizontal Drilling Bring More Oil out of the Wilmington Oil Field of Southern California

Donald D. Clarke

City of Long Beach, Department of Oil Properties
Long Beach, California, U.S.A.

Christopher C. Phillips

Tidelands Oil Production Company
Long Beach, California, U.S.A.

ABSTRACT

The giant Wilmington oil field of Los Angeles County, California, on production since 1932, has produced more than 2.6 billion barrels of oil from basin turbidite sandstones of the Pliocene and Miocene. To better define the actual hydrologic units, the seven productive zones were subdivided into 52 subzones through detailed reservoir characterization. The asymmetrical anticline is highly faulted, and development proceeded from west to east through each of the 10 fault blocks. In the western fault blocks, water cuts exceed 96%, and the reservoirs are near their economic limit. Several new technologies have been applied to specific areas to improve the production efficiencies and thus prolong the field life.

Tertiary and secondary recovery techniques utilizing steam have proved successful in the heavy oil reservoirs, but potential subsidence has limited their application. Case history 1 involves detailed reservoir characterization and optimization of a steamflood in the Tar zone of Fault Block II. Lessons learned were successfully applied in the Tar zone, of Fault Block V (4000 m to the east). Case history 2 focuses on 3-D reservoir property and geologic modeling to define and exploit bypassed oil. Case history 3 describes how this technology is brought deeper into the formation to capture bypassed oil with a tight-radius horizontal well.

INTRODUCTION

This paper describes three drilling projects that used computerized mapping, modeling, and simulation programs in conjunction with detailed reservoir characterization and advanced geosteering technology to successfully help tap bypassed, heavy oil in California's mature supergiant, the Wilmington oil field. To date, the Wilmington oil field has yielded more than 2.6 billion barrels of oil of the original 9 billion barrels of oil in place. This paper focuses on how detailed reservoir characterization and 3-D visualization tools applied to horizontal drilling have improved Tidelands Oil Production Company's recovery factor in the "Old Wilmington" or western portion of the field. Background data from THUMS' Long Beach Unit (the field contractor for the eastern portion of the field) are included where needed to provide a thorough field overview. The methods and technologies described herein have the capability of increasing the ultimate reserves of the Wilmington oil field by hundreds of millions of barrels and will find immediate application in other mature oil fields.

BACKGROUND

The Wilmington oil field of southern California (Figure 1), the largest oil field in the Los Angeles Basin (Biddle, 1991), has produced more than 2.6 billion barrels of oil (California Department of Conservation, 1999). Discovered in 1932, it produces from semi- and unconsolidated Pliocene and Miocene clastic slope and basin turbidite sandstones (Henderson, 1987; Blake, 1991). The individual reservoirs are defined by graded sequences of sandstone interlayered with siltstones and shales (Slatt et al., 1993). The entire sequence is folded and faulted (Mayuga, 1970; Clarke, 1987; Wright, 1991). Even the typical rhythmically deposited sequences have lenticular, lobate shapes and are complicated by basal scour, amalgamation, onlapping, and channeling. The result is a sequence of rocks that often appears to be uniform but is not. These complexities also result in permeability variations that hinder the producibility of the sandstones, impact waterflooding, and result in a substantial amount of bypassed oil.

The Wilmington field has been divided stratigraphically into seven producing zones, 52 subzones and, locally, into even finer subsubzones (Henderson, 1987). A serious effort was made to establish stratigraphic continuity in detail as fine as possible. The finer subdivisions are defined as hydrologic bodies or depositional sequences. Many techniques and tools were applied to characterize the thinner sand bodies into unique units, including core description combined with log-rock typing, detailed log correlation, production/injection history matching, bypassed pay-saturation analysis on recent pass-through wells, and reservoir simulation (Otott, 1996; Davies and Vessell, 1997; Davies et al., 1997). Six geologists spent the better part of one year working with thousands of old logs and assorted base maps to sort out a consistent and logical stratigraphic sequence. In addition to the authors, Keith Jones, Mike Henry, Linji An, Rick Strehle, and David K. Davies performed a significant portion of the characterization.

Although we are still learning about the intricacies of Wilmington field's reservoirs, today's computerized visualization tools, combined with advanced measurement-while-drilling (MWD) data, have contributed significantly to the collective knowledge base about this field. We thus confidently conclude that the field's poorly drained sandstones remain ideal targets for horizontal drilling.

HISTORY OF THE LONG BEACH UNIT

The Wilmington oil field is a faulted asymmetrical anticline. The reservoirs of the 10 larger fault blocks in

FIGURE 1. Location map showing oil fields in the Los Angeles Basin of southern California with the Wilmington oil field shown in light gray.

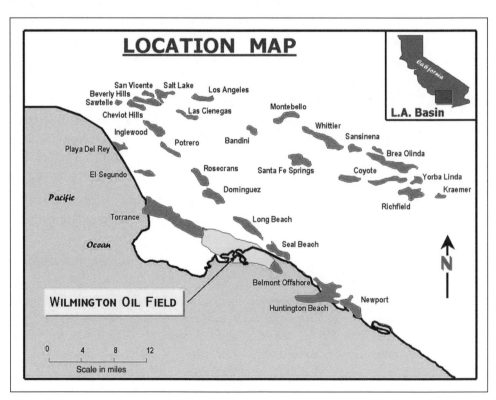

the field have been managed independently. The city of Long Beach, Department of Oil Properties, operates most of them (Figure 2). The Wilmington Townlot Unit (WTU), a portion of the westernmost Fault Block I, is operated by Magness Petroleum Company. Pacific Energy Resources operates a portion of Fault Block II. Tidelands Oil Production Company is the field contractor for most of the western portion (Fault Blocks I–V). THUMS Long Beach Company is the field contractor for the eastern part of Wilmington oil field (Fault Blocks VI–90N), which is called the Long Beach Unit (LBU). The LBU was originally produced with more than 1000 wells drilled between 1965 and 1982 (Otott and Clarke, 1996). Despite the very long completion intervals used and water injection for pressure support, oil remained in pockets of tight, thin sandstones, as well as in areas with poor injection support. Widely ranging permeabilities and faulting caused typically viscous oil (12.5° to 16° API) to be left behind in sandstone units 6 to 15 m thick.

An additional 460 wells were drilled in the Long Beach Unit from 1982 to 1986 using a subzone approach to improve sweep efficiency, which allowed another 25.4 million cubic meters (m^3), or 160 million barrels (bbl), to be produced. After 1986, bypassed sandstones were selectively perforated in even finer intervals. The first horizontal well was completed in November 1993 as part of the optimized waterflood project. Thirty-nine more horizontal wells have been drilled since then, using a combination of computerized mapping and digitized injection surveys to identify the dominant, unswept flow units. Reservoir exploitation was enhanced by geosteering supported by logging while drilling (LWD) and real-time analysis of MWD/LWD data to update cross sections.

Typically, the wells were placed 3 to 5 m (10 to 15 ft) below the top of the sandstone. Initial oil-production rates from the best horizontal wells exceeded 95.4 m^3/day (600 bbl/day) and about 47.7 m^3/day (300 bbl/day) from the average wells, at 80% water cut, stabilizing at about

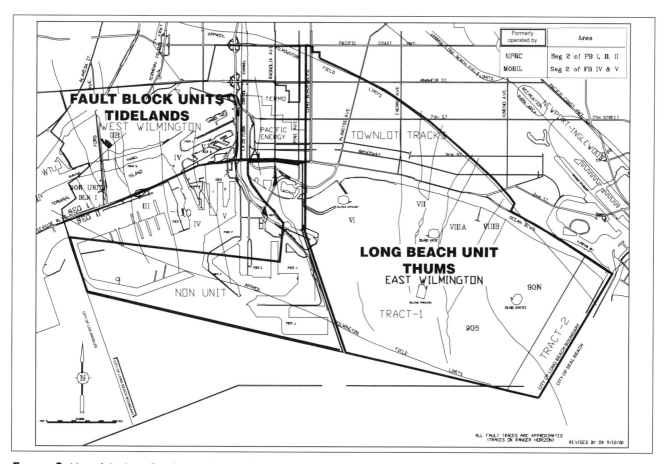

FIGURE 2. Map of the Long Beach area showing the areas of the Wilmington oil field where Tidelands Oil Production Company (western portion) and THUMS Long Beach Company (Long Beach Unit) are the field contractors. The city of Long Beach, Department of Oil Properties, operates both portions. The limits of the Wilmington field, along with those of the Long Beach field and the Seal Beach field, are shown. The coastline, harbor areas, breakwater, and oil-drilling islands are shown for reference. Pacific Energy Resources operates the polygonal area between the properties. The western notch is the Wilmington Townlot Unit (WTU), which is operated by Magness Petroleum Company. Exxon operates the southeast area (not delineated) as State Lease 186 Belmont Offshore. This portion is under abandonment.

15.9 m³/day after 300 days. Total unit oil production is 6042 m³/day (38,000 bbl/day). Blesener and Henderson (1996) describe several of the new engineering technologies that have been applied to the Long Beach Unit. These include coiled-tubing drilling, drill-cuttings injection, and reclaimed-water injection. In 1995, the LBU ran a 3-D seismic survey to help define the subsurface in greater detail (Otott et al., 1996). The survey did not provide the desired results, but it did serve as a valuable tool for deep work. Several exploratory prospects were identified. One or more of these may be drilled in 2003.

THUMS was purchased from ARCO by Occidental Oil and Gas Corporation in May 2000. In 2002, THUMS conducted a 3-D vertical seismic profile (VSP) in the area between Islands Freeman and Chaffee. These data are being processed.

HISTORY OF THE OLD WILMINGTON AREA

The history of the Wilmington oil field has been detailed by Mayuga (1970), Ames (1987), and Otott and Clarke (1996). More than 5000 wells have been drilled conventionally in the 70 years Old Wilmington has been on production. The entire field is on secondary recovery, and oil production is down to 1113 m³/day (7000 bbl/day) with an average water cut of 96.9%. Because of the steep, 14%-per-year decline, it was decided to investigate new ways to produce more oil. As part of this effort, Tidelands Oil Production Company has drilled 14 horizontal wells since 1993 in four heavily drilled (3000-plus wells) fault blocks (Phillips and Clarke, 1998; Phillips et al., 1998). The first horizontal well project was a "Huff 'n' Puff" conducted in 1993 in Fault Block I Tar zone. Two 274.3 m- (900 ft-) long horizontal wells were drilled into the D_1 sandstone. The second project was a steamflood in Fault Block II Tar zone. In 1995 for project 2, four horizontal wells were drilled on average 488 m (1600 ft) within the D_1 sandstone. A Fault Block IV Terminal zone waterflood well was drilled 335.3 m (1100 ft) within the Hxb sandstone in 1995 as the third project. Again in 1995, five horizontal wells were drilled into the Fault Block V Tar zone as part of a steamflood. The wells were drilled on average 457.2-m (1500 ft) horizontally within the S_4 sandstone. In 1997, a 304.8-m (1000-ft) horizontal well was drilled into the Hx_0 sandstone of Fault Block V Terminal zone to complete the fifth project.

For each project, the horizontal laterals for the waterflood wells were placed at the top of the sandstone to recover attic reserves. The laterals for the steamflood wells were placed at the bottom of the sandstone to maximize capture of oil through steam-assisted gravity drainage.

Except for the first project, 3-D modeling and visualization were used from planning through completion. To

be effective, horizontal wells require precision placement. The studied areas required significant geologic evaluation and characterization. The area then was modeled with software that provided 3-D visual displays of stratigraphic and structural relationships and enabled excellent error checking of data and grids in 3-D space. The geologic model was revised and modified in 3-D space. The 3-D model provided a visual reference for well planning and communicating the spatial relationships contained within the reservoir. Accurate 2-D and 3-D visualization was used for interpreting the LWD response and monitoring well progress while drilling. Maps and section plots brought to the rig site allowed the drilling team to relate to the geology, thus providing a strong confidence factor. Accurate and rapid postdrilling analysis for completion-interval selection and LWD analysis completed the process.

CASE HISTORIES

Three case histories are presented herein. Case history 1 describes a thermal-enhanced recovery project that expanded on an existing steamflood project. The expansion area was subjected to detailed characterization with 3-D modeling and visualization before completion of the development project. The technologies developed in the steamflood project were applied to the areas in Fault Block V and are described as case histories 2 and 3. Figure 3 shows the location of the three case histories.

CASE HISTORY 1: FAULT BLOCK II, TAR ZONE STEAMFLOOD PROJECT

Case history 1 is in the Tar zone of Fault Block II (Figure 4). The Tar zone of the lower Pliocene Repetto Formation (Figure 5) is the shallowest of the major oil-producing zones in the Wilmington field. It has been interpreted to consist of large, lobate, submarine-fan deposits (Redin, 1991), which are composed of interbedded siltstones, shales, and unconsolidated fine- to medium-grained arkosic sandstones. The sand bodies were deposited as a set of compensating turbidite lobes, as opposed to the sheet sandstones (or larger sheet lobes) that occur lower in the section. This section is composed of smaller sandstone lobes that are generally limited to less than two miles in lateral extent. The sequence is also complicated by onlap and channeling. In Fault Block II, the Tar zone is 76–91 m (250–300 ft) thick and occurs at depths of 697–848 m (2300–2800 ft) below sea level. The T and D sandstones (Figure 5) are the best developed and most productive. Oil gravity ranges from 12° to 15° API, with a viscosity of 260 cp at the ambient reservoir temperature of 51.7°C (125°F).

Fault Block IIA is located in the western portion of

the field between the Wilmington and the Ford faults (Figure 3) and is downplunge from the crest of the Wilmington structure. The fault block is bounded to the west by the Wilmington fault and to the east by the Cerritos fault, both of which are permeability barriers (Figure 3). The faults show normal displacement with vertical offsets that range from 15 to 30 m (50 to 100 ft), but they may have complex histories of movement. In addition, several smaller-scale faults (Ford, Ford A-1, Ford A-lB) exist in the southeastern portion of the block (Figure 4). These faults exhibit vertical offset on the order of 4.5–9.0 m (15–30 ft) and are only partially sealing. The north and

south limits of production are defined by oil-water contacts within the productive sandstones.

A Tar zone steamflood in Fault Block II was initiated in 1982 and expanded in 1989, 1990, 1991, and 1993. In 1995, a plan was created to expand the steamflood to the south (Figure 4). Instead of the inverted seven-spot pattern used in the earlier phases, it was decided to use horizontal wells, carefully laid out so that each horizontal well would replace three or four vertical wells. Five temperature-observation wells (OB2-1 to 5, Figure 4) would be interspersed to monitor distribution of the thermal energy.

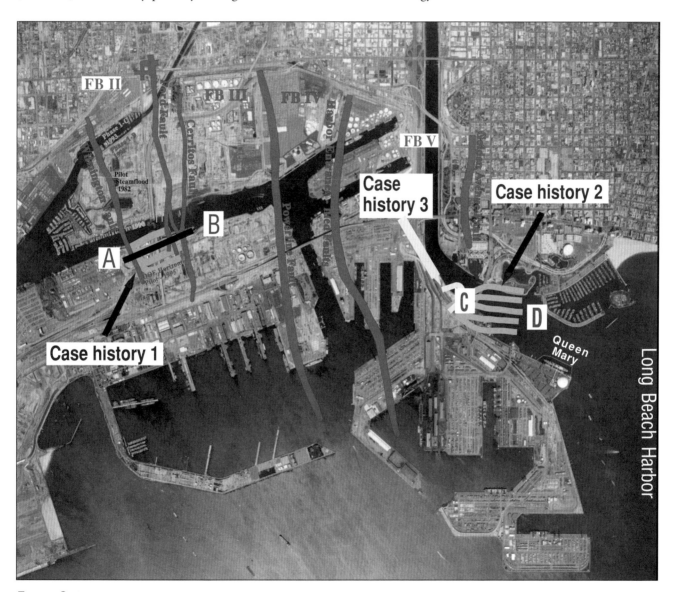

FIGURE 3. Aerial photograph of Long Beach/Los Angeles Harbor showing the location of the three horizontal-well projects. North is to the top. On the left (west) is the steamflood project (case history 1) with four horizontal wells placed in the Tar zone below Terminal Island. The location of cross section A-B is shown. Case history 2 consists of five horizontal wells that cross under the Los Angeles River channel and are part of a Tar zone steamflood project in Fault Block V (FB V). The location of cross section CD is shown. Case history 3 involves a tight-radius horizontal well that lies stratigraphically below case history 2 in Fault Block V in the upper Terminal zone. Oil activities coexist with the busiest harbor in the country, and the Fault Block V steamflood is below Long Beach's new $180 million aquarium.

The four horizontal wells were drilled into the bottom of the 18.3 m- (60 ft-) thick D_1 sandstone. Two steam injectors and two producers were placed about 122 m (400 ft) apart horizontally as part of a pseudosteam-assisted, gravity-drainage project. This innovative Fault Block II steamflood project received partial funding from the U.S. Department of Energy as part of a class III mid-term project (Koerner et al, 1997; U.S. Department of Energy, 1999).

The existing maps had insufficient detail for the planned development. The only way to obtain success was to perform a detailed geologic analysis of Fault Block II. The well tops, coordinates, and fault data were entered into a computer modeling package and, after a rough 3-D model was constructed to assess the problems, it was clear that a complete revision of the geology was necessary.

The existing six subzone intervals were further divided into 18 subsubzones, and the faults were reevaluated. A team of geologists spent months on detailed log work to define the 18 horizons and six faults. The log data ranged from electric logs from the 1930s through complete log suites of the 1980s. Each subsubzone was hand-mapped

FIGURE 4. Map of Fault Block II area showing the location of the Wilmington oil field Tar zone, Fault Block II-A steamflood project. Each phase of the project is shown in color code. The project described here focuses on the southern area where the four horizontal wells were drilled. Cross section A-B follows the well course of UP955.

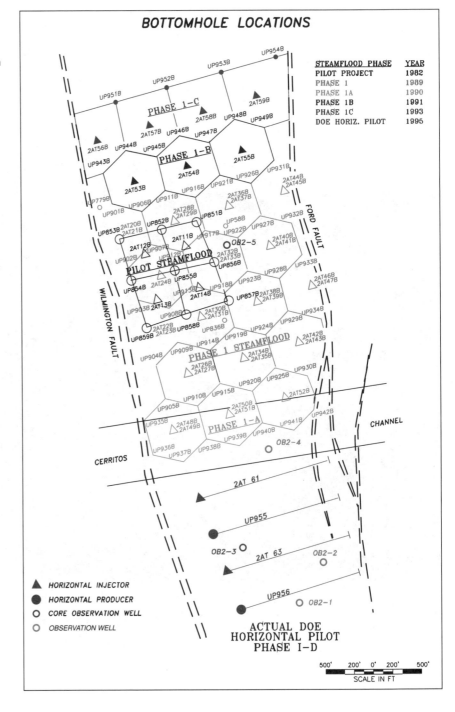

for lateral extent. The faults and horizons were then modeled three-dimensionally and compared to the original interpretation.

A significant amount of the well planning was performed using this detailed, 3-D working model, which made visualization of the inconsistent data very easy. The data inconsistencies came from differentially subsiding horizons caused by intraformational compaction from oil withdrawal over a 60-year period and an assortment of data entry and coordinate conversion errors. These errors were rapidly identified and corrected.

Subsidence was probably the most difficult problem to solve. The intraformational compaction of the producing reservoirs varied over time and directly impacted the surface (and the distance to the producing horizons). From 3.7 to 6.7 m (12 to 22 ft) of surface elevation was

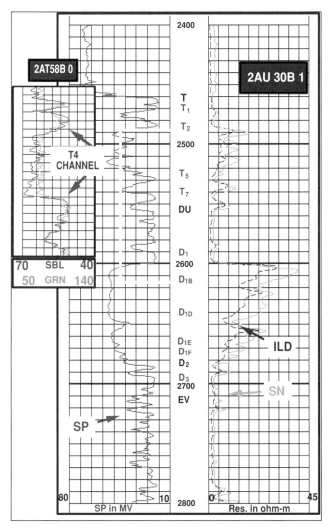

FIGURE 5. Type logs for Tar zone, Fault Block II (Well 2AU 30B 1). Original markers are shown in black, and the newly picked markers in red. The inset shows the T4 channel from well 2AT58B. Note the good saturation between D_1 and D_{1E}. The location of these two wells is shown in Figure 10.

lost above the proposed horizontal lateral locations (Figure 6). The subsidence varies, increasing from west to east toward the center of an elliptical subsidence bowl, where the maximum subsidence to date is 8.8 m (29 ft).

To compensate for the errors, data were adjusted for ground-level change and internal compaction. These adjustments are time dependent. For example, a well drilled in 1940 could have been drilled to 762 m (2500 ft) below sea level to reach the T marker. The same well drilled today to the same X, Y position might require drilling to 768 m (2520 ft) below sea level to reach the T marker (Phillips, 1996). The ground level is lower now because of subsidence, and the depths to the other markers also are different (intraformational compaction). The stratigraphic section has been compressed. Figure 7 illustrates the corrections that are applied.

After data were modified, the mapping software facilitated the rapidly generated new geologic models by using the predefined geologic criteria. This data was quickly integrated into a more comprehensive structural model (Figure 8), which was edited and modified where necessary. The 3-D model was recalculated many times during this iterative process. Not only was the resulting model excellent at revealing subtle differences in the geology, it also was an invaluable tool for finding data errors.

When the acceptable 3-D deterministic model was established, cross sections along the well courses were constructed and used for geosteering (Figure 9). The cross sections derived from the model proved extremely accurate and were used extensively. The combination of detailed sequence characterization and 3-D modeling allowed us to accurately map a previously unrecognized channel (Figure 10) and onlap (Figure 11).

The computerized 3-D displays greatly enhanced communication among the geologist, the petroleum engineer, and the driller. The geologist could rotate, slice, and change the look of the model to improve the visualization. The geologist also displayed the offset log information on a cross section along a well course that had been scaled up to match the real-time LWD logs. This was invaluable during drilling because the geologist could accurately follow the drill bit by plotting the MWD data directly onto the computer-generated cross section.

In Fault Block II, the bottom of the D_1 sandstone was targeted. Instantaneous drilling rates as great as 183 m/hr (600 ft/hr) were achieved because the accurate geologic model enabled the well site team to bypass slow-drilling, problematic shales and otherwise to modify the drilling program for improved efficiency. In the end, steam-assisted, gravity-drainage horizontal wells UP-955, UP-956, 2AT-61, and 2AT-63 were successfully drilled within a 4.5-m (15-ft) target window (Figure 12).

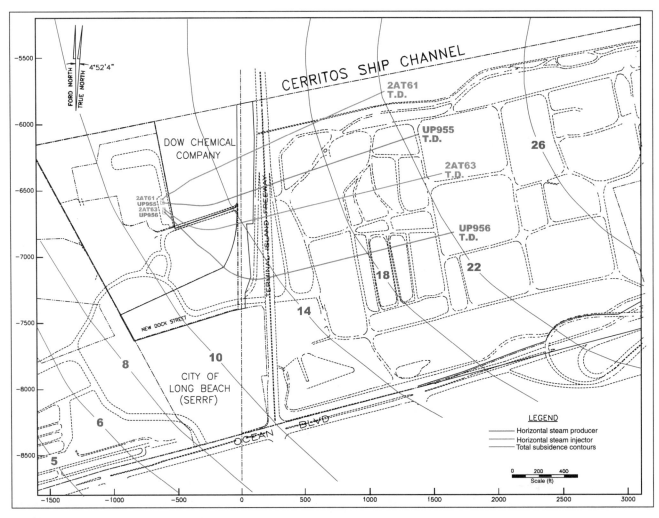

FIGURE 6. Map showing location of Fault Block II horizontal wells in relation to the subsidence bowl. Red contour lines are total elevation loss in feet. The four horizontal wells were completed in an area of 14 to 22 ft (4.3 to 6.7 m) of subsidence. Producing well courses are shown in blue, and steam-injection well courses are shown in green. The line of cross section A-B parallels well UP 955.

FIGURE 7. Three components of subsidence correction. (A) Adjustments must be made for rock compaction that occurs after a well is drilled. (B) The kelly bushing must be corrected for subsidence that occurred prior to drilling. (C) Finally, an adjustment must be made within the formation to correct the overlying sediments for compaction that has occurred below.

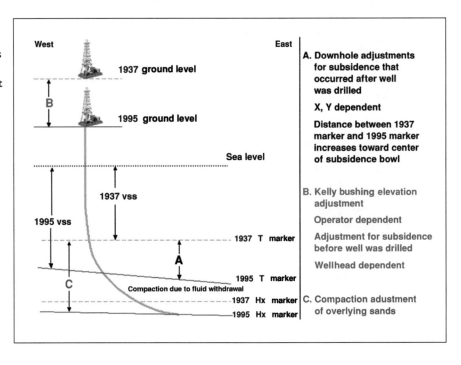

The steamflood project had to be terminated in January 1999 because ground elevations had dropped nearly one foot. Subsidence has been a historical problem for the city of Long Beach, and the continuation of activities that may cause subsidence is not permitted by the city.

Although this project was marginally economic, we consider it a technical success. The project started with a steam/oil ratio of 7; by the time the steam project was shut down, the steam/oil ratio was 14. More drilling would have helped greatly, but expansion was not possible at the time. In October 1999, flank wells were converted to cold-water injection. A 3-D deterministic reservoir simulation model that calculated mass balance and heat balance was used for injection conversion. Subsidence was halted, and by September 2002, the area was very profitable, producing 179 m^3/day (1130 bbl/day) net with a gross of 4515 m^3/day (28,400 bbl/ day).

FAULT BLOCK V PROJECTS (CASE HISTORIES 2 AND 3)

The next step was to see if these techniques could be applied to Fault Block V. There are two horizontal-well projects in this block. The first is in the Tar zone, where five horizontal wells were drilled. The second is in the Upper Terminal zone, where a single well was drilled into the thin, shaley Hx$_0$ sandstone. The accuracy of the 3-D geologic model and the usefulness of the computerized tools used to extract information from the model greatly enhanced the success of both projects.

Case History 2 (Tar Zone)

As with the Fault Block II project, the more than 60-year-old electric logs were reviewed and recorrelated, dividing the Tar zone into 14 subsubzones. The log (Figure 13) shows a portion of the stratigraphic section from probe-hole well FJ-204. The S$_4$ sandstone was chosen as the target because it shows the highest resistivity (oil saturation) and is the thickest, continuous, clean sandstone across the fault block. A probe hole was drilled to verify reserves, not for horizontal placement.

A deterministic geologic model was created, from which the maps and cross sections were extracted and used to geosteer the horizontal wells. The modeling was more straightforward than the earlier project, because the area where the horizontal wells were planned is unfaulted (Figure 14).

The experience gained in Fault Block II and improvements to the software made modeling still easier. Areas of no data were controlled by adding interpretive "ghost" points through the 3-D viewer, then reconstructing the model. This interpretative technique cut modeling time significantly.

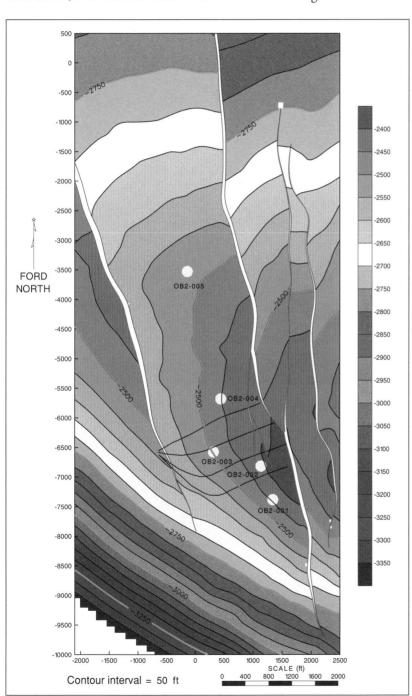

FIGURE 8. Structure map of the T marker in Fault Block IIA. Observation and horizontal wells are shown. Contour intervals shown are 50 ft (15.2 m) from −2400 to −3400 ft (−731.5 to −1036 m) below sea level.

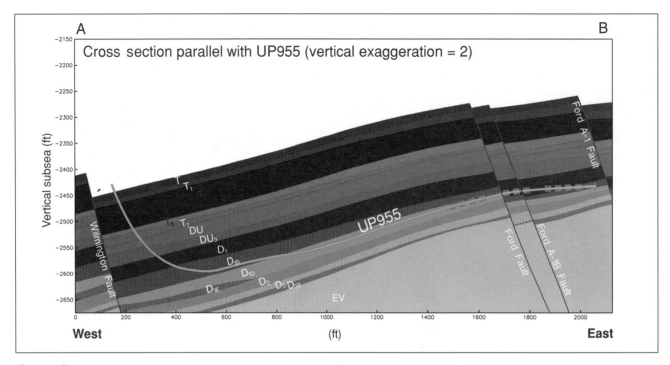

FIGURE 9. Cross section A-B, which follows the well course of UP 955. Perforations are shown on well course. The onlap of the D_{1E} is shown; no detail below EV is shown. The section is scaled in feet and has a 2x vertical exaggeration. The locations of the cross section and well UP 955 are also shown in Figures 3, 4, 6, and 8.

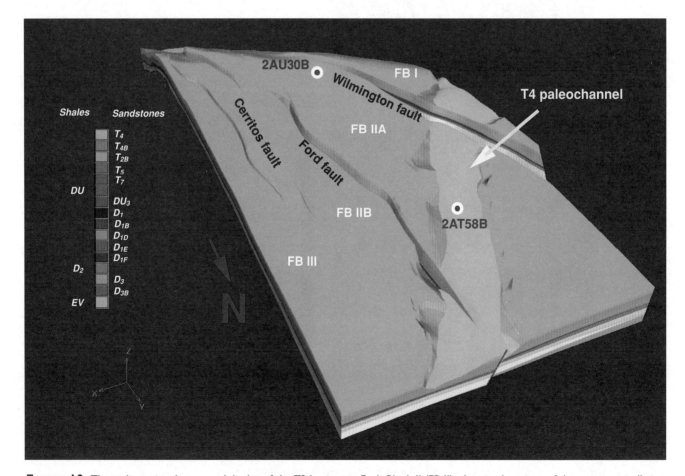

FIGURE 10. Three-dimensional structural display of the T2 horizon in Fault Block II (FB II), showing locations of the two type wells in Figure 5. The T4 paleochannel cuts through several horizons.

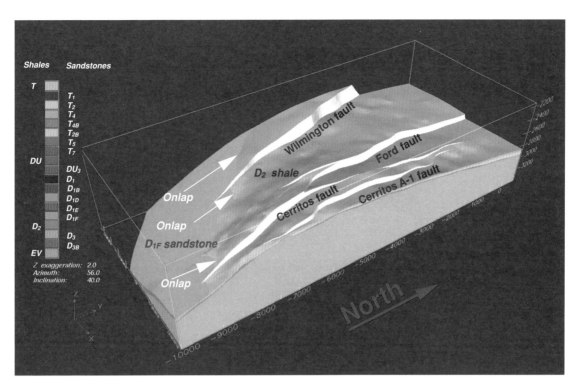

FIGURE 11. Three-dimensional display of the D_{1F} onlap onto the D_2 shale in Fault Block II. The figure has a 2x vertical exaggeration, and the units displayed are in feet.

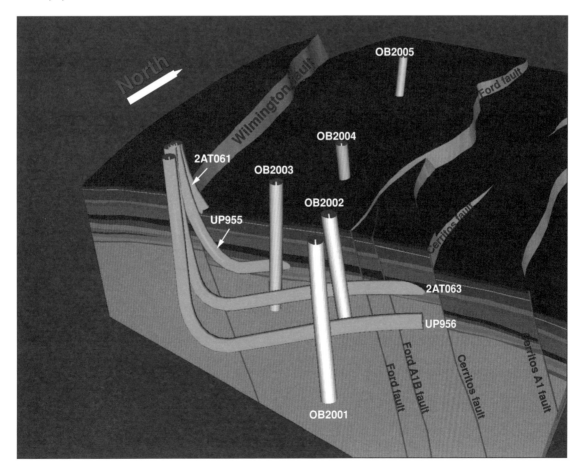

FIGURE 12. Three-dimensional display of Fault Block II showing locations of wells drilled for the steamflood project. The figure has a 2x vertical exaggeration, and the wellbores are greatly exaggerated to enhance the visual impact of the well pattern.

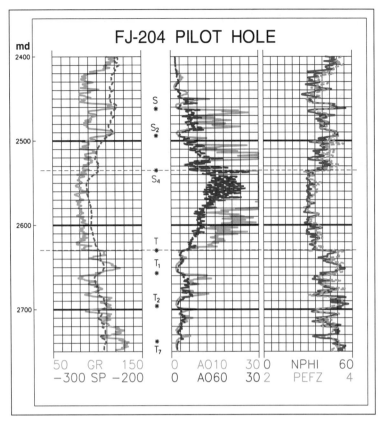

FIGURE 13. Well log for FJ-204 pilot hole in Fault Block V Tar zone. The S and T section is shown. The S₄ sandstone is the target. The location for this well is shown in Figure 14.

Data from one area of the model indicated an anomalous structural low. The survey and log picks appeared to be correct for a well located in the area of this low. The data point was honored, and horizontal well J-201 was drilled into the area. It was apparent from the LWD curve separation and bed boundary intersections that the T shale was shallower than the model indicated. The offending well data were removed, and the model was rebuilt based on the horizon picks from well J-201. Because this remodeling can now be done in almost real time, the geologist revises the model as drilling proceeds. An improved model is built if needed as each new well is completed. Well J-201 did not go as planned; it was difficult to determine the completion interval until the other horizontal wells and their perforations were displayed in 3-D (Figure 14).

The 3-D model in Figure 14 is bench cut and shows the five horizontal wells and their perforations. The goal was to keep the wells parallel to the top of the T shale to maximize recoverable reserves from the superjacent S₄ sandstone. The maps, cross sections, and geologic model all were used to place the horizontal wells accurately. Figure 15 shows the cross section for well J-203.

Overall, the Tar V drilling project (case history 2) was a major technical and economic success. Based on what was learned in Fault Block II and the accuracy of the 3-D model, the drilling team was able to plan and drill with confidence. It was easy to anticipate the highs and lows of the horizons and the locations of bed boundaries. No wells were plugged back for geologic reasons, and drilling time was reduced by spreading out survey lengths, using less time for correctional sets, and rotating the tool string while drilling a large percentage of the horizontal section. Roller reaming prior to running casing was eliminated by avoiding shales, thus allowing reaming with the bit already in the hole. In addition, no pilot holes (except for FJ-204) were necessary. As a result, time and money were saved.

The drilling team appreciated having visuals from 3-D modeling at the rig site because they stimulated better feedback and established a clearer understanding of the geology encountered. The team could see what a particular directional tool set accomplished and thus refine drilling techniques for added efficiency. Previously, drillers had relied only on numbers, which were much less intuitive and informative.

The Tar V horizontal well budget was based on the Fault Block II wells. An average savings per well was realized of U.S. $12,400 on directional costs and U.S. $18,000 as a result of fewer drilling days. In total, U.S. $152,000 was saved on the five horizontal wells drilled. Because of the monetary savings and the drilling team's confidence in the 3-D model, all of the laterals were extended an extra 12% on average, effectively increasing the producible area and adding 60,734 stock tank m³ (STCM), or 382,000 stock tank barrels (STB), of oil.

The five horizontal wells were steam cycled and placed on production (Figures 16, 17, and 18). Two of them, FJ-204 and FJ-202, were placed on permanent steam injection. A-186 3, A-195 0, and A-320 0, each well more than 30 years old, remained on production within the steam-project boundaries. As of March 1996, they had averaged 2.5 m³/day (16 bbl/day) net with 31.8 m³/day (200 bbl/day) gross, at an average water cut of 92%.

When the horizontal project was initiated, this area had only about five years of remaining economic life under waterflood, and recoverable reserves were estimated at 11,924 m³ (75,000 bbl). The average pool-water cut prior to steaming was 95%. The water cut in the project area was 81% in 1998 and 92% in July 2000; another 270,283 m³ (1,700,000 bbl) of reserves has been added to the Tar V pool.

Steam communication to the existing waterflood wells, from cyclic steam injection into wells FJ-202 and FJ-204, resulted in a six- to tenfold net production increase in the old waterflood wells (Figure 17). Peak annual production rates under steam drive were forecast at 93.8 m³/day (590 bbl/day) for the horizontal project. During the first four months of 1998, the average oil production was 111 m³/day (698 bbl/ day). In July 2000, the average oil production was 38.5 m³/day (242 bbl/ day). The production rates should be several times greater, but the performance of each well has been hindered by fluid levels exceeding 457.2 m (1500 ft). The high fluid level suppresses oil production and cools the produced fluids, resulting in lower recoveries. The success of the program is reflected in Figures 16, 17, and 18, which show how the project area has changed over time. Note in Figure 16 that prior to steaming, the average net was about 2.5 m³ day (16 bbl/day). By January 1998 (Figure 17). the average net was more than 23.9 m³/day (150 bbl/day). In August 2000, the average net was still more than 15.9 m³/day (100 bbl/day).

Three-dimensional techniques contributed significantly to the success of the Tar zone horizontal project. Assuming a 50% recovery factor, every foot above the target is equivalent to 2524 STCM (15,876 STB) in lost reserves (Phillips, 1996). At U.S. $14/bbl oil, an error of as much as 1.5 m (5 ft) vertically would equate to U.S. $ 1.1 million in lost revenue.

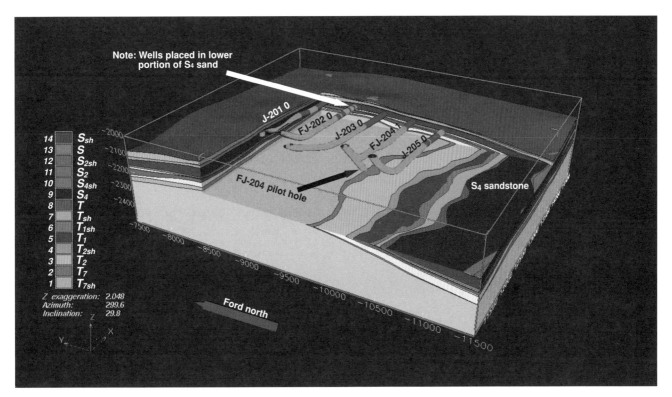

FIGURE 14. Three-dimensional bench cut of the Fault Block V Tar zone showing the steamflood project in the S_4 sandstone. Perforation intervals are shown in red. This nearly horizontal area has a 2x vertical exaggeration.

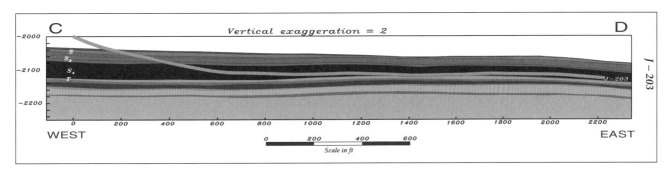

FIGURE 15. Cross section C-D along well course J-203. This well was placed as close to the bottom of the S_4 sandstone as possible to increase recovery. See Figure 14 for location of well J-203.

FIGURE 16. Fault Block V Tar zone water-flood configuration. Production and cut data are shown prior to steaming operations. Each production well is labeled in green with the name, gross production, net oil production, and water cut. The injection wells are labeled in red with well name, injection rate in barrels per day, and surface pressure. Structural contours are vertical subsea on the S sandstone with a 20-ft contour interval. Ford north refers to a local Cartesian coordinate system.

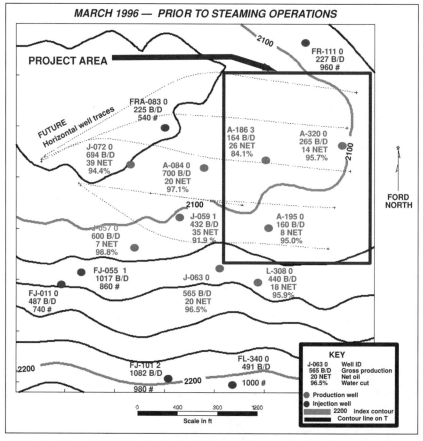

FIGURE 17. Fault Block V Tar zone water-flood configuration, with production and cut data for January 1998.

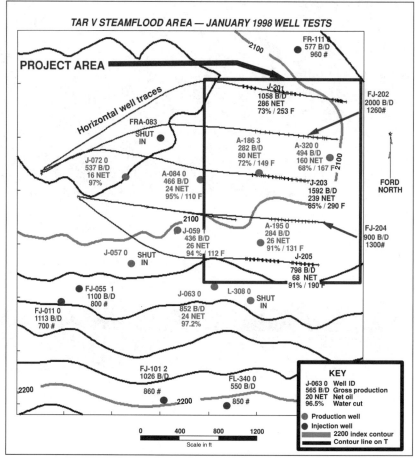

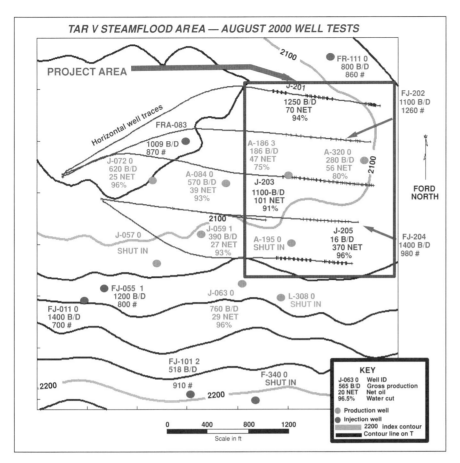

FIGURE 18. Fault Block V Tar zone waterflood configuration, with production and cut data for August 2000.

CASE HISTORY 3: UPPER TERMINAL ZONE: Hx_0 THIN-SANDSTONE SEQUENCE

The Hx_0 sandstones of Fault Blocks V and VI were reviewed as part of a U.S. Department of Energy (DOE) class III short-term project (Phillips, 1998). The project proposed using new reservoir characterization tools and techniques to exploit bypassed oil. The new technologies included detailed reservoir characterization; 3-D geologic modeling; geosteering in thin, heterogeneous beds; and modeling the LWD responses (MacCallum et al., 1998).

A deterministic geologic model was created to define the Hx_0 layer and the horizons above and below it (Hx_1 above, Hx_2 and Hx below). The sandstone percentage was calculated for each data point. A 3-D property model was created by gridding the sandstone percentage in 3-D space using the top and bottom of the Hx_0 as confining surfaces (Figure 19). The original oil saturation (So) was property modeled similarly to identify target areas for exploitation (Figure 20).

A display of the So model and wells drilled in the 1980s clearly showed that Fault Block VI was effectively drained, but Fault Block V still had reserves. The difference between the original So and that indicated by the 1980s wells was quantifiable. This is easily seen in Figure 20 for wells A-160 and A-189. The calculated So is less

than 40%, whereas the property model shows the original So to be 60%. The So calculated from the old wells was decremented, the two data sets were combined, and Fault Block V was again property modeled (Figure 21). The sandstone percentage model and the So model were combined, and the original oil in place was calculated to be 540,566 STCM (3.4 million STB). The current oil in place was calculated, and the reserves were reduced to 445,172 STCM (2.8 million STB) (Phillips and Clarke, 1998). Obviously, significant reserves remain.

Based on the geologic model, the block engineer proposed that a horizontal well be drilled within and adjacent to the modeled area along the structural high. An existing wellbore was sidetracked with a horizontal lateral to capture hydrocarbon reserves not economically recoverable with conventional methods. Idle well J-017 was selected for drilling the high dogleg horizontal well, and a production rig was configured for drilling to keep costs to a minimum. By investigating the area west of the original Hx_0 project area, it was determined that the target sandstone thins and shales out to the west, thus reducing oil saturation. Electric logs from wells penetrating the area as far as 305 m (1000 ft) to the west were correlated, and a second 3-D geologic model was created.

A facies boundary was delineated to constrain the planned well course within the higher water-saturation

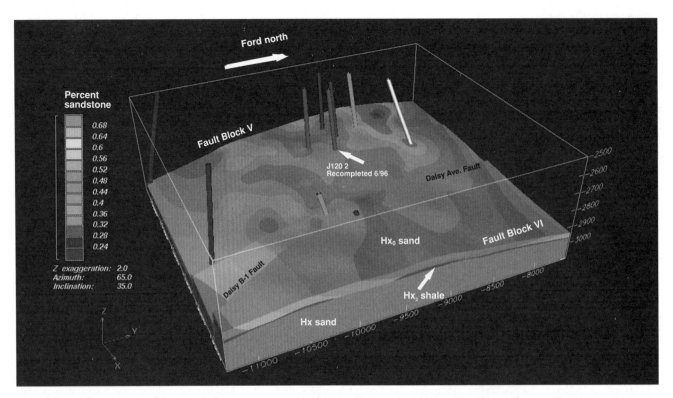

FIGURE 19. Three-dimensional display of the percentage of sandstone in the Hx_0 to Hx_2 interval of the Terminal zone in Fault Block V. Cartesian coordinate system is in feet, and vertical exaggeration is 2x.

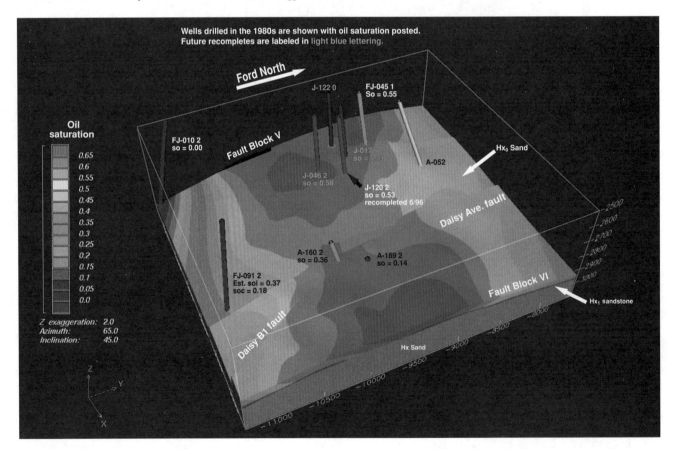

FIGURE 20. Three-dimensional display of qualitative oil saturation (So) for the Hx_0 sandstone of the Terminal zone in Fault Block V. Cartesian coordinate system is in feet, and vertical exaggeration is 2x. Oil saturation is posted with each well. Soc = current oil saturation; soi = initial oil saturation; So = oil saturation calculated from 1980s logs.

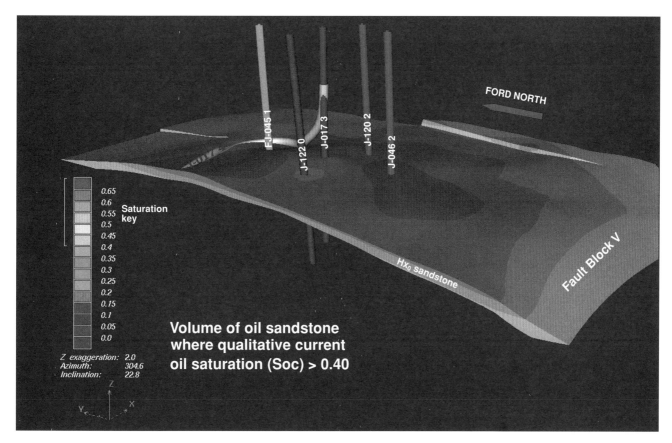

FIGURE 21. Three-dimensional display of the Fault Block V Terminal zone showing the J-017 well and surrounding wells. The oil saturation is mapped within the Hx_0 sandstone. This is essentially a net-pay map for the Hx_0 sandstone.

(So) target. The Hx_0 layer was subdivided, and two sandstone lobes were identified within the Hx_0 layer. The Hx_{0J} and Hx_{0B} horizons were defined and added to the 3-D model. Maps and cross sections were extracted from the 3-D model and used for well planning (Figures 21 and 22).

A cross section along the well course was created for the geologist. The directional vendor required three linear cross sections for drilling because the well plan showed a U-turn (Figure 22). Stratigraphic sections consisting of adjacent wells also were created to help in geosteering. Both pasteup and digital varieties were used.

Structure maps were created on the Hx_1, Hx_0, and newly defined Hx_{0J} (Figure 23) and Hx_{0B} sandstones. These plots, as well as the 3-D model, were used for geosteering. Ultimately, they also were used for directional control because of rig-site problems.

A recently introduced, probe-based multiple-propagation-resistivity (MPR) device was used to provide LWD geosteering as well as directional information. This resistivity sensor was part of a slim-hole, positive-pulse-type MWD/LWD system that was used instead of carrier wave tools because of its smaller size (the tool diameter is $4\frac{3}{4}$ in). These newer tools have well-integrated surface equipment, are battery-powered, and provide more reli-

able telemetry signals. The MPR tool is a four-transmitter, two-receiver array that provides a total of eight compensated resistivities at 2 MHz and the deeper-reading 400 kHz in boreholes as small as $5\frac{7}{8}$ in. For additional geosteering control, an inclinometer and gamma-ray scintillation detector are placed below the MPR sub (MacCallum et al., 1998). The well was successfully placed within two sandstone lobes of the thin sequence by geosteering using the LWD data.

The interval is thin and shaley (total thickness of 5.2 m [17 ft]), and the LWD showed the anisotropic effect throughout the log (Figure 24). The resistivity response in anisotropic conditions is similar to conductive invasion in that the short-spaced measurement reads less than the long-spaced for both the phase difference and attenuation resistivity measurements. However, the shallow-reading, phase-difference resistivity curves measure a higher resistivity than the deep-reading attenuation curves for both frequencies and spacings. This curve order is not indicative of conductive invasion but of anisotropy (Meyer et al., 1996). The presence of anisotropy plus formation heterogeneity complicated the interpretation of the LWD data so that the geosteering team had to rely significantly on the geologic model.

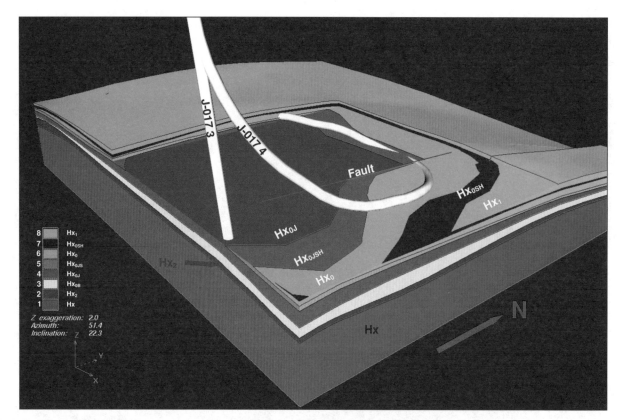

FIGURE 22. Three-dimensional bench-cut display showing the J-017 4 tight-radius sidetrack into the Hx_0 sandstones.

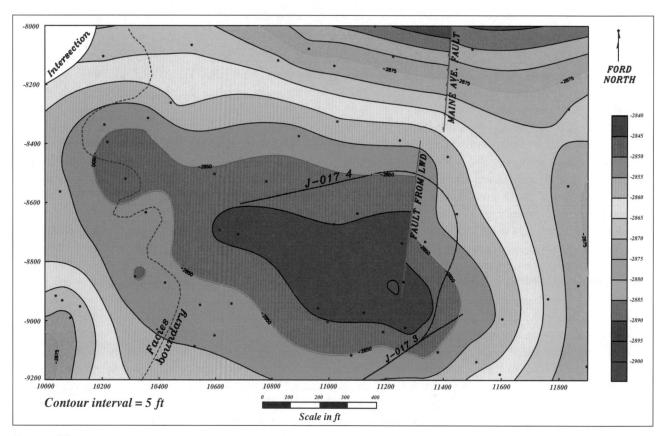

FIGURE 23. Structure map of the Hx_{0J} sandstone showing the well course for J-017 4. The sandstone shales out and merges with the Hx_{0J} shale in the western portion of the area. Five-feet contour intervals are in feet vertical subsea. A small fault was detected during drilling and was incorporated into the model.

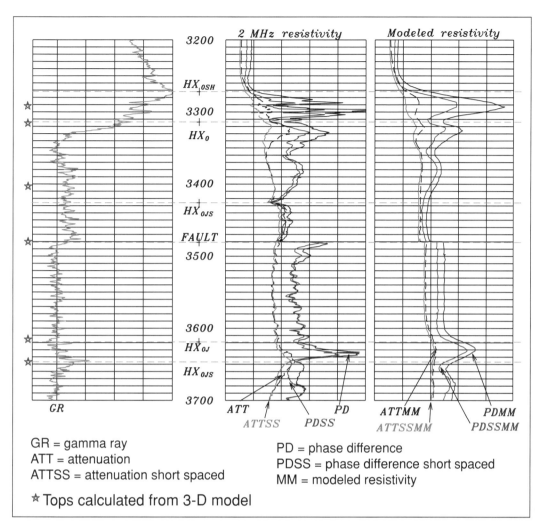

FIGURE 24. J-017 4 horizontal-well log, showing logging-while-drilling (LWD) and modeled logs. An excellent match was obtained. The location of this well is shown in Figure 23.

GR = gamma ray
ATT = attenuation
ATTSS = attenuation short spaced

PD = phase difference
PDSS = phase difference short spaced
MM = modeled resistivity

★ Tops calculated from 3-D model

A simple layer model was used for previous horizontal-well projects. The sandstone package was thick enough that the LWD gave a unique, easily interpretable response. The Hx_0 sandstone is divided by a continuous shale. The upper sandstone, referred to as the Hx_0, is 1.8 m (6 ft) thick; the lower sandstone, Hx_{0J}, is 2.4 m (8 ft) thick. Again, the horizontal well was successfully placed into each of these sandstones.

Postwell analysis and support were excellent. The LWD analyst spent significant time studying the LWD data and explaining the results. For wells drilled parallel to bedding, adjacent beds and formation anisotropy were significant factors in the log response. The anisotropy was quantified, the horizontal and vertical resistivity was determined, and a mathematical model of the LWD response was created (MacCallum et al., 1998).

The 3-D model was refined based on conclusions reached by collaboration between the LWD analyst and the geologist. The shales above the Hx_0 and Hx_{0J} sandstones were modeled, which helped signifigantly in data interpretation. A new anisotropy inversion algorithm and the inclusion of shales in the geologic model allowed for a

clearer understanding of the 2-MHz resistivity responses to the formation and their boundaries.

The fault geometry of a previously unidentified fault also was determined during this process using the 3-D model and further mathematical modeling of the LWD. There was a good correlation between the tops calculated from the 3-D geologic model and the tops selected from the LWD log. The average vertical distance between the bed boundary calculated from the 3-D geologic model and the well, as determined by the LWD log, is less than 0.08 m (0.25 ft).

CONCLUSIONS

A geologist working with carefully characterized rock data and 3-D modeling and visualization techniques adds greatly to a horizontal drilling team. The highly accurate 3-D visualizations of the reservoir greatly increase the confidence factor of the team, thus enabling Wilmington field reserves to be maximized.

To be effective, horizontal wells require precision placement. Three-dimensional models help isolate data inconsistencies, and 3-D viewers are good for adding data to correct the geologic model. Once the final geologic model is created, the drilling team can use the resulting 3-D visuals with confidence to improve drilling techniques and directional control. Postwell analysis of the LWD data also is facilitated using 3-D geologic models.

ACKNOWLEDGMENTS

We would like to acknowledge the help and support of Mike Domanski, president of Tidelands Oil Production Company; Jim Quay, Steve Siegwein, Scott Walker, Scott Hara, Rudy "Bud" Payan, and Chris Parmelee, technical engineering staff at Tidelands Oil Production Company; Dennis Sullivan, director of the Department of Oil Properties, city of Long Beach; Donald McCallum of Baker-Hughes INTEQ; and Art and Tamara Paradis and Heather Kelley of Dynamic Graphics Inc. Computer modeling was done with DGI EarthVision on an SGI Iris Indigo workstation.

REFERENCES CITED

Ames, L. C., 1987, Long Beach oil operation—A history, *in* D. D. Clarke and C. P. Henderson, eds., Geologic field guide to the Long Beach area: Pacific Section AAPG, p. 31–36.

Biddle, K. T., 1991, The Los Angeles basin: An overview, *in* K. T. Huddle, ed., Active margin basins: AAPG Memoir 52, p. 5–24.

Blake, G. H., 1991, Review of the Neogene biostratigraphy and stratigraphy of the Los Angeles Basin and implications for basin evolution, *in* K. T. Biddle, ed., Active margin basins: AAPG Memoir 52, p. 135–184.

Blesener, J. A., and C. P. Henderson, 1996, New technologies in the Long Beach Unit, *in* D. D. Clarke, G. E. Otott Jr., and C. C. Phillips, eds., Old oil fields and new life: A visit to the giants of the Los Angeles Basin: Pacific Section AAPG, p. 45–50.

California Department of Conservation, Division of Oil, Gas, and Geothermal Resources, 1999, 1998 annual report of the state oil and gas supervisor: California Department of Conservation, Sacramento, 269 p.

Clarke, D. D., 1987, The structure of the Wilmington oil field, *in* D. D. Clarke and C. P. Henderson, eds., Geologic field guide to the Long Beach area: Pacific Section AAPG, p. 43–56.

Davies, D. K., and R. K. Vessell, 1997, Improved prediction of permeability and reservoir quality through integrated analysis of pore geometry and openhole logs: Tar zone, Wilmington field, California: Society of Petroleum Engineers, SPE Paper 38262, 9 p.

Davies, D. K., R. K. Vessell, and J. B. Auman, 1997, Improved prediction of reservoir behavior through integration of quantitative geological and petrophysical data: Society of Petroleum Engineers, SPE Paper 38914, 16 p.

Henderson, C. P., 1987, The stratigraphy of the Wilmington

oil field, *in* D. D. Clarke and C. P. Henderson, eds., Geologic field guide to the Long Beach area: Pacific Section AAPG, p. 57–68.

Koerner, R. K., D. D. Clarke, S. Walker, C. C. Phillips, J. Nguyen, D. Moos, and K. Tagbor, 1997, Increasing waterflood reserves in the Wilmington oil field through improved reservoir characterization and reservoir management: Annual report submitted to the U.S. Department of Energy, Cooperative Agreement Number DE-FC22-95BC14934, unpaginated.

MacCallum, D., M. Pactel, and C. C. Phillips, 1998, Determination and application of formation anisotropy using multiple frequency, multiple spacing propagation resistivity tool from a horizontal well, onshore California: Society of Professional Well Log Analysts 39th Annual Logging Symposium, Keystone, Colorado, May 1998.

Mayuga, M. N., 1970, Geology and development of California's giant Wilmington oil field, *in* Geology of giant petroleum fields—Symposium: AAPG Memoir 14, p. 158–184.

Meyer, W. H., T. Maher, and P. J. McLean, 1996, New methods improve interpretation of propagation resistivity data: Presented at the Society of Professional Well Log Analysts 37th Annual Symposium, Taos, New Mexico.

Otott Jr., G. E., 1996, History of advanced recovery technologies in the Wilmington field, *in* D. D. Clarke, G. E. Otott Jr., and C. C. Phillips, eds., Old oil fields and new life: A visit to the giants of the Los Angeles Basin: Pacific Section AAPG, p. 37–44.

Otott Jr., G. E., and D. D. Clarke, 1996, History of the Wilmington field: 1986–1996, *in* D. D. Clarke, G. E. Otott Jr., and C. C. Phillips, eds., Old oil fields and new life: A visit to the giants of the Los Angeles Basin: Pacific Section AAPG, p. 17–22.

Otott, Jr., G. E., D. D. Clarke, and T. A. Buikema, 1996, Long Beach Unit 3-D survey, *in* D. D. Clarke, G. E. Otott Jr., and C. C. Phillips, eds., Old oil fields and new life: A visit to the giants of the Los Angeles Basin: Pacific Section AAPG, p. 51–55.

Phillips, C. C., 1996, Enhanced thermal recovery and reservoir characterization, *in* D. D. Clarke, G. E. Otott Jr., and C. C. Phillips, eds., Old oil fields and new life: A visit to the giants of the Los Angeles Basin: Pacific Section AAPG, p. 65–82.

Phillips, C. C., 1998, Geological review of Hx_0 Sands, *in* U.S. Department of Energy, Increasing waterflood reserves in the Wilmington oil field through improved reservoir characterization and reservoir management: 1997 Annual Report, Contract No. DE-FC22-95BC7 4934, Appendix 1, 12 p.

Phillips, C. C., and D. D. Clarke, 1998, 3D modeling/visualization guides horizontal well program in Wilmington field: Journal of Canadian Petroleum Technology, v. 37, no. 10, p. 7–15.

Phillips, C. C., D. D. Clarke, and L. Y. An, 1998, Give new life to aging fields: Oil and Gas Investor, v. 39, no. 9, p. 106–115.

Redin, T., 1991, Oil and gas production from submarine fans of the Los Angeles Basin, *in* K. T. Biddle, ed., Active margin basins: AAPG Memoir 52, p. 239–259.

Slatt, R. M., S. Phillips, J. M. Boak, and M. B. Lagoe, 1993,

Scales of geologic heterogeneity of a deep water sand giant oil field, Long Beach Unit, Wilmington field, California, *in* E. G. Rhodes and T. F. Moslow, eds., Frontiers in sedimentary geology, marine clastic reservoirs, examples and analogs: New York, Springer-Verlag, p. 263–292.

U.S. Department of Energy, 1999, Increasing heavy oil reserves in the Wilmington oil field through advanced reservoir characterization and thermal production technologies, partners: The city of Long Beach, Tidelands Oil Production Company (Tidelands), University of Southern California, and David K. Davies and Associates: 1996 Annual Report, Contract No. DE-FC22-95BC1 493, 85 p.

Wright, T. L., 1991, Structural geology and tectonic evolution of the Los Angeles Basin, California, *in* K. T. Biddle, ed., Active margin basins: AAPG Memoir 52, p. 35–134.

4

Mason, E. P., M. J. Bastian, R. Detomo, M. N. Hashem, and A. J. Hildebrandt, 2003, Results and conclusions of a horizontal-drilling program at South Pass 62 salt-dome field, *in* T. R. Carr, E. P. Mason, and C. T. Feazel, eds., Horizontal wells: Focus on the reservoir: AAPG Methods in Exploration No. 14, p. 49–65.

Results and Conclusions of a Horizontal-drilling Program at South Pass 62 Salt-dome Field

E. P. Mason

Shell Exploration & Production Company
New Orleans, Louisiana, U.S.A.

M. J. Bastian

Shell Exploration & Production Company
New Orleans, Louisiana, U.S.A.

R. Detomo

Shell Exploration & Production Company
New Orleans, Louisiana, U.S.A.

M. N. Hashem

Shell Exploration & Production Company
New Orleans, Louisiana, U.S.A.

A. J. Hildebrandt

Shell Exploration & Production Company
New Orleans, Louisiana, U.S.A.

ABSTRACT

A horizontal-well redevelopment drilling program around the flanks of the South Pass 62 salt-dome field resulted in significant successes and costly failures. Successful wells exploited thin, oil-filled shoreface sandstones; partially depleted zones; and massive, sand-filled channels. Failures were those wells that attempted to connect multiple fault blocks and drain low-resistivity/laminated-sandstone reservoirs. This paper reviews the field history; describes the geologic setting, including a summary of significant structural features and producing-sandstone depositional environments; discusses the horizontal-well strategy; and examines successful and unsuccessful wells.

South Pass 62 field lies 50 km (30 mi) east of the Mississippi River delta in 104 m (300 ft) of water. The field was discovered in 1965, developed with 61 directionally drilled wells from three platforms in the late 1960s, redeveloped from 1986 to 1988 with 31 wells from a fourth platform, and redeveloped again from 1994 to the present with horizontal and directionally drilled slim-hole sidetracks. A 3-D seismic-based field study completed in 1994 identified reservoir targets for the horizontal-drilling program.

Nearly 60 stacked, variable pay sandstones combine with steep formation dips and extensive faulting to create a complex field with hundreds of reservoirs. The field lies on the north flank of a mushroom-shaped, south-leaning salt dome that rises from below 8000 m (25,000 ft) to within 200 m (656 ft) of the seafloor. Typical formation structural dips decrease from 70° adjacent to the salt to 10° off structure. Several generations of faults exist, with throws ranging from centimeters to more than 100 m. Approximately 60 Pliocene and Miocene deltaic and turbidite pay sandstones ranging in depth from 1158 to 5791 m (3800 to 19,000 ft) onlap the salt.

INTRODUCTION

Between 1994 and 1997, Shell Exploration and Production Company redeveloped South Pass 62 field (SP62) in the Gulf of Mexico in an attempt to restore the aging giant's profitability. Daily production had declined from 30,000 bbl of oil to 2000 bbl, and operating costs had soared. Although large volumes of oil and gas remained in the field, most were producible only through low-rate, conventional, gravel-pack recompletions of partly depleted zones and thin, laminated sandstones. Nearly 200 faults dissecting 60 pay sandstones created hundreds of reservoirs in the field. Thirty years of production had resulted in many depleted reservoirs and many others in various stages of depletion. Many small, unpenetrated fault blocks with reserves too low to justify a well also existed. These conditions made SP62 an extraordinarily complicated field and difficult to redevelop profitably. Horizontal wells appeared to be the solution.

HORIZONTAL DRILLING IN THE GULF OF MEXICO

By 1990, horizontal wells had become commonplace nearly everywhere except in the Gulf of Mexico. The benefits of recovery and rate in unconsolidated sandstones were thought to be small, although dramatic improvements had been observed in similar environments elsewhere (Stark, 1992).

Texaco U.S.A. initiated horizontal drilling in the Gulf of Mexico when it spudded well B-12 in East Cameron Block 278 in May 1990 (Fisher and French, 1991; Cochrane and Reynolds, 1992). This well was drilled successfully to deplete a gas sandstone 442 m below the mud line; it proved that horizontal wells were a viable alternative to extended-reach development wells. The company followed up with horizontal wells in other fields. At South Marsh Island 239, a horizontal well was drilled to a then record true vertical depth of 3076 m (Jenkins and Patrickis, 1992). At the "Teal" Prospect (Eugene Island 338), a horizontal well was drilled to improve recovery from a depleted oil reservoir (World Oil, 1992). At Eugene Island Block 228, a 431-m lateral section was drilled and completed (Pardo and Patrickis, 1992).

Other operators, large and small, also began to have success with horizontal wells in the Gulf of Mexico. In 1992, Chevron U.S.A., Inc., experienced a sixfold increase in production rates from three horizontal wells drilled in shallow, Pleistocene gas sandstones at two fields (Gidman et al, 1995). Kerr-McGee Corp. exploited a thin oil column, reducing the effect of coning and realizing an eightfold increase in initial rates at Breton Sound Block 20 field, also in 1992 (Paull, 1993). In 1993, Pogo Producing Company completed a five-well horizontal program at Eugene Island Block 295 "B" in zones that would have been uneconomical to develop with vertical or directional wells. Combined rates reached 101 MMCFGD (Schroeder et al., 1995). Amoco successfully redeveloped thin oil reservoirs with a four-well horizontal-drilling program in bottom-water-drive reservoirs with 7-m oil columns.

Reentry sidetracks were also being attempted successfully, including Shell's first horizontal wells in the Gulf of Mexico. At Mississippi Canyon 194 field ("Cognac") in 1994, oil rims 18 to 30 m thick were successfully exploited using slim-hole sidetracks. Productivity improvements of three to ten times were observed (Danahy and Scheibal, 1997). These very profitable sidetracks were drilled with workover-class platform rigs to accelerate recovery of oil reserves prior to gas-cap blowdown. They opened up a new avenue for increasing production rates at many of Shell's older, declining fields on the Gulf of Mexico shelf. More recently, Exxon Company U.S.A. has applied this same strategy in redeveloping mature fields in the Gulf of Mexico (Batchelor and Moyer, 1997).

At South Pass 62, we decided to apply this "new" technology for Shell in the Gulf of Mexico to redevelop the field. Ultimately, nine horizontal sidetracks, two new horizontal wells, and five directional sidetracks were drilled at SP62. Production rates increased dramatically, peaking at 15,000 BOPD, largely because of the productivity improvement from the horizontal wells. However, the phenomenal successes were tempered by several costly failures.

This paper describes the types of horizontal wells drilled, examines the successes and failures, and summarizes the conclusions and best practices.

SOUTH PASS 62 FIELD HISTORY

South Pass 62 field, lying 50 km southeast of the Mississippi River delta in 100 m of water (Figure 1), was discovered in 1965. It was developed with 61 directionally drilled wells from three platforms in the late 1960s, redeveloped from 1986 to 1988 with 31 wells from a fourth platform, and redeveloped again from 1994 to 1997 with horizontal and directionally drilled slim-hole sidetracks (Figures 2 and 3). A 3-D seismic-based field study completed in 1994 identified reservoir targets for the horizontal-drilling program. Cumulative production from block 62 is 108 MMBO and 170 BCFG. Peak daily production rates reached 30,000 bbl and 50 MMCFG in 1970 before declining to 2000 bbl in 1994.

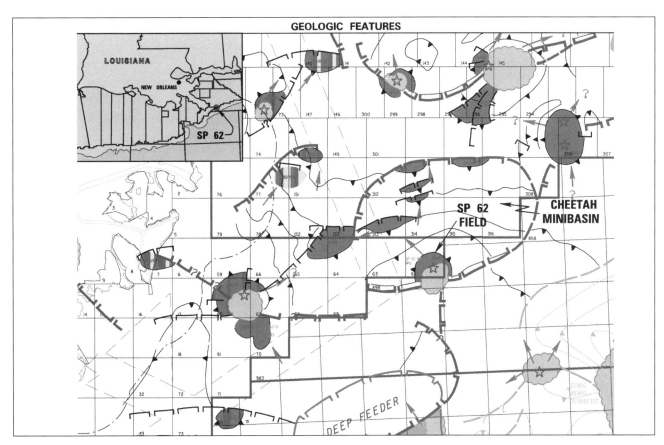

FIGURE 1. South Pass 62 field is located 50 km southeast of the mouth of the Mississippi River in 100 m of water. Dark gray indicates oil fields. Light gray refers to salt domes or tabular salt.

GEOLOGIC SETTING

SP62 field lies on the south rim of a large, salt-withdrawal minibasin (Cheetah minibasin, Figure 1). The south rim is defined by a large, pressure-changing, north-dipping, counterregional fault. The north flank is defined by a series of arcuate, regional, south-dipping faults.

Basin formation began in the early Miocene in response to sediment loading. As the ancestral Mississippi River deposited sediment in the north, mobile salt was forced southward into a salt ridge. This ridge provided a backstop for early Miocene turbidites cascading down the continental slope. Continued deposition resulted in further loading, continued salt-ridge growth, and eventual rupture of the overlying Eocene chalk along the ridge crest. Salt, squeezed out of the ridge, localized into a dome at SP62.

The Cheetah minibasin is extraordinarily rich; almost every closure within the basin contains oil.

STRUCTURE

Salt and salt-related faulting dominate the structure and form the trap at South Pass 62 field. Salt rises from a depth of approximately 8000 m through Eocene chalk and Miocene, Pliocene, and Pleistocene clastics to within

200 m of the seafloor (Figure 4). The salt body leans to the south and is mushroom shaped, with a thin root and massive head. The north face is concave, sculpted by a north-dipping arcuate slump. Numerous thin salt lenses as much as several hundred meters thick are found on the west flank of the dome. These lenses are defined by well penetrations but are unseen on 3-D seismic data.

Formation structure is characterized by dips in excess of 70° adjacent to the salt, and decreasing to less than 10° away from the structure (Figure 5).

Faults are prevalent at SP62. Approximately 200 seismically resolvable faults with throws ranging from five to hundreds of meters dissect and compartmentalize the producing horizons. The field is separated into four megacompartments by three major faults: (1) a large, counterregional fault/salt evacuation surface separating the productive north flank from the nonproductive south flank; (2) an arcuate slump defining the north salt face; and (3) a fault system antithetic to the slump. These faults have always been barriers to flow, despite sandstone-on-sandstone juxtaposition across each respective fault.

The mushroom-shaped salt overhang provides an excellent trap at SP62. Hydrocarbon column heights in excess of 900 m are common.

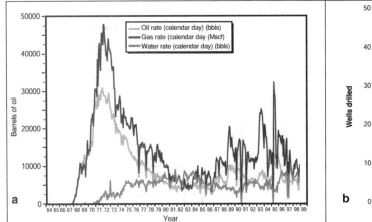

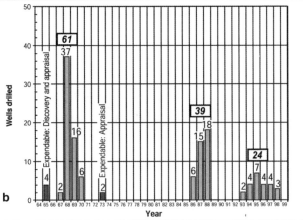

FIGURE 2. (a) Daily production of oil, gas, and water in the South Pass 62 field, plotted by year. Production decreased from peak rates in 1971 to a low of 2000 BOPD in 1994. (b) Well activity by year in the South Pass 62 field. Dark gray bars indicate exploratory and appraisal wells. Light gray bars indicate development and redevelopment wells. Primary field development occurred between 1967 and 1970, redevelopment in the mid-1980s, and the horizontal-drilling program in the mid-1990s.

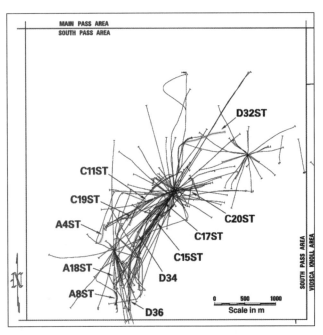

FIGURE 3. South Pass 62 field base map. Eleven horizontal wells and sidetracks were drilled between 1993 and 1997.

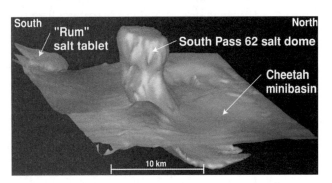

FIGURE 4. Three-dimensional rendering of South Pass 62 salt dome piercing an early Miocene horizon.

STRATIGRAPHY AND DEPOSITIONAL ENVIRONMENT

Approximately 60 Piocene and Miocene deltaic and turbidite pay sandstones range in depth from 1000 to 6000 m. Horizontal wells focused on seven late Miocene deltaic and shoreface sandstones (Figure 6). Typically, these sandstones were fine grained, well sorted, uncemented, and normally pressured.

Upper-shoreface sandstones (e.g., M4) typically were 3 to 6 m thick, continuous, upward coarsening, well connected, and easy to correlate. Horizontal wells in these reservoirs were the program's most productive and profitable (C11 and D32). Excellent connectivity and continuity resulted in high rates and good drainage.

Lower-shoreface intervals were characterized by thin beds, low resistivity, and high continuity (N9 and U-U1). These zones consisted of 30- to 90-m bundles of centimeter- to meter-thick sandstone members. Individual sandstone members were difficult to correlate, but the gross intervals were not. Historically, at SP62, these intervals had produced small volumes at low rates using conventional gravel packs. Laterals drilled through lower-shoreface deposits were disappointing in that they sanded up after producing small amounts of oil. The use of preperforated screens might have alleviated this problem.

Distributary channels, typified by the L10, M8, and V series, were characterized by rapid changes in thickness, variable connectivity and water contacts, and inconsistent drainage. Thicknesses ranged from a few meters to more than 30 m. Channel geometries and drainage were difficult to predict, despite integration of all well, production, and seismic data. In nearly every well penetrating channelized reservoirs, sandstone thickness and hydrocarbon

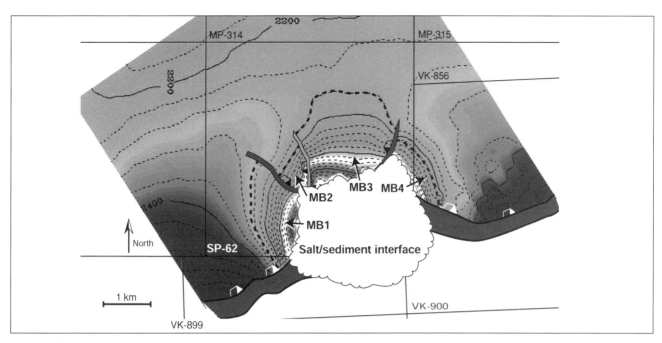

FIGURE 5. N6N8 sandstone time-structure map in milliseconds showing structural dips in the South Pass 62 field increasing adjacent to the salt. The four megacompartments (MB1, MB2, MB3, and MB4) are defined by large faults.

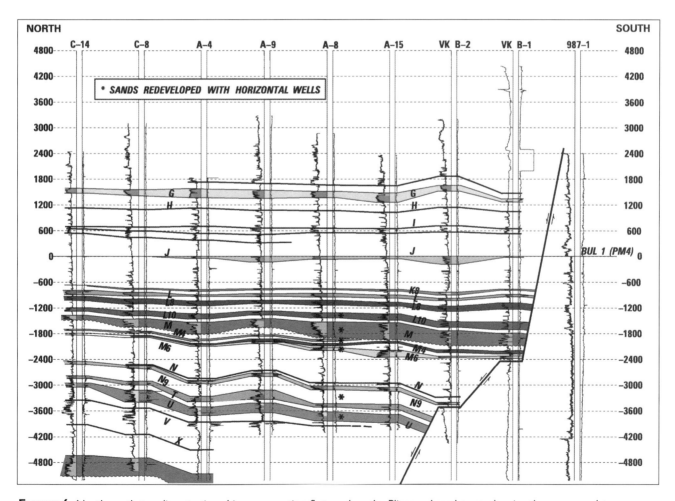

FIGURE 6. North-south-trending stratigraphic cross section flattened on the Pliocene J sandstone, showing the seven sandstones exploited with horizontal wells in the South Pass 62 field.

distribution were significantly different than predicted. Laterals placed in channelized intervals yielded both extremely profitable and productive wells (C20st, A8st), as well as the most costly failure (D34).

One horizontal well was drilled into the M sandstone, an incised valley deposit. The M sandstone is a massive, high-net/gross-ratio interval characterized by extreme variations in thickness. Thicknesses ocassionally exceed 100 m. On gamma-ray logs, the M is blocky, with a sharp base and top. On seismic data and in cross section, the base cuts through different stratigraphic intervals, characteristic of an incised valley. The M sandstone exhibited excellent connectivity and productivity, and it produced with a strong water drive.

DRIVE MECHANISM

Reservoir drive mechanisms varied from strong water drive to pure depletion drive. Thick, higher net/gross sandstones (M and M8) exhibited strong water drives where sandstone-on-sandstone juxtaposition across faults combined with exceptionally good stratigraphic connectivity to allow communication with downdip aquifers. The strong water drive seen in the M4 upper-shoreface deposit is attributed to excellent continuity and connectivity.

With few exceptions, reservoirs adjacent to or near the salt face were depletion, or partial, water drive. Increased faulting and more variable stratigraphy combined to create small compartments disconnected from downdip aquifers. Deeper reservoirs also tended to be depletion drive.

HORIZONTAL WELL STRATEGY

SP62 was one of Shell's first horizontal programs in the Gulf of Mexico; therefore, the company wanted not only to increase daily production and recover oil from low-rate/low-profitability reservoirs, but also to experiment with designs and try new concepts. Very little had been published in 1994 on horizontal-well applications in the Gulf of Mexico, and very few wells had been drilled. Successful techniques from SP62 could therefore be used in other fields.

The strategy was to test play concepts on early wells, then prosecute the more successful plays with later wells, use slim-hole sidetracks (4¾-in. holes) from shut-in or temporarily abandoned wellbores as a low-cost alternative to new holes, and leverage service-company expertise in drilling horizontal wells.

PLAY CONCEPTS

Plays prosecuted at SP62 included using horizontal wells to drain underpressured reservoirs (two wells) and

TABLE 1. OVERVIEW OF HORIZONTAL WELLS DRILLED IN THE SOUTH PASS 62 FIELD REDEVELOPMENT PROGRAM.

Well	Success?	Sandstone name	Cumulative production (MBO)	Maximum daily rate (BOPD)	Why drilled	Pilot?
A4	No	V1	121	400	Acceleration, high rate, eliminate recompletions	Yes
A8	Yes	M8	330	3300	High rate, better drainage	Yes
A18	No	U/U1	52	500	High rate, better drainage, laminated sandstones	No
C11	Yes	M4	390	2400	Acceleration, high rate, eliminate recompletions	No
C15	Yes	M	432	3000	High rate, combine fault blocks, eliminate recompletions	No
C17	Marginal	U/U1	248	500	Low pressure, combine fault blocks, laminated sandstones	Yes
C19	No	N9N10	170	1900	Laminated sandstones, combine fault blocks	No
C20	Yes	L10	596	2200	Low pressure, combine fault blocks, high rate	Yes
D32	Yes	M4	933	3400	Acceleration, high rate, eliminate recompletions	No
D34	No	L10	36	1500	Recover attic oil, high rates, combine fault blocks	Yes
D36	Marginal	L10	274	350	High rates, recover attic oil	Yes

laminated/thin-bedded intervals (two wells), to connect multiple fault blocks with a single lateral (two wells), and to eliminate multiple recompletions and accelerate production (five wells). Frequently, several plays were tested with the same well. For example, well C17st connected three partly depleted fault blocks in a laminated/thin-bedded sandstone interval. The most successful application at SP62 was the use of horizontal wells to eliminate recompletions in multiple wells and accelerate production, and to drain depleted sandstones.

SLIM-HOLE SIDETRACKS

Slim-hole sidetracks were seen as a low-cost alternative to drilling new wells at SP62. Many old boreholes were temporarily abandoned and ready to be recompleted or sidetracked. A typical procedure was to mill a window through $7\frac{5}{8}$-in. casing, drill a $6\frac{1}{2}$-in. medium- or long-radius build section into the top of the objective sandstone, and then set a $5\frac{1}{2}$-in. liner. A $4\frac{3}{4}$-in. lateral section followed; $3\frac{3}{8}$-in. wire-wrapped, prepacked screen was then run to total depth (TD). An oil-based mud system was used on most wells. Initial attempts using a salt-water-based mud system resulted in frequent and expensive hole problems.

SERVICE-COMPANY EXPERTISE

The expertise of contracting companies Baker Hughes and Halliburton in drilling horizontal wells was used to the maximum extent possible in designing the horizontal program. Shell provided all available data to both companies and asked them to design the most efficient sidetracks. We provided target x and y coordinates and depths for the lateral section, directional data for all 130 wells from the four platforms, and mechanical sketches of all boreholes available for sidetrack. Both companies then developed a drilling program.

We believe this resulted ultimately in a better-designed program and reduced cycle time, because Shell had insufficient drilling staff available to design this program.

RESULTS

Of the 11 horizontal wells drilled, five (D32st, C11st, A8st, C15st, and C20st) were very profitable, two (C17st, D36) generated a small profit, and four (D34, A18st, A4st, C19st) were commercial failures (Table 1). Well D32st far exceeded expectations and more than compensated for the unprofitable wells. Failures resulted either from mechanical failure, large cost overruns in

Depositional environment	Thickness (m)		Lateral length (m)	Dip	Side-tracks	Drive mechanism	Comments
	Gross	Net					
River-mouth bar	9	7.5	187	10	0	H$_2$O	May not have penetrated entire sandstone
Channel	18	16.5	132	35	2	H$_2$O	Sidetrack 1 = steering failure (missed sandstone)
Lower shoreface	18	9	303	40	0	Depletion	Sanded up, depleted quickly
Upper shoreface	5	4	731	15	0	H$_2$O	
Incised valley fill	30	27	449	55	2	H$_2$O	Stuck liner, initial sidetrack sanded up
Lower shoreface	18	11	196/457	35	1	Depletion	Stuck drill pipe
Lower shoreface	30	12	432	20	0	H$_2$O	Cut interval twice, screen failure/sanded up; workover unsuccessful
Stacked channels	23	21	164/316	20	0	Partial H$_2$O	Stuck screen at fault halfway into the lateral
Upper shoreface	5	4	761	15	3	H$_2$O	Sidetrack 1 = stuck liner, Sidetrack 2 = stuck liner, Sidetrack 3 = stuck drillpipe
Nested channels	25	18	201/776	30	2	Depletion and H$_2$O	Geology poorly understood; completion too complicated
Nested channels	18	14	207	45	0	Depletion	Reservoir partly depleted

drilling the well, or geologic "surprises" (e.g., undetected faults, perched water, unexpectedly low pressure).

The two most successful wells and two significant failures are discussed in detail to highlight some of the most important information gleaned from this program.

SUCCESSES

D32ST (M4 THIN SHOREFACE SANDSTONE)

Well D32st was drilled to accelerate production and improve drainage from a large, unproduced M4 sandstone reservoir. Several directionally drilled wells that had penetrated the reservoir were to be recompleted eventually in the M4 sandstone. Well D32st effectively drained much of the reservoir, making recompletions in several nearby directionally drilled wells unecessary.

In the reservoir traversed by D32st, the M4 shoreface sandstone averaged 5-m gross (4-m net) thickness, coarsened upward with a sharp upper contact, and was highly continuous (Figures 7 and 8). The sandstone pinched out updip toward the salt.

The structural dip of the M4 reservoir averaged 15° (Figure 9). Well D32st was drilled along structural strike, traversing the fault block downdip of the salt face. Lateral length was 761 m, of which 427 m consisted of clean sandstone (Figure 10).

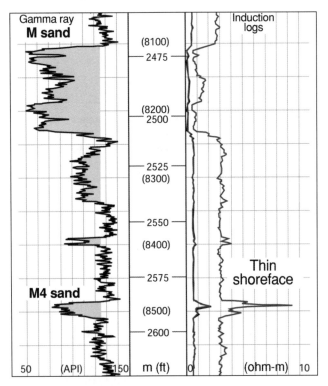

FIGURE 7. Well D32 (original hole) gamma-ray and resistivity curves through the M4 sandstone, upper-shoreface deposit, South Pass 62 field. The interval typically coarsened upward, exhibiting a sharp upper contact.

Well D32st has produced nearly 940,000 bbl of oil with a peak daily rate of 3400 bbl of oil, nearly 10 times that for conventional M4 completions (Figure 11). Similar success was achieved with the C11st horizontal well completed in the M4 sandstone in an adjacent fault block.

C20ST (L10 PARTIALLY DEPLETED DISTRIBUTARY CHANNEL SANDSTONE)

C20st, the field's first horizontal sidetrack, was drilled to improve recovery and accelerate production from a partially depleted L10 reservoir. By recovering M4 reserves that would have had to be recovered by recompleting nearby wells, those recompletions became unnecessary.

The L10 consists of variable, channelized sandstones exhibiting sharp upper and lower contacts on well logs (Figure 12). In the objective fault blocks, the L10 was nearly 30 m thick.

Well C20st was drilled along structural strike, crossing a fault thought to separate two L10 reservoirs (Figures 13, 14, and 15). Substantial production from downdip wells had depleted the reservoirs from initial pressures of 15,859 to 11,032 kPa (2300 to 1600 psi), as determined by Repeat Formation Tester (RFT) pressure measurements taken in the pilot hole and bottom-hole pressures taken in nearby wells. Because adjacent fault blocks exhibited similar bottom-hole pressures, we believed the fault blocks to be in communication through the downdip acquifer, if not across the faults.

A 535-m, 4¾-in. lateral was drilled. While running the completion assembly into the borehole, the prepacked screen became stuck at about 275 m into the lateral section, leaving the last 260 m unscreened (Figure 16).

Well C20st has produced 600,000 bbl of oil with peak daily rates exceeding 2200 bbl of oil, far surpassing expectations (Figure 17). Rate and pressure increases in 1998 and 1999 indicated that either additional sandstone stringers had begun to produce or that the last 260 m of the borehole had begun to contribute production.

FAILURES

C19ST (N9 LOWER-SHOREFACE LAMINATED SANDSTONE)

Well C19st was drilled to recover reserves from a low-resistivity/laminated-sandstone reservoir (Figures 18 and 19). The N9N10 zone consisted of a 37-m gross (12-m net) sandstone package over a large fault block. Numerous N9N10 penetrations existed that could have been recompleted as conventional gravel packs, but unfortunately, these completions produced at noncommercial rates. Peak rates from conventional completions reached only

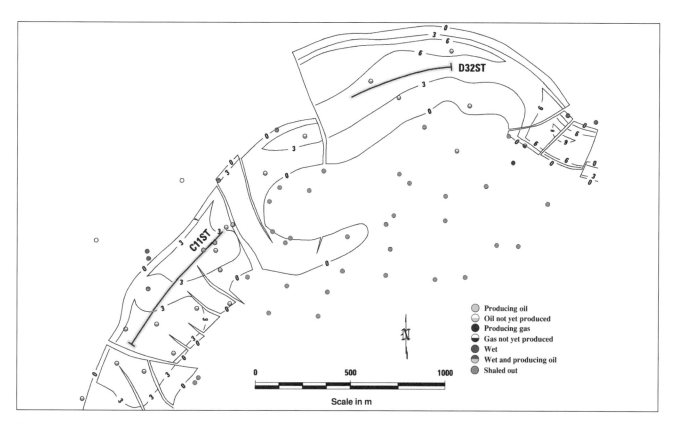

FIGURE 8. M4 sandstone net-oil isopach at original conditions, showing D32st and C11st lateral wells, South Pass 62 field.

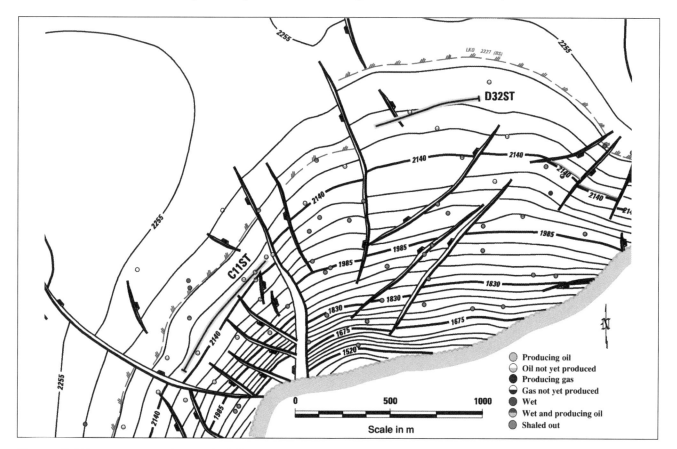

FIGURE 9. M4 structure map, showing the D32st lateral, which traversed a large fault block along structural strike, South Pass 62 field. Because a strong water drive was expected, the lateral was placed as far updip as possible. Depths are in feet.

about 200 BOPD. We had hoped for rates of as much as 1500 BOPD.

Well C19st was drilled across structure, so that the front of the lateral cut the interval from top to bottom, and the end of the lateral cut the interval from bottom to top (Figure 20). We at Shell believed that better drainage and higher rates would result. Particular care was taken in steering the lateral section so that the overlying, wet, N6N8 sandstone was not penetrated at the end of the lateral section. We had hoped that production rates and cumulative volumes produced would be large enough to make horizontal wells profitable in laminated sandstones across the field.

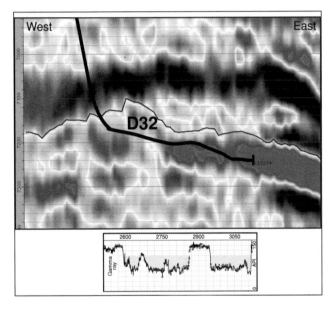

FIGURE 10. Well D32st gamma-ray curve (bottom) along the well path (top), South Pass 62 field. Approximately 450 m of the 750 m lateral consisted of clean sandstone.

Well C19st produced 170 MBO before sanding up after nine months of production. Initial peak rates reached 1900 BOPD, surpassing expectations (Figure 21). However, from the beginning of production, water was produced along with the oil, leading us to believe that communication between the lateral and the overlying N6N8 sandstone existed. Coiled tubing was used to wash sandstone from the lateral, but the well quickly sanded up again. Further workovers were deemed unlikely to bring the N9N10 lateral section back on production; therefore, the horizontal section was abandoned, and C19st was completed to another sandstone uphole. Preferential high-velocity fluid flow through more permeable sandstone laminae is thought to have eroded the screen at those points, allowing sandstone into the borehole.

Failures in laminated sandstones at SP62 and other Shell fields were borne out in the laboratory (Fair et al., 1996). Other operators were experiencing similar screen failures (Pardo and Patrickis, 1992; Perdue, 1996; Foster et al., 1999). This convinced Shell to try "Frac'n'Pac" completions, which yielded high production rates without mechanical failures. They rapidly became the completion method of choice at SP62.

D-34 (L10 DISTRIBUTARY CHANNEL SANDSTONE CONNECTING MULTIPLE FAULT BLOCKS)

Well D34 was drilled to recover attic reserves in the L10 sandstone. The lateral was intended to connect three separate fault blocks and drain oil updip of producing wells (Figures 22 and 23). A pilot hole drilled near the front of the lateral section indicated well-developed oil-bearing sandstones (Figure 24) at close to original pressure. The lateral section was drilled successfully (after one sidetrack) and completed with two tubing strings (Figure 25). A short tubing string was set in the front of the later-

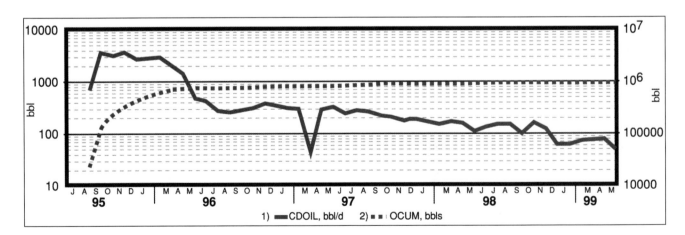

FIGURE 11. Plot of production at South Pass 62 field versus time, showing peak rates of 3400 BOPD, nearly 10 times the rate seen from conventional M4 completions. CDOIL = daily oil; OCUM = cumulative oil.

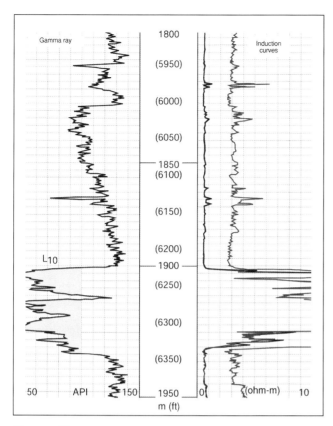

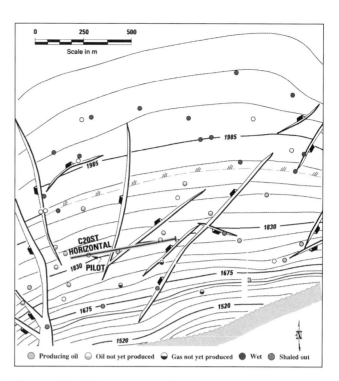

FIGURE 12. Well C-20 pilot-hole gamma-ray and resistivity curves, South Pass 62 field. The L10 sandstone was 30 m thick.

FIGURE 13. L10 structure map, South Pass 62 field. Well C20st traversed three fault blocks along a 316-m lateral strike to structural dip. Depths are in feet.

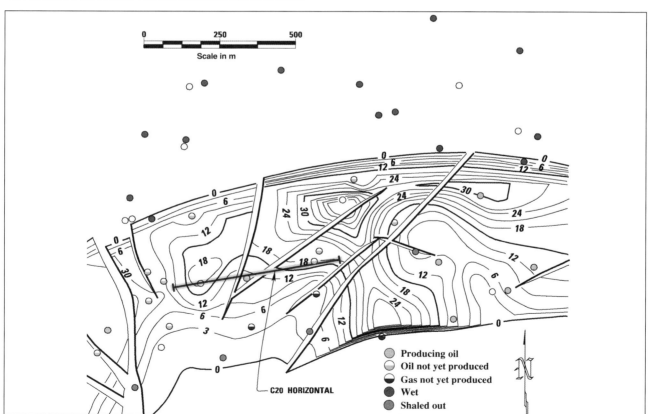

FIGURE 14. L10 net-oil isopach at original conditions prior to any production, showing C20 horizontal well, South Pass 62 field. Thicknesses are in feet.

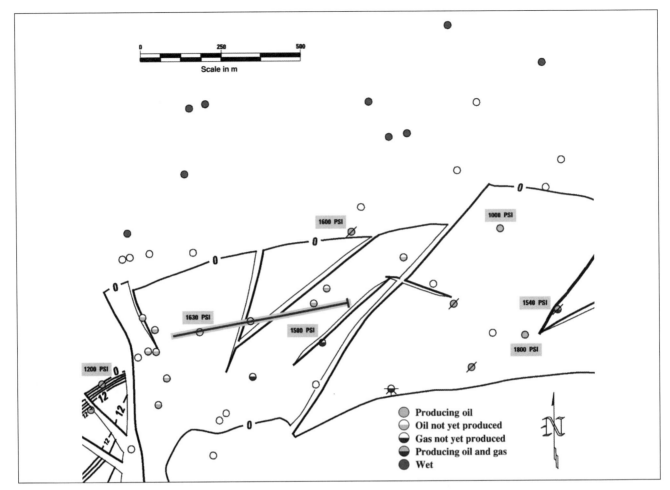

FIGURE 15. L10 net-oil isopach at the time C20st horizontal was drilled, South Pass 62 field. Several million bbl of oil had been produced, and reservoir pressures had been depleted to approximately 700 psi. Bottom-hole pressures indicated that the reservoirs were in communication either through the downdip aquifer or across faults.

al, and a long tubing string was set at the end of the lateral to allow production from separate fault blocks.

The well began to produce at high rates, but pressures and rates declined quickly from the short string after only 35 MBO had been produced. In addition, the long string sanded up, nullifying the dual-completion concept. The well was worked over but could not be brought back on production. Well D34 was later sidetracked when it encountered the L10 sandstone at unexpectedly low pressure. It was subsequently recompleted in a sandstone farther uphole.

Subsequent seismic reprocessing and reinterpretation revealed additional faults not recognized on the original 3-D seismic data set. These faults compartmentalized the reservoir much more than expected. In retrospect, it is clear that our original well plan had no chance of success.

Well D34 was the most expensive failure in the drilling program. We conclude that we attempted too much with one borehole in an area where the geology was more complicated than we had realized.

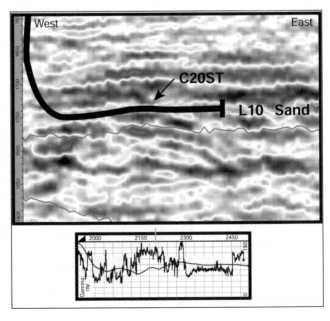

FIGURE 16. Well C20st gamma-ray curve (bottom) along the well path (top), South Pass 62 field. Approximately 381 m of the 540 m lateral consisted of clean sandstone.

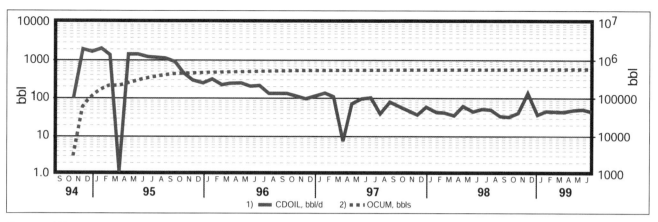

FIGURE 17. Plot of production versus time for well C20st, showing peak rates of 2200 BOPD, several times the rate seen from conventional L10 completions. CDOIL = daily oil; OCUM = cumulative oil.

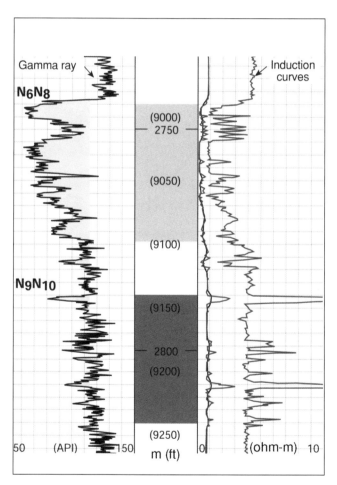

FIGURE 18. Typical N9N10 interval seen on gamma-ray and resistivity curves in well A4, South Pass 62 field. The massive N6N8 sandstone above the N9N10 was typically drained, taking care not to penetrate the base of the N6N8 at the toe of the lateral section.

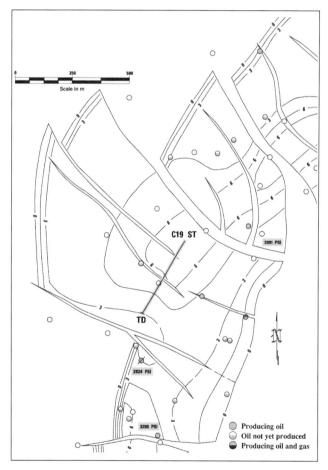

FIGURE 19. N9 net-oil isopach at the time well C19st was drilled, South Pass 62 field. The objective fault blocks had not previously been produced.

FIGURE 20. N9N10 structure map showing the C19st lateral drilled slightly oblique to strike and connecting two fault blocks, South Pass 62 field. Depths are in feet.

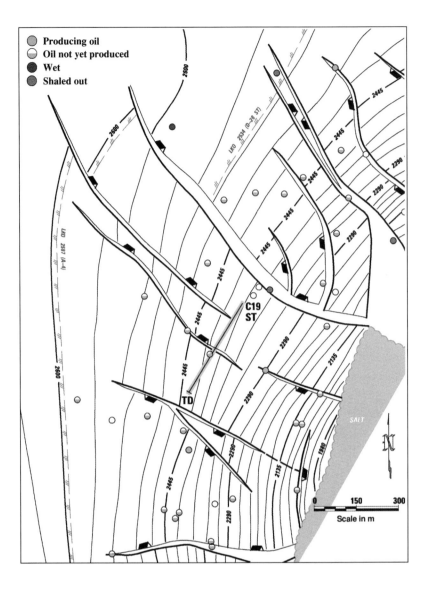

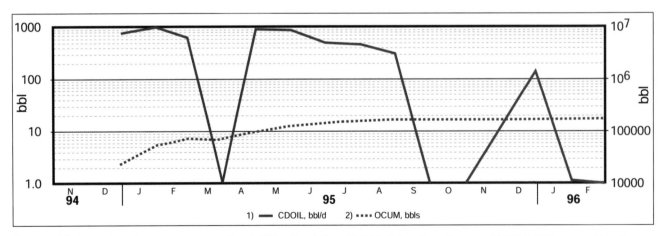

FIGURE 21. Plot of production versus time of well C19st, South Pass 62 field. Although peak rates of 1900 BOPD were achieved, nearly nine times the rate seen from conventional completions, the completion sanded up and went off production after only nine months and 170 MBO of production. Two workovers failed to bring the lateral back on line. CDOIL = daily oil; OCUM = cumulative oil.

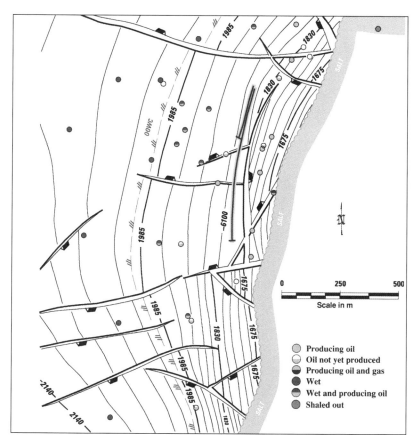

FIGURE 22. L10 structure map in the vicinity of well D34, South Pass 62 field.

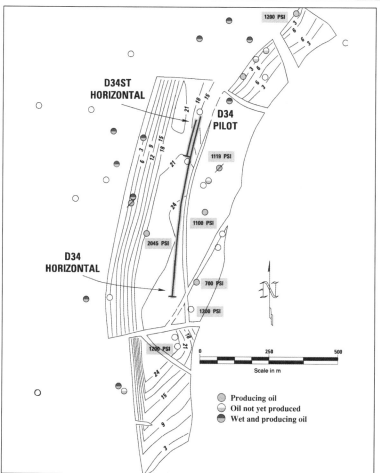

FIGURE 23. L10 net-oil isopach at the time well D34 was drilled, South Pass 62 field.

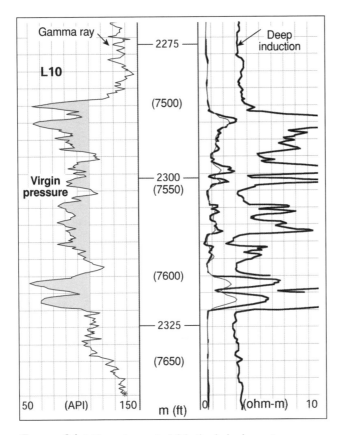

FIGURE 24. L10 sandstone in D34 pilot hole shown in gamma-ray and resistivity curves, South Pass 62 field. RFT pressures indicated near-virgin pressures, which encouraged us to proceed with the horizontal section.

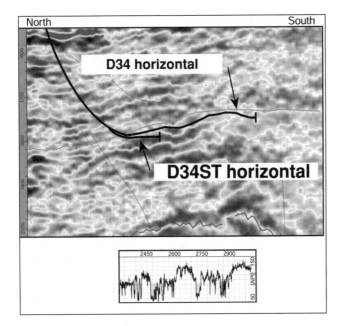

FIGURE 25. L10 sandstone as portrayed by gamma-ray curve (bottom) in the D34 lateral section (top), South Pass 62 field. Wet zones within the overall pay sandstone indicated more stratigraphic complexity than expected.

CONCLUSIONS

Conclusions derived from this drilling program include:

1) The most successful horizontal wells were drilled in thin, shoreface sandstones (M4 sandstone) and massive channels (M and M8 sandstones) that had extraordinarily good continuity and connectivity.
2) Horizontal wells drilled in/through nested channels (L10) and laminated lower-shoreface deposits (N9N10 sandstone) were generally unsuccessful. Nested channels with perched water, variable water contacts, and unpredictable connectivity and continuity—typical of the L10 sandstone where well D34 was drilled—were disappointing and unprofitable. Laminated, thin-bedded intervals frequently resulted in early screen failure, which we attributed to screen erosion at the points of preferential/concentrated production in the cleanest, most permeable zones.
3) Risk of reservoir compartmentalization increased adjacent to the salt structure because the degree of faulting increased with proximity to the salt face. Wells A18st, D34, and D36 were drilled near the salt, and all encountered small compartments attributed to faulting. At SP62, laterals placed in massive, high net/gross, well-connected sandstones (e.g., M sandstone) had the highest probability of success adjacent to salt because connection to downdip aquifers and production across faults was more likely.
4) At South Pass 62, downdip reservoirs were preferred, primarily because unseen structural complexity was less likely. These sandstones also tended to be more continuous and connected.
5) Connecting multiple, small-fault blocks was costly. Lateral sections were more likely to collapse at faults, increasing well costs and decreasing profitability.
6) Simple, short well paths less than 450 m long that cut across or through the zone of interest were preferred. These boreholes provided a large increase in rates over vertical wells, without the high cost and drilling problems that accompanied nearly every long lateral drilled at SP62.
7) Low-pressure/depleted zones performed well. Three wells drilled through depleted reservoirs all produced at higher rates and for greater lengths of time than expected.
8) Prepacked, wire-wrapped screens frequently sanded up. Today we would use perforated screens.
9) The increased cost of pilot holes was more than offset by the value of knowledge gained from them. Pilot

holes provide reservoir pressure information and a preview of the geology to be encountered at the start of the lateral section. This is critical information in a field with complicated geology and extensive production history.

ACKNOWLEDGMENTS

The authors would like to thank Shell Exploration and Production Company for permission to publish and Tom Wilson and Vu Cung for their constructive comments.

REFERENCES CITED

Batchelor, B. J., and M. C. Moyer, 1997, Selection and drilling of recent Gulf of Mexico horizontal wells: Proceedings, 29th Offshore Technology Conference, Houston, Texas, OTC-8462, v. 3, p. 235–245.

Cochrane, J. F., and K. C. Reynolds, 1992, Pressure transient analysis of a shallow horizontal gas well: Test objectives, design, procedure, interpretation, and application of results: Proceedings, Society of Petroleum Engineers, European Petroleum Conference, Cannes, France, SPE-25052, v. 2, p. 349–361.

Danahy, M. A., and J. R. Scheibal, 1997, Exploitation of thin oil rims using horizontal sidetracks at Mississippi Canyon 194 (Cognac) Field (abs.): Annual AAPG Convention, Dallas, Texas.

Fair, P. S., J. Kikani, C. D. White, 1996, Modeling high angle wells in laminated pay reservoirs: Proceedings, Annual Society of Petroleum Engineers Conference, Denver, Colorado, SPE-36027, p. 19–28.

Fisher, E. K., M. R. French, 1991, Drilling the first horizontal well in the Gulf of Mexico: A case history of East Cameron Block 278, well B-12: Proceedings, 66th Annual Society of Petroleum Engineers Technical Conference, Dallas, Texas, SPE-22545, p. 111–123.

Foster, J., T. Grigsby, and J. Lafontaine, 1999, The evolution of horizontal completion techniques for the Gulf of Mexico: Where have we been and where are we going?: Proceedings, Sixth Society of Petroleum Engineers/Latin American and Caribbean Petroleum Engineering Conference, Caracas, Venezuela, SPE-53926, 15 p.

Gidman, B., L. R. Hammons, M. D. Paulk, 1995, Horizontal wells enhance development of thin offshore gas reservoirs: Petroleum Engineer, v. 67, no. 3, p. 19, 22–24.

Jenkins, R. W., and A. N. Patrickis, 1992, Drilling the deepest horizontal well in the Gulf of Mexico, South Marsh Island 239 D-10: Proceedings, International Association of Drilling Contractors/Society of Petroleum Engineers Drilling Conference, New Orleans, Louisiana, IADC/SPE-23897, p. 317–329.

Pardo, C. W., and A. N. Patrickis, 1992, Completion techniques used in horizontal wells drilled in shallow gas sandstones in the Gulf of Mexico: Proceedings, 67th Annual Society of Petroleum Engineers Conference, Washington, D.C., SPE-24842, p. 797–804.

Paull, B. M., 1993, Horizontal well boosts mature Gulf of Mexico reservoir: Petroleum Engineering International, v. 65, no. 11, p. 16–19.

Perdue, J. M., 1996, Completion experts study Gulf of Mexico horizontal screen failures: Petroleum Engineering International, v. 69, p. 31–33.

Schroeder, T., D. Mathis, R. Howard, G. Williams, and J. Sun, 1995, Teamwork and geosteering pay off in horizontal project: Oil & Gas Journal, v. 93, p. 33–39.

Schroeder, T., D. E. Mathis, A. Mathis, W. A. Hill, A. Ellis, and R. G. Hea, 1996, Various techniques optimize horizontal, shallow gas well, synopsis of SPE Paper 36489, scheduled for presentation at the SPE Annual Technical Conference and Exhibition, Denver, Colorado, October 6–9, 1996: Journal of Petroleum Technology, v. 48, no. 8, p. 734–737.

Stark, P. H., 1992, Perspectives on horizontal drilling in western North America, in Geological studies relevant to horizontal drilling: Examples from western North America: Rocky Mountain Association of Geologists, p. 3–14.

World Oil, 1992, Texaco isolates gas cap at Eugene Island: Gulf Coast Oil World, v. 12, no. 6, p. 25–27.

5

Jolley, L., M. Nicol, A. Frankenbourg, A. Leonard, and J. Wreford, 2003, The use of horizontal wells to optimize the development of Andrew—A small oil and gas field in the UKCS North Sea, *in* T. R. Carr, E. P. Mason, and C. T. Feazel, eds., Horizontal wells: Focus on the reservoir: AAPG Methods in Exploration No. 14, p. 67–94.

The Use of Horizontal Wells to Optimize the Development of Andrew—A Small Oil and Gas Field in the UKCS North Sea

Liz Jolley

BP Exploration, Aberdeen, U.K.

Matt Nicol

BP Exploration, Aberdeen, U.K.

Amy Frankenbourg

BP Exploration, Aberdeen, U.K.

Andy Leonard

BP Exploration, Aberdeen, U.K.

John Wreford

BP Exploration, Aberdeen, U.K.

ABSTRACT

The Andrew field is a small oil and gas field with a 58-m oil column, a 66-m gas cap, and a simple dome structure, producing entirely from horizontal wells. It has been a successful development for BP and the Andrew field partners, with plateau oil production extending 18 months beyond the predicted onset of field decline. Development success has been helped substantially by focusing presanction activity on key reservoir uncertainties and business decisions. The decisions that resulted were to drill all horizontal producers to optimize low gas-oil-ratio (GOR) oil recovery, to closely manage the reservoir under production, to delay gas coning and water breakthrough, and to collect sufficient surveillance data to allow regular updating of the reservoir management plan. The objective of the Andrew development is to maximize oil recovery before going to gas-cap blowdown. The challenge is to manage the GOR throughout the life of the field. Central to this are well design, location, numbers, and the drawdown strategy. The horizontal wells produce at higher rates (average 10 MBOPD) and at relatively lower drawdown pressures (100 psi). They recover increased reserves per well (13 MMBO per well), compared with a conventional well. Project economics were improved as well numbers were reduced from 24 in a conventional well case to 10 horizontal producers. Low GOR oil production has been maximized by well positioning relative to the gas-oil contact (GOC) and oil-water contact (OWC); by drilling long wells that enter the reservoir on the crest and exit through the flank of the field; and by completion design, perforation strategy, careful well management, and drilling two additional infill wells. As a result, the recovery factor has risen from 45% at sanction in July 1996 to 49% by the end of 2002. The final field recovery factor is expected to rise to 53% by sidetracking low-rate producers and continuing to manage the reservoir drawdown. Oil reserves also have increased from 132 to 154 MMBO from 1996 to 2002 as the result of an increase in field STOOIP (stock tank original oil in place) and better-than-expected reservoir and horizontal-well performance.

INTRODUCTION

Andrew field was discovered in 1974 in UKCS blocks 16/27a and 16/28, 230 km from Aberdeen, Scotland, by a crestal exploration well 16/28-1 (Figure 1). The Andrew field development was not sanctioned, however, until 1994, with first oil produced in June 1996, 22 years after field discovery. During this time, five appraisal wells were drilled, three of which were field-delineation wells, one a crestal equity well (16/27-1A), and finally, a central pilot hole with horizontal sidetrack drilled in 1992 to appraise nonconventional well performance with an extended well test (EWT).

Twelve years from discovery to first oil reflects the unfavorable economics of field-development plans using conventional well technology and traditional North Sea

FIGURE 1. Location map of Andrew field in the central North Sea.

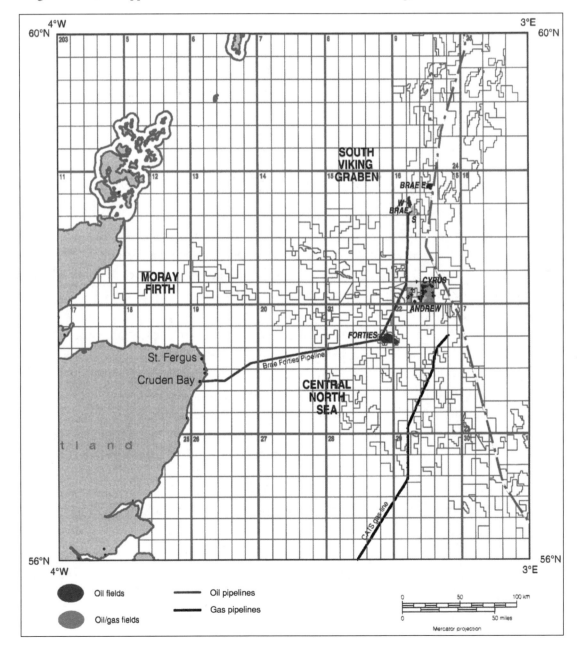

facility design. The field is small by 1980s North Sea standards, with high STOOIP uncertainty. It is located away from the main North Sea development hubs, in a fairway characterized by small Tertiary oil discoveries. It is too far from the Forties (in the south) and the Brae (in the north) fields to be a viable tieback opportunity and, therefore, requires a stand-alone facility. The split between gas-cap and oil-rim reserves increased uncertainty about the oil-reserves profile and ultimate oil-recovery factor.

The transformation of the Andrew discovery to an attractive development occurred as a result of challenging traditional North Sea development scenarios. The sanctioned facility design was small and simple, requiring low manning and shorter construction times. Platform commissioning times were accelerated. Uncertainties were identified, ranked, and managed by building in design contingencies and acquiring critical data. Horizontal-well design replaced conventional wells (Steele et al., 1993), and the Andrew Operating Alliance was set up to maximize teamwork and cost benefits. Figure 2 illustrates the Andrew development scheme.

This paper focuses on the impact of using horizontal wells to successfully maximize oil rates and ultimate recovery factors from the Andrew field.

BACKGROUND

Andrew field data and reservoir properties are listed in Table 1. The reservoir has been divided into seven field-wide zones, based on systematic changes in depositional style and biostratigraphy (Figure 3). Production performance on a reservoir scale is influenced by four categories of heterogeneities: (1) variations in channel geometry, (2) shale barriers and baffles, (3) high-permeability streaks, and (4) faults.

Variations in channel geometry are a function of changes in sandstone supply. The changes in channel distribution reflect a changing basin-floor topography as the Zechstein salt dome, which underlies the field, changed shape. The sandstone distribution will affect sweep efficiencies and coning rates in different regions of the field.

Shale barriers and baffles occur on a thickness scale of 25 cm to 5 m, but volumetrically account for less than 10% of the reservoir. Of concern is the lateral extent of any shale. Laterally restricted shales are associated with channel-abandonment fills and are considered low impact. Unconfined sheet shales, on the other hand, may be extensive on the field scale (e.g., A3 shale) (Clark and Pickering, 1996). They could provide barriers to gas or water movement or could encourage overrun/underrun from aquifer/gas cap into the oil column. They are modeled as intralayer transmissibility barriers in the simulation model.

High-permeability sandstones (400 m–800 md) occur at the base of channels as coarse lags or as loosely packed sandstones toward the top of channel fills. These provide preferential routes for water and gas to move into the oil column, speeding up time to breakthrough.

Intrareservoir faults have been mapped from a 1992 3-D seismic survey and matched to mud-loss data acquired while drilling. Where a fault cross cuts the wellbore, the faulted zone has not been perforated, because it is believed that gas and water may preferentially move along a fault plane into the oil column.

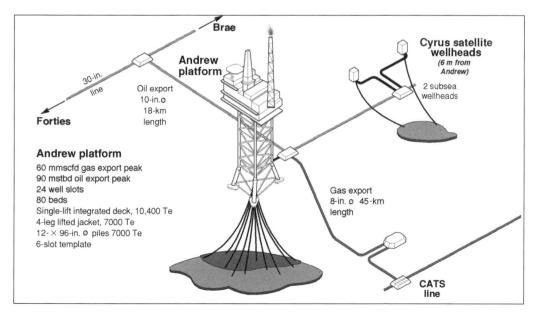

FIGURE 2. Andrew field-development scheme.

TABLE 1. ANDREW FIELD PARAMETERS.

Parameter	Description
Location	UKCS blocks 16/27a and 16/28; 230 km from Aberdeen, Scotland
Timing	Discovered in 1974 (16/28-1); 3-D seismic shot in 1991; sanction in 1994; first oil June 26, 1996
Partnership	BP (62.75%); Lasmo; MOEX; Talisman
Development scheme	Ten horizontal oil wells and a vertical gas-management well. Completion cemented and perforated liners. Water injection a retrofit option. Gas lift available.
Facilities	24-slot platform, gross fluid handling 115 MBOPD, planned peak oil rate 58 MBOPD, actual 75 MBOPD. Gas handling 95 MMSCFD debottlenecked to 125 MMSCFD, reinjecting up to 70 MMSCFD, gas lift of 55 MMSCFD, exporting up to 42 MMSCFD, fuel of 8 MMSCFD
Export	Oil through Brae Forties system via 16-km connector. Gas to Scottish power through CATS pipeline via 43-km connector
Production	Oil planned plateau of 58 MBOPD, actual of 75 MBOPD
Geologic setting	Paleocene submarine fan sandstones (Andrew Formation); four-way dip closure over salt dome. Andrew shales provide cap rock.
Pay parameters	97% net:gross for main reservoir zone, 20% porosity, 200-md average permeability; Kv/Kh ratios vary from 0.8 to 0.00001.
Pay cutoffs	In oil zones A2, AB, and AC = 9% phi; A1 and A3 = 15% phi (1md). In gas: 9% phi (0.1 md)
STOOIP/Reserves	Oil 315 million bbl, gas 280 bcf, sales gas 135 bcf
Offtake rate	Reserves per annum on plateau = 17%
Depth/area	About 2430 m TVDss, vertical relief of 140 m/13 km^2
GOC	2496 m TVDss (maximum gas column 66 m)
OWC	Variable 2576–2558 m TVDss tilt with dead oil zone above
FWL (free-water level)	2554 m TVDss (oil column maximum = 58 m); residual oil leg 0 to 22 m to northwest
Oil API/Viscosity	40 API/0.286 cp (at reservoir conditions)
Oil density/GOR/ formation volume factor	0.825 kg/m^3, 871 scf/stb, 1.52 rb/stb @ 3720 psia
Gas-cap data	No H_2S, CO_2 = 1.1%; C_1 = 85.66%; C_2 = 6.51%; C_3 = 2.46%. Expansion factor = 207 scf/rcf.
Reservoir conditions	Datum = 2496 m TVDss, temperature = 230°F, pressure = 3720 psia.
PI	130–160 stb/psi (vertical wells about 0.42 PI/m, 16Z = 0.32 PI/m)

FIGURE 3. Andrew field reservoir zones.

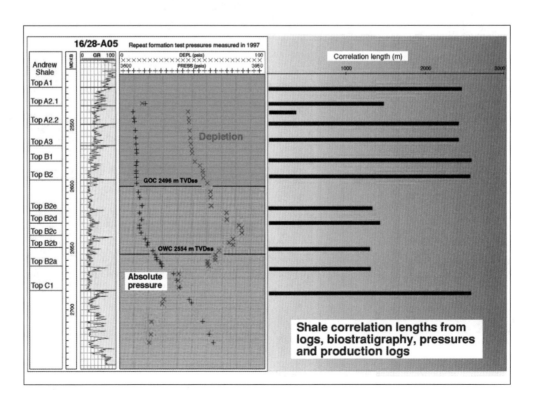

PRODUCTION OPTIMIZATION WITH HORIZONTAL WELLS

The economic success of the Andrew development depended on:

- maximizing ultimate oil recovery before going to gas-cap blowdown. For this reason, no net depletion of the gas cap was planned during the first three or four years of production.
- maximizing cumulative oil production before water or gas breakthrough. Maintaining a low and even drawdown and managing the GOR throughout the life of the field while achieving oil rates as high as possible were central to success.

ANDREW DEVELOPMENT CONCEPT

A development concept using only horizontal producers was key to achieving these objectives, despite considerable uncertainty at the time of sanction about their performance in this type of reservoir. In their favor, horizontal wells were believed to produce at higher rates (on average, 10 MBOPD) at relatively lower drawdown pressures (100 psi) and to recover increased reserves per well

(13 million bbl per well), compared with conventional wells. Unwanted water and gas production during early years of production could be reduced by well positioning, completion strategy, and well management (Tehrani, 1991).

The uncertainty about this development plan was reduced by performing an extended well-flow test on well 16/28-16z(A01), a horizontal-appraisal well drilled in 1993. Its location across the center of the field is illustrated in Figure 4. The position of the wellbore relative to gas-oil and oil-water contacts is illustrated in Figure 5. It demonstrates the greatly increased productivity of the horizontal-appraisal well compared with vertical-appraisal wells drilled into the same structure. The test data from well 16/28-16z(A01), particularly the relatively high flow rates at low drawdowns and lack of pressure depletion during flow, indicated that the Andrew reservoir has a generally high kV/kHz (vertical to horizontal permeability ratio) and is conducive to drainage by horizontal wells. The success of this appraisal well allowed the well numbers required to drain Andrew to be more than halved from the previous development model, which involved 24 conventional vertical wells. This breakthrough was one of the key factors that allowed the project to be sanctioned.

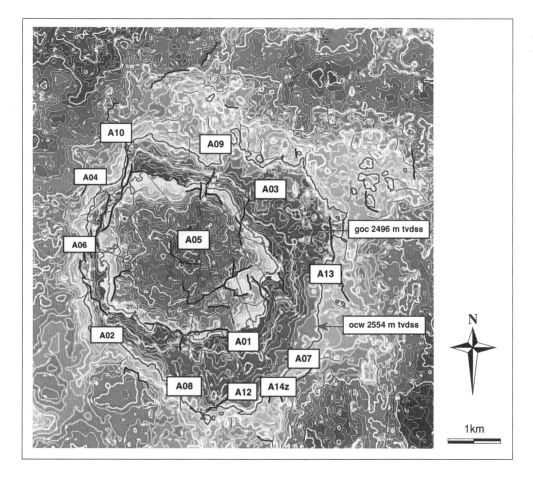

FIGURE 4. Andrew Field top-reservoir depth map with wells.

CRITICAL UNCERTAINTIES BEFORE DEVELOPMENT

Confidence in the sanction case was increased by creating a reservoir-uncertainty statement, which openly expressed the reservoir and development risks and their impact on project value (in reserves and development costs) (Smith and Buckee, 1985). The uncertainties identified, their impact and possible solutions are listed in Table 2 and Figure 6. Much of the analysis was done by running multiple sensitivities on the full-field reservoir model (Eclipse, a trademarked reservoir-simulation program of Geoquest). The model relied on an integrated reservoir and fluids description of the Andrew field and strongly influenced the location, standoff, perforation strategy, and drawdown management of the resulting oil producers.

TRUE-VERTICAL-DEPTH STANDOFF OF HORIZONTAL WELLS

At sanction, there was some uncertainty about the appropriate standoff of the horizontal well between the GOC and the mobile OWC. The choice of standoff is critical in order to limit the rate of gas/water breakthrough and optimize cumulative oil production. Various methods, assuming different standoff heights from

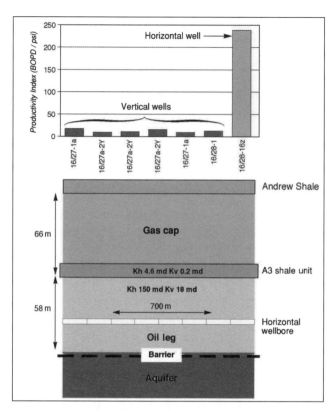

FIGURE 5. Viability of horizontal wells in Andrew field, and position of 16/28-16z(A01) horizontal wellbore in hydrocarbon column.

TABLE 2. KEY UNCERTAINTIES IN ANDREW RESERVOIR DESCRIPTION AND DEVELOPMENT PLAN.

Uncertainty	Impact	Contingencies/solutions
STOOIP Depth conversion and top reservoir pick	• wide reserves range, especially on flanks • project economics	• multiple depth conversion • parametric analysis of HCIIP (hydrocarbon initially in place)
Aquifer Lateral extent and permeability	• no water injection versus water injection needed • degree and timing of aquifer pressure support	• facility design with extra well slots for injectors (+12) and space to retrofit water-injection unit
Residual oil leg Reduced permeability to water (Krw) near free-water level	• degree and timing of aquifer pressure support	• surveillance program to monitor aquifer; gauge in A05 (gas injector) water leg
Lateral extent of key shales, specifically A3 shale	• degree of pressure support from gas-cap expansion	• gas reinjection to manage gas-cap pressures
Saturation distribution in transition zone	• oil reserves	• SCAL tests • use of analog data, e.g., Forties field
Lateral extent of barriers to flow Shales and faults	• ultimate recovery factor • rate of gas and water coning	• sensitivities run and P90-P10 trends defined • surveillance data collection plan to monitor well trends
Horizontal well design Reservoir Kh/Kz	• oil rates and reserves per well	• extended-well test on 16/28-16z(A01) horizontal appraisal well
Rate of water and gas coning	• oil rates and reserves per well	• optimize standoff • modeling to understand likely mechanisms
Inflow performance	• rates and reserves per well • drawdown	• PLTs to measure inflow • completion and perforation strategy to maximize inflow
Sanding risk	• erosion of flow lines • oil-rate reduction	• perforation and completion strategy

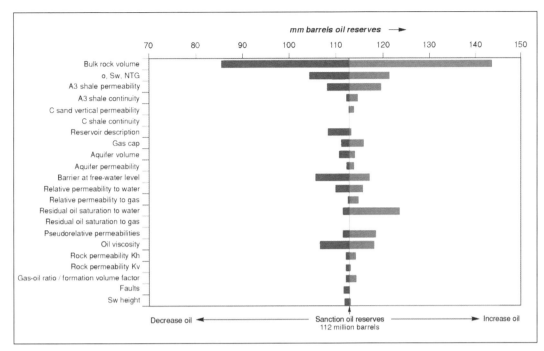

FIGURE 6. Andrew reservoir uncertainties and impact on oil reserves.

the OWC and GOC, were applied to predict time to gas and water breakthrough. An analytical approach predicted that breakthrough would occur 11 months after start-up. A homogenous fluid-simulation model predicted breakthrough after seven months at a 55:45 standoff. Model runs using the integrated reservoir and fluids description, including the expected suppression of water-relative permeability, predicted gas and water to cone within 12 months of first oil. This model indicated that a standoff of 67:33 is optimal for Andrew.

Dynamic simulation indicated that a greater level of heterogeneity can cause profound effects on the movement of water and gas through the oil column. The effect of the A3 shale, which straddles the GOC at the field crest, was investigated in detail by running two sensitivities on a single well-simulation model (Eclipse). Two cross sections along well A03 from Eclipse runs are presented in Figure 7. The first sensitivity assumes a 67:33 standoff and a leaky A3 shale that is not a barrier to gas movement. In this case, there is rapid gas and water breakthrough at the heel. In the second sensitivity, the A3 shale is a substantial barrier to the movement of gas from the gas cap down into the heel of the well. Consequently, gas coning at the heel is delayed. Water breakthrough can be delayed by optimizing the well profile at the heel by raising the heel 29 m up (+10 m) from the free-water level (FWL) to a 50:50 stand-off in the underlying B sandstones.

Results from the modeling sensitivities were applied to well-profile design. The 16/28-16Z appraisal well (A01 producer) and the two subsequent wells (A02 and A03) were drilled with a 67:33 standoff from gas and water.

Supported by modeling work and new data from A02 and A03, the next three wells, A04, A06, and A07, were drilled with optimized profiles with a 50:50 standoff at the heel, dropping to a 67:33 standoff beyond the shelter of the A3 shale. This change in profile was made in response to increased confidence that the A3 shale would act as an extensive gas umbrella, as well as the desire to experiment with optimizing the profile. Increased confidence resulted from the integration of biostratigraphic and sedimentologic data, and penetration of the shale unit in $12\frac{1}{4}$-in. and $8\frac{1}{2}$-in. sections of A02 and A03.

The final six wells all returned to the 67:33 standoff trajectory. Modeling, surveillance data (well-test data), and a rising GOR trend increased the desire to delay gas coning, making this the preferred profile.

COMPLETIONS

It was originally intended, for cost reasons, to complete Andrew wells with predrilled or slotted liners. However, after completing A01, it was decided to change to cemented and perforated liners to minimize risk of formation damage, reduce the risk of sand production, and provide options for future interventions (e.g., shutoff). The ability to select the intervals, spacing, density, and orientation of the perforations has tremendous advantages for sand control and water and gas shutoff. All Andrew producers, with the exception of A01 and A13, have a cased, cemented, and perforated $5\frac{1}{2}$-in. liner, producing into $5\frac{1}{2}$-in. tubing. The vertical-gas injection well, A05, has a cased, cemented, and perforated $9\frac{5}{8}$-in. liner and $5\frac{1}{2}$-in. tubing.

A formation isolation valve (FIV) run in the tubing is used for completion. It isolates the formation from kill-weight drilling and completion fluids. After perforations are shot, the FIV is closed by means of a shifting tool at the bottom of the perforating guns, as they are pulled out of the hole. When the well is hooked up, the FIV valve is opened and the well is produced. Utilization of the FIV is thought to limit reservoir damage, accounting for the high percentage (>90%) of perforations that are contributing to flow (Anderson et al., 1988).

PERFORATION STRATEGY

The production profile along the horizontal section is controlled by perforation density (Figure 8). The optimal production strategy is to uniformly deplete the reservoir and avoid creating a pressure sink in the center of the field where the distance between wells is smallest. The strategies put in place to manage risks are as follows:

FIGURE 7. Andrew field cross sections through Well A03 in Eclipse simulation model illustrating the potential controls on gas movement exerted by the A3 shale unit.

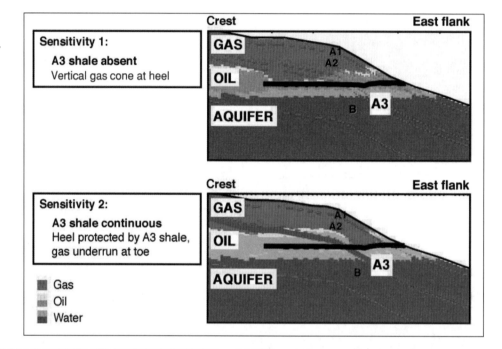

FIGURE 8. Perforation strategy for Andrew field horizontal wells.

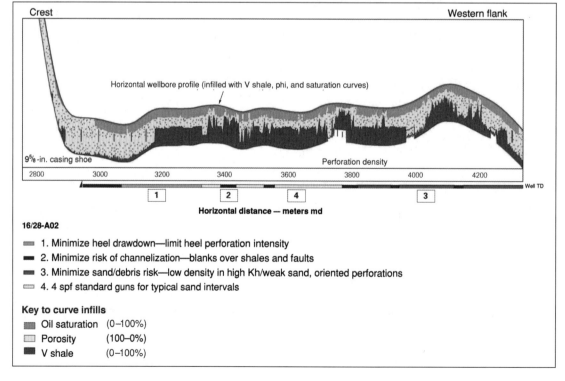

16/28-A02

1. Minimize heel drawdown—limit heel perforation intensity
2. Minimize risk of channelization—blanks over shales and faults
3. Minimize sand/debris risk—low density in high Kh/weak sand, oriented perforations
4. 4 spf standard guns for typical sand intervals

Key to curve infills

Oil saturation (0–100%)
Porosity (100–0%)
V shale (0–100%)

- Limit the perforation density at the heel so as to minimize heel drawdown, delaying early breakthrough.
- Limit the perforation density in weaker lithologies (e.g., highly permeable sandstones), thus minimizing the risk of sand production without materially impacting the productivity or ultimate recovery from each well.
- Maximize the inflow distribution along the horizontal section so as to maximize production and ultimate recovery and extend time to first gas and water breakthrough.
- Place blank sections over shales or faults to minimize the risk of channelization. This provides the option to shut off sections in the well, thus managing the decline of oil productivity after gas/water breakthrough.

SAND PRODUCTION

The risk of sand production is regarded as low (Masie et al., 1987), but it is a high-impact hazard if it occurs. When it occurs, the flow lines would be at risk from erosion by formation sand carried in the produced fluids. If sand were to be produced in quantities greater than 1 lb per 1000 barrels (PPTB), current guidelines dictate that maximum well rates will be limited to less than 7000 BOPD. It is therefore beneficial to reduce the risk of sanding. Most uniaxial core-strength measurements imply that Andrew sandstones are significantly stronger than sandstones in other fields (e.g., Forties) that are known to produce sand (Aadnoy, 1987). The reservoir units most at risk from sanding are those containing coarser-grained sandstones in high-density turbidite channels or associated facies. These units are especially prevalent in A2 and B1 zones (Figure 3). In the predrilled template wells, these sandstones were perforated using oriented tubing-conveyed perforation (TCP) guns in which charges were oriented to penetrate only the top 60° and base 60° of the wellbore. This is believed to reduce the possibility of wellbore instability caused by the perforation. The lowest-strength lithologies are shales, silts, and faulted rocks because of their ductile, rich, compacted framework, e.g., the A3 shale. These thin intervals were isolated with blank pipe. Sand production was first recorded in 2000, after four years of production.

INFLOW

The perforation strategy aims to manage inflow of oil into the wellbore such that the recovery from all the sandstones penetrated by the well is optimized. In the horizontal well, a pressure drop caused by friction between the flowing oil and the wellbore wall causes the drawdown to be greater at the heel of the horizontal well than at the toe. This tendency for greater inflow at the heel has to be managed carefully if the toe sandstones are to make their full contribution to production. In wells where higher-permeability units are more likely, inflow is managed by reducing the relative length of the perforated interval. Conversely, lower-permeability sandstones have increased perforation density. High-risk zones for preferential gas or water breakthrough—for example, higher-permeability units immediately overlying or underlying low-permeability units—are isolated with blank pipe.

A further advantage of cemented and perforated production liners is the potential in the future for selective shutoff (especially at the toe end). Shutoff of part of the wellbore requires that the reservoir around the well be isolated; otherwise, the unwanted phase will make its way through the reservoir around the shutoff zone and break through somewhere else. Hence, shales that are expected to be laterally extensive (hundred of meters and more) away from the well can isolate reservoir sands, and the zone of their intersection with the well is a suitable place for setting plugs. These shales—e.g., the A3 shale—have been selected preferentially as zones for blank pipe to protect them from perforation.

DRAWDOWN STRATEGY

The reservoir description also has influenced the drawdown strategy. Before drilling began, the wells were ranked in order of potential to cone gas. In Figure 9, gas and water movements in three wells are investigated using the Eclipse model. Once on production, shorter wells (A04 and A03) and/or those with a greater level of heterogeneity (e.g., A03) are preferentially flowed at lower drawdowns than those that are longer and/or have a lesser degree of heterogeneity (e.g., A02). This reduces the risk of early gas/water breakthrough to individual wells.

HORIZONTAL-WELL DESIGN AND IMPLEMENTATION

WELL DESIGN

All 12 Andrew oil producers follow the same basis of design represented in Figure 10. Wells A01 to A10 drill top reservoir just off the crest of the field in $12\frac{1}{4}$-in. holes to maximize the horizontal distance available for the $8\frac{1}{2}$-in. section. Wells A12 to A14Z enter the reservoir further downdip to avoid the more depleted B sandstone oil in the center of the field. In all cases, the $12\frac{1}{4}$-in. section begins at a near-vertical inclination and builds angle at 3–4°/30 m over an 80-m-thick interval, reaching total depth of 15–20 m beneath the GOC at 2496 m TVDSS in the oil-bearing reservoir. A $9\frac{5}{8}$-in. casing was then run to ensure isolation of the gas column from the producing wellbore.

FIGURE 9. Cross sections through a simulation model (Eclipse) illustrating gas and water movements into the wellbore after two years of oil production.

A02 Well length/standoff modeling results

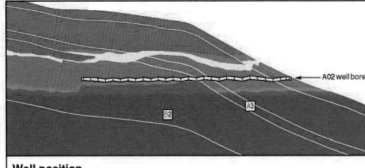

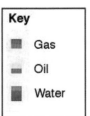

Key

- Gas
- Oil
- Water

A02 well bore

B2   A3

Well position
- 67/33 standoff heel to toe

Breakthrough predictions
- toe gas breakthrough is drawdown sensitive
- heel gas breakthrough is managed with perforation strategy
- early B-sandstone water breakthrough if A3 acts as barrier
- least risk of early gas breakthrough
- longest well with high PI

A03 Well length/standoff modeling results

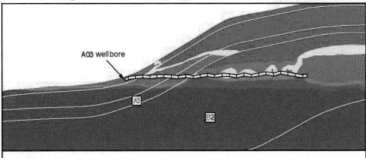

A03 well bore

A3   B2

Well position
- 67/33 standoff heel to toe

Breakthrough predictions
- most sensitive to drawdown
- early gas and water breakthrough in B-sandstones
- least protection of wellbore from gas cap
- B1 high permeability
- shortest well

A04 Well length/standoff modeling results

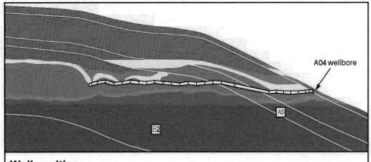

A04 wellbore

B2   A3

Well position
- 50/50 standoff in B-sandstone drop to 67/33 at toe from A3 shale

Breakthrough predictions
- heel gas breakthrough managed with perforation strategy
- early B sandstone water breakthrough if A3 acts as barrier
- good PI offset by wellbore pressure drop
- greater gas breakthrough risk than A02

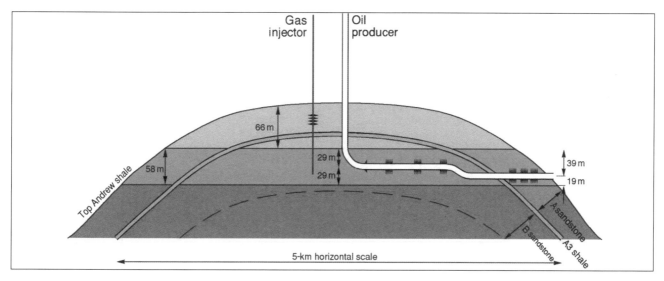

FIGURE 10. Basis of horizontal well design on Andrew field.

The GOC position was logged with logging-while-drilling (LWD) gamma-ray (GR)/resistivity logs and pipe-conveyed density/neutron/GR logs. The deviation survey data was then tied back to the GOC, allowing the error to be zeroed and allowing more accurate recalculation of the 8½-in. section and landing of the horizontal section. The final build angle to horizontal in the 8½-in. hole was 6–7° per 30 m, which landed the wells at 2535-m TVDss or 2516-m TVDss, respectively, for wells with a 67:33 or 50:50 standoff. Once horizontal, the trajectory is maintained until the well exits top reservoir on the flank of the field. The 8½-in. hole was drilled within a true-vertical-depth (TVD) tolerance of +/-5 m at the heel, increasing linearly to the toe of the well by +/-3 m per 1000 m. This tolerance assumes that the GOC had been successfully picked from the 12¼-in. logs. If the GOC was not successfully logged, the TVD tolerance at the heel increases to +/-8 m.

Measurements while drilling (MWD) and well-site biostratigraphy allowed real-time geosteering and allowed the trajectory to be monitored and adjusted and progression through the stratigraphy to be charted. The horizontal section drills through progressively younger beds from heel to toe, cutting across key intrareservoir heterogeneities (e.g., shale barriers). Drill-pipe conveyed resistivity/sonic/neutron density/GR/RFT (repeat formation test) logs were run in the 13 8½-in.-hole sections of nine wells.

DEVELOPMENT DRILLING ORDER

The three template wells (A02, A03, and A04) and eight Andrew platform wells are in a radial pattern (Figure 11). For the first nine wells, the heels were moved toward the crest of the field, resulting in more than half the perforations being made in the B sandstone. The heels of the last three wells stepped out to target reserves in the A sandstone and to avoid the more depleted B sandstone, which is prone to early water or gas coning. After A01 (16/28-16Z EWT), the wells were drilled in opposing quadrants of the field, then gaps between each were progressively infilled to ensure even production of oil.

WELL LENGTHS

Table 3 summarizes horizontal-well lengths, meters perforated, and standoff by well. A01 and A02 are the only development wells that do not crosscut all reservoir zones. Success in drilling and completing the first four Andrew horizontal wells brought confidence to drill longer horizontal sections than originally envisioned. The remaining eight wells all reached total depth in the Andrew Shale. Longer horizontal lengths allowed production at higher rates and lower drawdowns.

HORIZONTAL-WELL PROFILES

The horizontal-well profiles continuously improved during development drilling from 1996 to 1999 (Fine et al., 1993), in part because of the buildup of experience gained from a continuous drilling program executed by the same drilling crew and onshore-wells team. It also reflects improvements in directional drilling technology during three years of development drilling. Figure 12 illustrates three profiles from wells A02, A07, and A14Z drilled in 1996, 1997, and 1998, respectively. Wells A01 through A12 were drilled using a variable-gauge stabilized assembly. The final two infill wells, A13 and A14Z, were drilled using Baker Hughes Intec's latest phase of directional tool, "the Autotrak," which achieved a smoother borehole, with a reduced TVD error, and halved the drilling time in the horizontal section. Because standoff is critical to Andrew well performance, the ability to drill within a narrow TVD window has a high value in terms of reserves and rate. This will be especially true for any future infill wells targeting remaining oil targets with column heights of 10 m TVT.

Because knowing the absolute TVD of the well is critical for simulation modeling (where a few meters difference in wellbore vertical depth can affect predicted water and gas rates), directional survey data from each well has been analyzed to understand the depth un-

certainty about each horizontal wellbore. Absolute uncertainty at the well heel varies from −0/+4.6 m to −2.4/+3.9 m for two standard deviations in the best and worst wells. This increases at the toe, when it varies from −1.5/+1.6 m to −7.1/+7.3 m.

SURVEILLANCE STRATEGY
SURVEILLANCE DATA

Gathering surveillance data is one of the most cost-effective and useful uncertainty management tools available to improve reservoir management during a field's

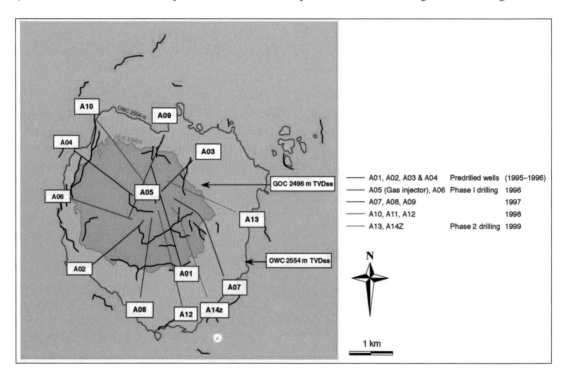

FIGURE 11. Andrew field development-well drilling order from predrilled wells (1995) to end of phase 2 infill drilling (1999).

TABLE 3: ANDREW HORIZONTAL WELLS—FINAL DRILLED LENGTHS, PERFORATED RESERVOIR METERS, STANDOFFS, AND TOTAL DEPTH FORMATION.

Oil producer	Horizontal section (drilled m)	Gross open reservoir (m)	Perforated reservoir (m)	% Wellbore perforated	Standoff GOC: OWC	TD reservoir unit
A01	1500	860	860	Slotted liner 100	67:33	B1 sandstone
A02	1279	1090	762	70	67:33	A2.2 sandstone
A03	917	642	370	58	67:33	Andrew Shale
A04	1090	991	599	61	50:50 heel to 67:33 toe	Andrew Shale
A06	1391	1299	605	47	50:50 heel to 67:33 toe	Andrew Shale
A07	1963	1348	842	62	50:50 heel to 67:33 toe	Andrew Shale
A08	1052	1020	537	53	67:33	Andrew Shale
A09	1316	1283	506	39	67:33	Andrew Shale
A10	1168	1135	617	54	67:33	Andrew Shale
A12	1316	749	506	68	67:33	Andrew Shale
A13	1280	482	462	Slotted liner 96	67:33	Andrew Shale
A14Z	1759	1245	646	57	67:33	Andrew Shale

life. At Andrew, data has been collected since first oil production to

- fill in gaps in static reservoir description, both areally and vertically
- allow the dynamic behavior of fluids to be monitored closely
- enable the optimization of the drilling plan
- evaluate intervention options

Well and reservoir performance is monitored by means of individual well tests. It is planned to test each well at least twice per month (conditions permitting) or at any time that well conditions change dramatically. Oil, gas, and water-rate data are plotted and compared with P10-P50-P90 trends modeled in 1996. It is through careful well-test monitoring that changes in reservoir conditions have been caught early and intervention, when appropriate, has been implemented rapidly.

Reservoir pressure is monitored by use of RFT logs in 8½-in. and 12¼-in. holes of all but two of the producing wells. These pressures are compared to predictions and are used to define reservoir behavior. On a single-well scale, reservoir pressure is monitored by means of downhole pressure gauges in wells A02, A03, A04, and A07, as well as aquifer pressure in A05. Downhole pressure is used to ensure that drawdown limits are not exceeded. Pressures are monitored daily by a software package that links live downhole data with the Aberdeen office. The result is the daily management of offtake strategy on a well-by-well basis.

Detailed surveillance is being integrated into the reservoir description, allowing reservoir management decisions to be made which will optimize the longer-term oil recovery.

ANDREW FIELD PERFORMANCE, 1996 THROUGH 2000

STOOIP, reserves, and recovery factor have been adjusted since field sanction; Table 4 records these changes. The field was sanctioned in 1993 with a STOOIP of 262

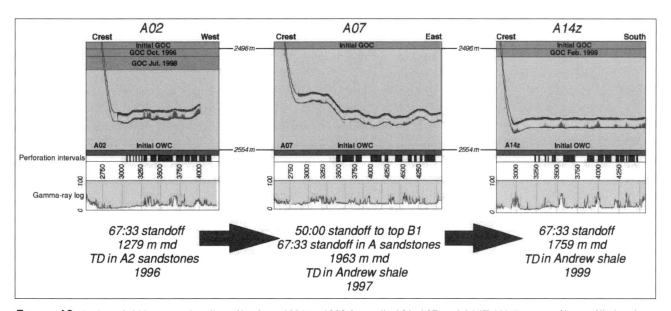

FIGURE 12. Andrew field horizontal-well profiles from 1996 to 1998 for wells A01, A07, and A14Z. Wellbore profile is infilled with V shale curve. The change in GOC from initial is also indicated.

TABLE 4. CHANGES IN STOOIP, RESERVES, AND OIL-RECOVERY FACTOR FROM SANCTION TO 2000.

Key dates	STOOIP (MMBO)	Gas initially in place (bcf)	Oil reserves (STBO)	Sales gas reserves (bcf)	Recovery factor (%)
1993 • sanction	262	182	111.6	134.9	45
1997 • results from six oil producers • 3-D seismic remapped	292	296	131.2	241	45
August 2000 • 2 additional infills (A13 and A14z) drilled	315	296	154	241	49

MMBO, reserves of 111.6 MMBO (a recovery factor of 45%), and sales gas reserves of 134.9 bcf. In 1997, after new seismic mapping and integrating the results of six producing wells, the STOOIP was increased to 292 MMBO with reserves of 131.2 MMBO. Sales gas reserves were increased to 241 bcf. In August 1998, reserves were increased from 131.2 to 140 MMBO, with an increase in recovery factor to 48%. This increase was supported by identification and sanction to drill two additional infill wells and better-than-predicted performance of the 10 producing wells. The field continued to outperform predictions into 2000, with more than 50% field reserves produced while on plateau from 12 wells. STOOIP was increased to 315 MMBO after a remapping exercise, in response to better-than-predicted oil-production rates. The ultimate recovery factor was increased to 49% (with an aspirational target of 53%), and field oil reserves were increased to 154 MMBO. Ultimate recovery factor and field reserves were evaluated by integrating field surveillance data with seismic data in the full-field reservoir simulator. These have not been adjusted since 2000.

PRODUCTION

Andrew's production history is summarized in Table 5. Total cumulative oil production one year after first oil was 16.26 MMBO (6% OOIP). The A05 gas injector came on line in 1997. Cumulative gas injected into Andrew as of June 30, 1997, was 2.237 bcf (0.1863 MMBOE) at a maximum rate of 55 MMSCFD. By August 31, 1998, 44.09 MMBO (16% of OOIP) had been produced from nine wells. In addition, Andrew gas injector A05 had injected a total of 16.28 bcf (2.8 MMBOE) at a maximum rate of 55 MMSCFD. Total cumulative oil production to August 31, 1999 was 68.4 MMSTB from 12 wells. This has risen to 120 MMSTB at the end of 2002.

Oil production has been distributed among all the Andrew producers, with more oil produced from low GOR wells and less oil produced from higher GOR wells (Table 6).

The wells are performing as expected and better than expected in several cases. Their productivity indices range from 130 to 160 BOPD/psi. This is attributed in part to the reservoir, which appears to have a larger STOOIP, to be well connected, and to have pressure support from the gas cap and the aquifer. However, optimized horizontal-well design is also a major contributing factor to this success story—specifically, the well standoff between the GOC and OWC, the length of horizontal sections (900–1700 m measured depth), evenly spaced perforations targeting all reservoir units, and equal production contribution from the length of the wells.

At sanction, wells were predicted to achieve maximum rates of 15,000 BOPD and produce on average 12 million bbl reserves per well. The additional two wells (A14Z and A13) drilled in 1998–1999 were targeting 4 MMBO new reserves each. Initial well rates varied from 10,000 to 17,000 BOPD; however, in compliance with the field-drawdown strategy, wells are produced at rates below their maximum potential. Average daily oil, gas, and water-production rates on plateau are listed in Table 7. Wells which have low GOR (below 800) are produced at 10,000–12,000 BOPD (A02, A06, A07, A08, A10, A12, A13, and A14z). Wells in which GOR is higher than 800 (A03, A04, and A09) are produced at a reduced rate of 5000–6000 BOPD.

Cumulative oil, water, and gas production to the end of 1999 by well is displayed in Table 6. Wells affected by higher GOR are rate constrained, which accounts for the reduced cumulative production from A03 and A04 compared with the other two template wells, A01 and A02. Generally, wells penetrating the steeper-dipping areas of the field (A01, A02, A06, and A08) have access to greater STOOIP and have produced cumulatively more oil. The exceptions are A04 and A09, where the well rate is constrained to control gas production. This compares with wells drilled into the shallow-dipping eastern-flank areas with a thinner oil column (A13z, A03, and A07). These wells are surrounded by smaller STOOIP volumes and are predicted to have lower per-well reserves and to experience shorter times to gas and/or water breakthrough.

TABLE 5. ANDREW FIELD PRODUCTION FROM JULY 1996 TO AUGUST 1999.

Production	1996–1997 (6 wells)	1997–1998 (9 wells)	1998–1999 (12 wells)	Total (1996–1999)
Oil production (million bbl)	16.26	26.91	25.25	68.42
Water production (million bbl)	0.18	1.3	5.8	7.1
Water cut (%)	0–>4	4–>11	11–>25	—
Gas production (bcf)	14.6	21.9	24.7	61.2
GOR (scf/bbl)	745–862	800–931	800–>1100	—
Gas injection (bcf)	2.237	14.04	10.55	26.83

ANDREW OIL PLATEAU

Figure 13 plots the rise in Andrew field plateau oil rates. The sanction plateau oil rate was 54.5 MSTBOPD from 10 producers and one gas injector. This was to be reached six months after first oil (July 1996) and maintained for two years (end of 1998). By January 1997, six wells had been drilled and were producing, and the average oil rate had risen to 59.5 MSTBOPD. By early 1998, rates of 66.5 MSTBOPD were possible from nine producing wells. In September 1998, the Andrew 75 MSTBOPD production strategy was implemented to take advantage of spare oil-plant capacity. This was

achieved from 10 producing wells. Two additional producers were drilled in late 1998–early 1999 in response to identification of unswept oil reserves. They enabled extension of field plateau rates at 75 MBOPD to the first quarter of 2000. Rates declined (from second quarter 2000) gradually to an average oil rate in 2002 of 32.3 MPOBD.

During Andrew Plateau production, this production strategy of increasing oil rate in line with capacity was evaluated constantly to ensure that the elevated drawdown had no long-term detrimental effect on Andrew reserves. The drawdown was maintained strictly through communication of instantaneous and daily average oil constraints. This provided both management assurance and opportunities for production optimization via excess potential. No negative impact on well or field performance has been observed, and the Andrew reservoir has continued to deliver better-than-predicted performance in gas and water breakthrough and GOR and water-cut development.

TABLE 6. ANDREW CUMULATIVE PRODUCTION BY WELL TO END OF 1999.

Well	First oil	Oil production (MMBO)	Gas production (bcf)
A01	September 1996	9.53	8.17
A02	June 1996	11.22	8.93
A03	August 1996	6.68	7.40
A04	July 1996	7.87	8.60
A06	January 1997	7.75	7.21
A07	June 1997	8.04	6.68
A08	August 1997	7.72	7.34
A09	October 1997	4.84	4.93
A10	March 1998	5.18	4.65
A12	September 1998	3.64	3.03
A13	March 1999	2.18	1.92
A14Z	April 1999	2.19	1.71
A05*	April 1997	Total injected gas	32.84

*Gas-injector well.

ANDREW RESERVOIR PRESSURE

The Andrew field is being produced under primary reservoir depletion (gas-cap expansion) with gas injection for primary pressure support and the underlying aquifer providing additional pressure support. The initial reservoir pressure is considered to be 3698 psia at 2496-m TVDss. This is consistent with the RFTs from the appraisal well 16/28-16. There is no evidence of pressure decline from other fields connected to the regional aquifer.

TABLE 7. ANDREW WELL-TEST SUMMARY.

Well	Target oil (produced liquids BOPD)	*Test oil (produced liquids BOPD)	*Test GOR	*Test gas (mscf/d)	*Test water cut (produced liquids BOPD)	*Test water
A01	9000	8614	957	8.2	22.1	2450
A02	5000	8258	951	7.9	44.9	6307
A03	9000	9038	1847	16.7	12.3	1264
A04	10,000	9605	1762	16.9	0.4	35
A06	9000	9351	710	6.6	26.7	3405
A07	6000	9360	957	9.0	31.2	4250
A08	10,000	10,559	1250	13.2	12.0	1439
A09	5000	5707	1155	6.6	4.0	234
A10	6000	7717	1091	8.4	20.6	2004
A12	10,000	10,549	767	8.1	1.3	139
A13	10,000	10,115	701	7.1	0.3	30
A14Z	10,000	9838	771	7.6	0.4	40

Well cycle pairs.
* At last well test, April 14, 1999.

Field data indicate high early-depletion rates (1996–1997) declining to levels of about 5 psi per MMB liquid production (1998). The reservoir pressure measured in 1999 had dropped from 3546 to 3477 psi. The initial steeper, overall pressure decline is thought to be caused by weaker-than-expected aquifer support at start-up. This is attributed to lower water mobility in the residual oil leg below the free-water level and in the transition zone. The reduced rate of pressure decline from 1998 to 2000 was attributed to increased levels of gas injection, along with improved aquifer response.

The horizontal-well design has enhanced connectiv-ity in the Andrew reservoir. The 8½-in. sections of all wells, with the exception of A01, A02, and A13, pene-trate every reservoir layer, which dips across the well tra-jectory. RFT pressure measurements indicate that all Andrew wells are in pressure communication and that there is pressure communication across reservoir zones. Figure 14 plots RFT data from wells A06 through A12. Depletion is greater at the heel than at the toe by about 25 psi. In addition, the data indicate that each of the zones within the Andrew reservoir is being depleted at a similar rate. It is interesting to note that although there are small pressure breaks across faults, or key shales, none

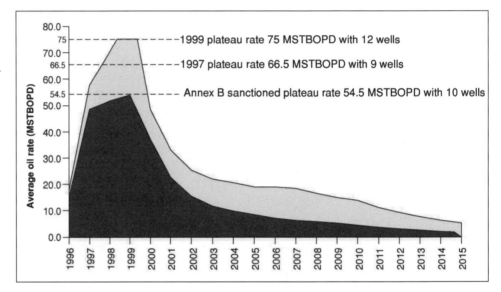

FIGURE 13. Andrew field oil plateau lengths. Annex B sanction case compared with 1997 and 1999 actual plateau oil rates.

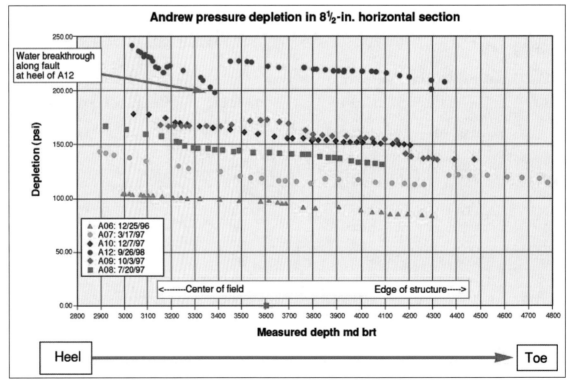

FIGURE 14. Andrew field changing reservoir pressure depletion measured from field crest (well heel) to flank (well toe) from 1996 to 1998.

of them results in zone isolation. From Figure 14, it also is apparent that with increased cumulative production and pressure decline, more breaks in along-wellbore pressures occur. The most extreme example of this is A12, discussed in the section "Water-breakthrough Trends." When integrated with reservoir data, each pressure jump occurs either at a shale baffle or across a fault. These features appear to have more impact later in field life.

INFLOW PROFILES

Inflow from all 12 oil producers is assumed to be greater than 90%, with flow contributions from the whole length of the well. Baseline production logs (spinner/pressure/temperature) were run in October 1996 in A02, A03, and A04 to measure inflow performance. Figure 15 displays the results from well A03. The average-inflow profiles suggested flow contribution from 91% of the wellbore. Although percentage flow contribution does vary by reservoir unit according to sandstone quality, all wells demonstrated flow contribution from all perforations heel to toe. A second production logging campaign was conducted in July 1998 on wells A06, A03, A02, and A01 to investigate inflow performance, gas and water entry detection, slotted liner performance, and GOC movements in a 12¼-in. hole. The production logging-tool (PLT) operation was run using coiled tubing, after the failure to gain access to the horizontal section of the wells with the tool on wireline tractors. Again, for both predrilled liner (A01) and for wells with cemented and perforated liners, it is possible to confirm that there is

- good inflow (heel to toe) in all wells
- no degradation in inflow performance resulting from water production (confirmed on repeat PLT in A02)
- no degradation resulting from gas production (confirmed by repeat PLT in A03)
- no significant reduction in wellbore diameter by scale seen in A02, A03, and A06
- no decrease in productivity seen in A06 as a result of well-kill activities during completion

As a result of the 1998 PLT campaign, a slotted liner was run in A13, which reduced well costs by £1million (U.S. $1.5 million).

GAS AND WATER MANAGEMENT

GAS AND WATER CONING

Maintaining low GOR oil production is critical to Andrew plateau production, because the facilities are gas constrained. The better-than-expected field GOR and water performance has been linked in part to horizontal-well design. Significant are:

- drilling pattern—even well spacing around the dome structure
- wellbore standoff and the careful positioning of the wellbore relative to the A3 shale layer, where the A3 acts as a barrier to early heel gas breakthrough
- well length—the longer the producing wellbore and the more even the offtake along it, the lower the drawdown and risk of gas coning
- selective perforation, in which shales and faults are isolated by blank cemented pipe
- drawdown management

Water breakthrough and the speed of gas coning are expected to be drawdown sensitive. Modeling and analytical work confirm that cumulative oil to breakthrough increases as drawdown decreases. Drawdown strategy has been implemented since 1996 by ranking producers in the order of their probable risk of gas coning or water breakthrough and setting drawdown limits accordingly on a well-by-well basis. Overall, the reservoir management plan and careful monitoring of well performance have optimized low GOR production.

GAS-OIL RATIO TRENDS

The trends in field GOR and water relative to average daily oil production are illustrated in Figure 16. A rise in GOR above solution, indicating gas-cap breakthrough, was first measured in early 1997 in wells A04 and A03, where GOR increased from 770 to 1000 MMSCF test gas/test stbd. By August 1998, seven wells (A01, A02, A03, A04, A06, A08, and A09) had experienced a rise in GOR from 800 to above 900 MMSCFD platform/stbd, and by August 1999, eight of twelve producing wells had gas breakthrough. The arrest in the rising GOR trend in 1999 was the result of three new wells coming on stream—A12 in October 1998, A13 in March 1999, and A14Z in April 1999. The cumulative oil-to-gas breakthrough is in line with preproduction expectations, and the rate of GOR buildup still falls within the P10-to-P90 range of the 1996 (before field start-up) reservoir model predictions. The majority of wells (A03, A04, A06, A08, and A10) has GOR behavior that follows the most likely 1996 model trends. Three wells (A01, A02, and A07) have performed better than expected and have a GOR rise that has followed the P10; one well, A09, has had early gas breakthrough and follows the P90 1996 model trend. Figure 17 plots well test data against the expected P90, P50, and P10 trends in gas and water.

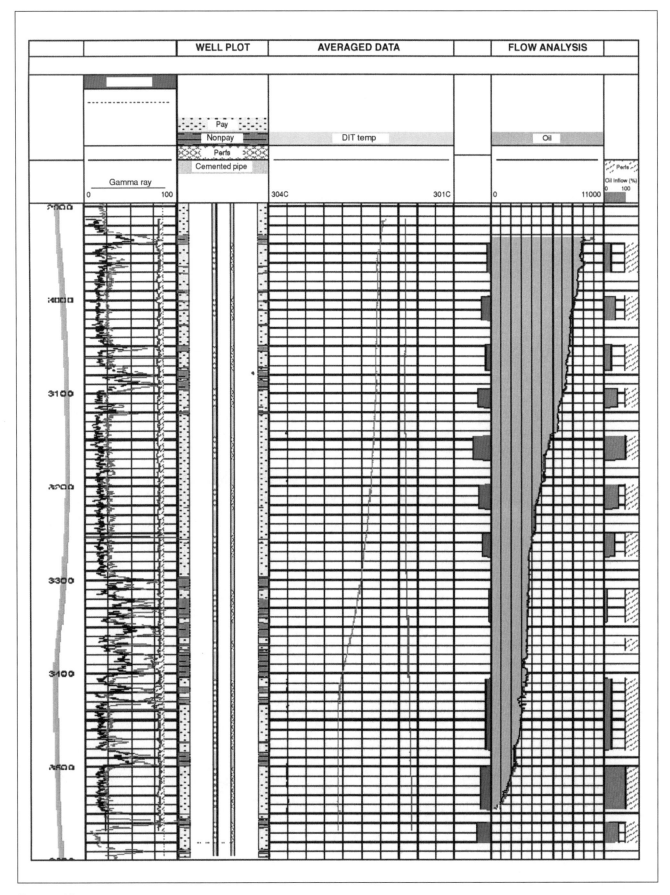

FIGURE 15. Andrew field inflow profile from A03 PLT run in July 1996, showing flow contribution from heel to toe.

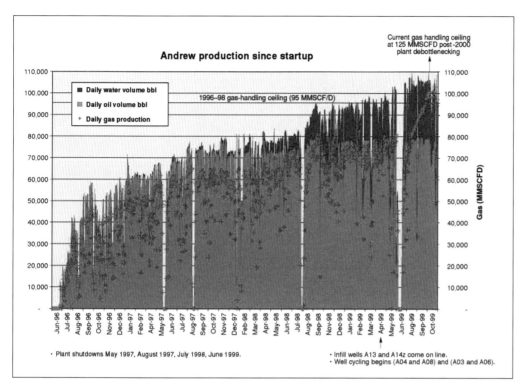

FIGURE 16. Andrew field gas-oil ratio trends from mid-1996 to October 1999, relative to average daily oil production and Andrew gas-plant capacity.

GAS MANAGEMENT

GOR is so critical to plateau oil production that its rise has been managed carefully. For the first 18 months of production, it was not necessary to set maximum GOR limits per well, because there were no gas-handling constraints, and oil rate was paramount. As GOR increased, a gas-management strategy has evolved and been implemented.

Tools used to delay buildup in gas production include increasing the producing length and GOC standoff of wells; using known reservoir shale baffles to reduce gas movement by not perforating these intervals; managing drawdown on a well-by-well basis; controlling oil rates; drilling infill wells to reduce well drawdown; and cycling production between high GOR wells.

The drawdown strategy permits a range of 60 to 125 psi, with well A02 (which has the lowest risk of gas coning) produced at the highest drawdown. There is no evidence that the higher drawdown on A02 has accelerated gas production. In fact, the well remains at solution GOR and has the highest cumulative oil without gas breakthrough (Figure 17). High GOR wells (800-plus) have their drawdown reduced to a maximum of 80 psi.

Once gas-cap breakthrough has occurred in a well, the rate of GOR increase is recognized to be rate dependent. From 1997 to 1998, well GOR rates were reduced by cutting oil rates to 5–6 MSTBOPD for high GOR wells. The early wells to be cut back (A03 and A04) remained at low gas rates for a limited time (months) before the GOR began to rise again. Subsequently, all wells with elevated

GOR are flowed at reduced oil rate, where possible, to reduce the gas rate. Table 7 lists well potential and actual oil rates for high GOR wells. To maintain oil rates, low GOR wells have been produced preferentially to minimize gas. There is some evidence that the wells produced at rates higher than recommended by the drawdown strategy (A06 and A08) experienced accelerated gas breakthrough.

Further evidence of rate dependence comes from fieldwide gas behavior observed after the planned platform shutdown in August 1998. When the Andrew field was put back on production after the seven-day shutdown, every well had a reduction in GOR. This gas behavior implies that the gas cones in the Andrew wells will "heal" when the well is shut in. This observation presented the opportunity to maximize low GOR production by well cycling. High GOR wells have been cycled on and off production to allow gas-cone healing and to reduce their gas rates long enough to produce an incremental oil rate. Cycling was carried out for six months, until the two latest infill wells became available and provided a new source of high-rate, low-GOR production.

Facility upgrades have formed a critical part of gas management. The Andrew platform oil and gas production trains have been "debottlenecked" several times since the Annex B submission. At field start-up, June 26, 1996, the maximum oil-production rate from Andrew and Cyrus was 70 MSTBOPD (wet pipeline). As a result of upgrade work carried out during the August 1998 platform shutdown, platform oil production has been increased from 85 to 90 MSTBOPD (wet pipeline) and

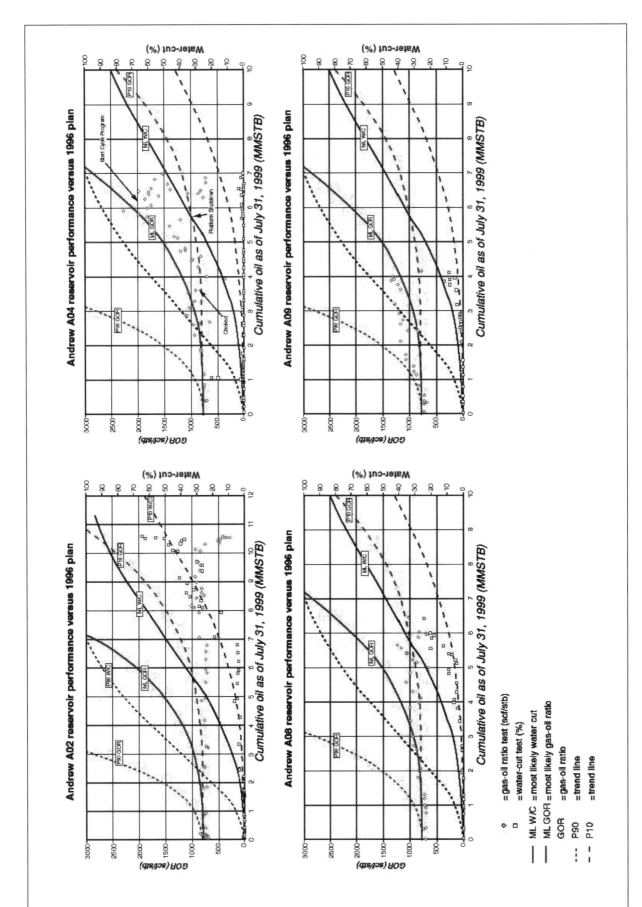

FIGURE 17. Andrew well performance against P90, P50, and P10 predictions from simulation model runs for wells A02, A04, A08, and A09.

platform gas handling from 90 to 110 MMSCFD. The capacity was increased to 125 MMSCFD post-2000 plant debottlenecking.

GAS ACCESS POINTS

Gas-cap gas will be produced from Andrew production wells as a result of local pressure reduction in the oil column, which induces gas to move from the gas cap preferentially toward the producing wells. This may take the form of a gas cone locally above the horizontal section or a gas underrun beneath a shale. Eventually, the field GOC will move down and enter the wellbores. Well-test data show that wells A03, A04, and A09 have the highest GORs. PLTs run in A03 and A04 in 1998 failed to record gas entry points. From a knowledge of field geometry, reservoir properties, and gas-cap distribution, it is as-

sumed that gas coning is more likely along high-permeability sandstones in the A zones toward the toe end of most wells. The B sandstones penetrated at the heel are protected from the gas cap by the A3 shale "umbrella." If these predictions are correct, gas shutoff would be an attractive option, because the toe of the well could be isolated by using a bridge plug.

GAS-OIL CONTACT MONITORING

Movements in the GOC may give useful information on gas-cap movements during continued reservoir pressure decline. The Andrew GOC in the 12¼-in. hole is monitored through LWD logging when a new well is drilled and through repeat logging with a resistivity tool in cased wells as part of an annual surveillance program. In Figure 18, where the data are plotted by well against

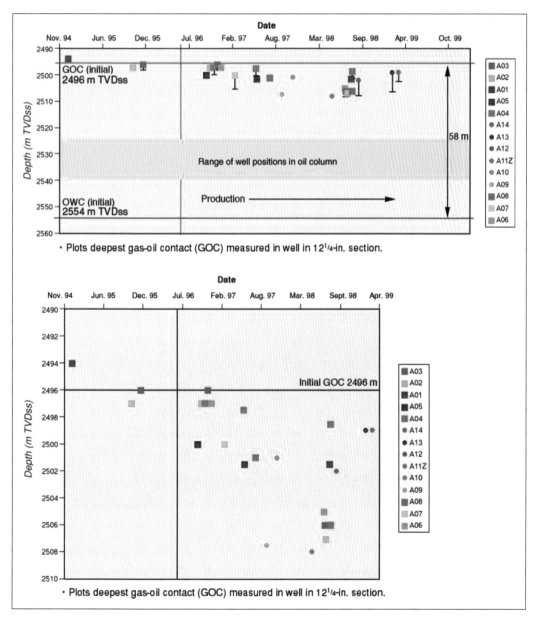

FIGURE 18. Andrew field crestal movements in gas-oil contact measured between 1995 and 1999 in producing wells.

time of measurement, a fieldwide deepening of GOC by 11 m TVDss is established. The measurements are biased to the center of the field and may reflect local gas coning; they are not representative of gas movement toward the flanks. There are also some wells which do not follow this trend, for example, A04, where the GOC, measured in August 1998, had moved only +2.5 m. Figure 19 shows the GOC hung above the B2e shale in this case. In several wells, the mobile GOC rests above intrareservoir shales, which adds support to the idea that shales influence gas movement (at least locally in the reservoir) and act as barriers to vertical gas movement and as baffles encouraging underrun or fingering.

WATER-BREAKTHROUGH TRENDS

Monthly tests have been carried out to measure changes in water cut. Water cut has followed the preproduction P10 predictions and is far lower than expected. The better-than-expected trend confirms that the 67:33 (gas: water) standoff is the optimal position for Andrew wells.

Water cut stayed below 5% for the first year of production (July 1996–August 1997). At that time, the only well with 5% or greater water breakthrough was A02

(5%), which started to cut water in April 1997 (Figure 20). During the second year of production (August 1997–July 1998), the water cut on six wells (A01, A02, A03, A06, A07, and A10) had risen to more than 5%, with field water cut rising to 11% by mid-1998. This trend continued into the third year of production, with the water cut rising from 11% to 25%; water breakthrough was experienced in 10 of the 12 wells (A04 and A14Z remain dry). In November 1999, water breakthrough was measured in A14z, followed by water breakthrough in A04 in December 1999. The field water cut in 2002 is approximately 60% (Figure 16). The slow rise in water rate has been better than originally modeled (Figure 17), with six wells (A01, A02, A03, A04, A06, and A08) following the P10 1996-model water trend. One well, A07, has followed the most likely (P50) 1996 water trend, and one well, A10, came on at a water cut of 4% and is now following the P90 water trend modeled in 1996.

WATER ENTRY POINTS

Water-cut rise from start-up to the end of 1999 in high water-cut wells (e.g., A02) displays a stepped charac-

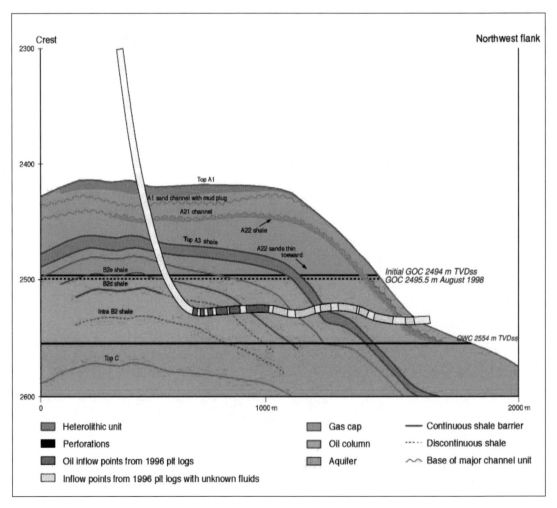

FIGURE 19. Andrew field cross section along A04 showing position of well relative to reservoir zones, oil-water contact, and gas-oil contact. Mobile gas-oil contact rests on intra-B2 shale. Production logging for 1996 recorded good inflow along well length.

ter rather than a smooth, constant trend. This is believed to reflect the influence of variable reservoir properties and faults on fluid movement. Initially, water is thought to run up a single, high-permeability sandstone and then jump to a second sandstone after a period of time, resulting in the stepped trend. The July 1998 PLT campaign successfully confirmed the effect of high-permeability sands and extensive (but thin) shales on water breakthrough in wells A02 (Figure 20) and A03 (Figure 21). Water-entry points were identified by comparing evaluations of the 1996 and 1998 RST porosity logs.

SHALE UNDERRUN

Water breakthrough preferentially occurs by underrun through high-permeability sandstones overlain by extensive, thin shales. Wells A02 (Figure 20) and A03 (Figure 21) record water entering midway along the wellbore, through the 10-m-thick, permeable B1 channel sandstone (mean 300 md, range 80–800 md), underrunning the fieldwide A3 shale. Both wells record heel water entering through a permeable B2 channel sandstone (mean 230 md, range 10–600 md) and underrunning the extensive B2e shale in A03 and the B2c shale in A02. The step-

like rise in water cut records breakthrough starting through one sandstone channel, then jumping to two, and so on. This pattern of focused water entry gives some confidence that water shutoff could work, if the water-wet sandstone can be isolated between two shales. The downside to this scenario is that water production would probably jump rapidly to the next-most-permeable sandstone along the wellbore.

FAULTING

Intrareservoir faults in some cases can act as conduits for early water breakthrough. It is likely that faults become conductive only as reservoir pressure is reduced to a critical point or if drawdown increases. This was only recognized in Andrew wells drilled two years after first oil. Well A12 was drilled and logged in September 1998, with a 67:33 gas-water standoff. The A12 oil target was located in the southeast corner of the field, flanked by well A07 to the east, well A08 to the west, and well A01 to the north. Well A01 had been producing for two years and wells A07 and A08 for one year before well A12 was drilled. MWD resistivity data, displayed against a seismic background in Figure 22, indicated the likely occurrence of water within 5 m

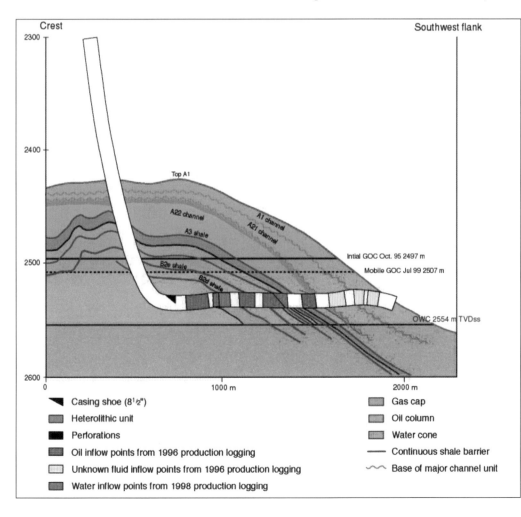

FIGURE 20. Andrew field cross section along well A02 showing water entry points along high-permeability (500-md) B1 and B2 sandstones, interpreted from an RST log run during the 1998 PLT campaign.

of the well heel. This anomaly was confirmed by RFT data, which measured reduced depletion at the well heel relative to the toe (Figure 14). The 350-m-wide anomalous zone coincided with a major low-amplitude anomaly and a mapped east-northeast–west-southwest reservoir fault. The situation was managed by cementing blank pipe across the faulted zone up to the top B2 shale, maximizing the chances of successfully isolating the water zone. A14Z was drilled through the same fault zone in December

1998. It was planned to step the well out flankward in a $12\frac{1}{4}$-in. hole to avoid drilling an $8\frac{1}{2}$-in. hole through it. However, the $12\frac{1}{4}$-in. hole jammed off at the fault, forcing a sidetrack. The same completion strategy was applied to the $8\frac{1}{2}$-in. hole, where a borehole through the faulted zone was completed with cemented blank pipe. Well A08 producing hole passes through the same fault and records a steep rise in water cut in September 1998, from 5% to 30% in six months (Figure 17). History-matching requires

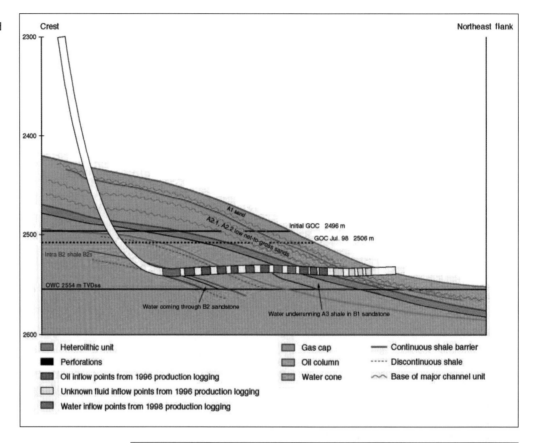

FIGURE 21. Andrew field cross section along well A03 showing water entry points along high-permeability (500-md) B1 sandstones, interpreted from an RST log run during the 1998 PLT campaign.

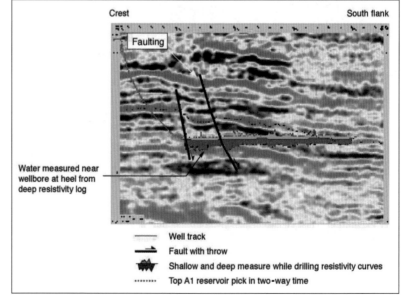

FIGURE 22. Andrew field water breakthrough along intrareservoir fault in well A12, where the resistivity log indicates water 5 m from the heel of the well.

the use of a fault to mimic this trend, and it seems the most likely explanation. Cemented liners and selective perforation allowed the field team to manage surprises in fluid type during drilling later in field life.

Increasing water cut does not seem to have negatively affected oil rate, even in the wettest wells (A02 and A07). However, a production impact has been realized during start-up after plant shutdowns. Gas lift is now required to bring on the wettest wells.

OIL-WATER CONTACT MONITORING

Based on appraisal well data, the original OWC was estimated at 2554 m TVDss. Well A05, a crestal gas injector drilled through the OWC in October 1996, recorded no apparent movement of the free-water level. A resistivity log run in May 1997 shows that the free-water level had moved up by 2.4 m. There is also evidence of higher-than-original water saturation as much as 7 m above the original free-water level in wells A01, A12, and A13. This could be a shale effect but also could be explained by water breaking through into the producing oil column.

INCREASED OIL RECOVERY

Andrew field came off plateau in late March 2000, when the GOR exceeded facility handling capacity. Well rates were therefore cut back until the next phase of plant debottlenecking was completed in May 2000. In March 2000, the field had produced 78 mmbo reserves with 52 mmbo remaining to be recovered from the 12 oil producers, assuming a 49% recovery factor. By the end of 2002, 120 MMBO reserves have been produced (78% of the total reserves) from 13 oil producers.

Many options to increase oil recovery have been investigated from 2000 to 2002 (Figure 23). The most effective way of increasing the final oil-recovery factor to 53% or higher is by drilling additional infill wells and continuing to manage drawdown on a well-by-well basis.

Full-field model runs have been used to predict size and location of unswept oil targets. The targets are generally small (0.5–4 MMBO reserves) with thin (10 m TVT) oil columns. Results of simulation (Figure 24) suggest that they occur between the producing wells, cut off from the producing wellbore by water or gas coning. To drain these targets, low-cost wells are essential, because the current well design is economically unattractive for targets smaller than reserves of 4 million bbl. A program of multilateral sidetracks has been appraised which would allow continued production from the original wellbore while accessing additional unswept oil by use of multiple arms (Figure 25). These would be drilled with coil tubing or through-tubing rotary drilling and completed barefoot, once the original wellbore had dropped below an oil rate

of 3 MBOPD. It is estimated that a program of eight multilateral sidetracks could increase reserves to 155 MMBO (53% recovery factor). In 2000, all Andrew wells were producing more than 5 MBOPD and did so until mid-2001. After comparing the risks and reserves from multilateral wells with new infill wells in 2000, it was decided to drill a new infill well (A15) in 2001 as part of the Andrew Phase 3 program. The multilateral sidetracks were considered too risky.

Gas and water shutoffs have been implemented successfully in other BP producing fields (Prudhoe Bay, Alaska), and their potential to reduce GOR or water rate in Andrew producers has been evaluated. Gas shutoff options being investigated include mechanical devices (straddles, patches, packers), cement plugs, chemical shutoff such as cross-linked polymers, inorganic gels, foams, or relative permeability blockers. Currently, two issues stand in the way of implementing this technology:

- mechanical risk of damaging the well and losing production/reserves associated with reentering a producing wellbore
- rapid movement of gas and water into the next-most-permeable sandstone once a gas- or water-producing zone is shut off, thus providing insufficient low GOR oil to support the activity

Gas entry points were identified from PLT data, collected in 2001 in high GOR wells.

Until some of these risks are resolved, it is unlikely that a shutoff will be programmed. However, a multilateral sidetrack program will look more attractive if gas-producing zones in the original producing wellbore can be shut off.

FLUID IDENTIFICATION FROM SEISMIC DATA

Three main uncertainties currently make the multilateral side-track program unattractive:

- the risk of losing production and reserves from the original wellbore through mechanical failure or wellbore damage during drilling
- the risk of successfully drilling and perforating the sidetracks
- the difficulty of accurately predicting target location and size

Large uncertainty exists about target size and location of remaining oil. In the Andrew field, this is not easily resolved by simulation modeling, because of the difficulties of history-matching a field with natural aquifer and gas-cap drive. Seismic data, however, do provide a useful uncertainty-reduction tool. Seismic rock-property modeling is a proven technology that, with the right data sets, can be used to identify fluids and lithologies directly from seismic data. If a time-lapse survey is shot after production has started (4-D seismic), it should be possible to identify the redistribution of oil, water, and gas in the reservoir at that time, compared with the preproduction picture. This technique has been used to identify unswept oil targets in the Forties field, where several targets have been drilled and are producing. The Andrew 3-D survey was reprocessed in 2000 to preferentially display fluid and lithology information. The appraisal phase was successful, and a second 3-D survey was acquired in 2001, along with three PLTS to calibrate fluid movements in the wellbores. Comparison of 2001 and 1992 preproduction 3-D surveys has enabled the team to image fluid (oil, water, and gas) movements in the reservoir with certainty. It has resulted in the mapping of four unswept oil targets with economic volumes and high certainty. Phase 3 was sanctioned in 2001 and the first infill well was drilled in 2001 between A02 and A06 (A16z). The second Phase 3 well was planned for late 2002.

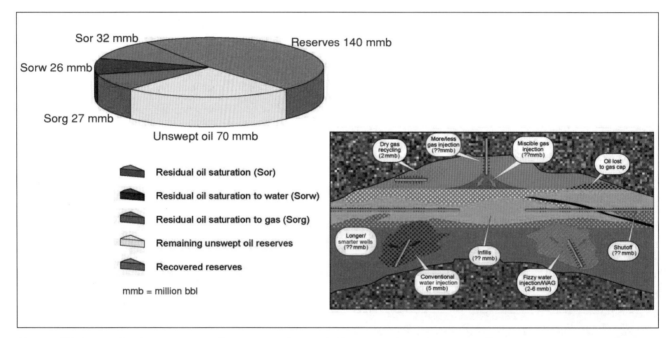

FIGURE 23. Future development options for increasing Andrew field recovery factor from 45% to 55%.

FIGURE 24. Andrew field remaining oil at 2002. Shown are infill and sidetrack locations.

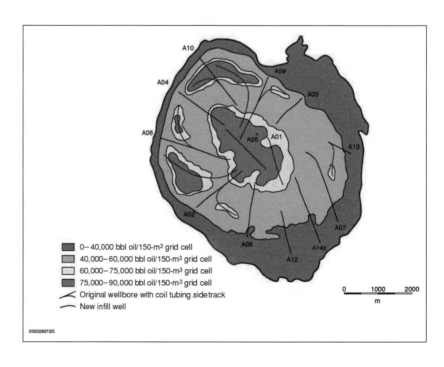

CONCLUSIONS

The Andrew field development has been a successful project for BP and the field partners. More than half the oil reserves (more than 70 million bbl) was produced in two-and-one-half years while at maximum oil rates. Plateau oil rates were increased from 54.5 to 75 MSTBOPD per day, and oil plateau was extended by 18 months (a 30% time increase).

This is partially because the reservoir has outperformed earlier expectations. STOOIP has increased from 262 million to 315 million barrels. The reservoir is well connected; at the same time, it has a sufficient level of heterogeneity to delay early coning of gas and water. Gas-cap expansion and aquifer support provide sufficient natural energy to drive fluids and maintain reservoir pressures without need for water injection.

The development plan and use of horizontal wells, however, also have made a significant contribution to Andrew field's performance. Identification and evaluation of critical reservoir uncertainties before development plan-

ning allowed the team to optimize well design, develop a drawdown strategy and reservoir management plan, and design contingencies to deal with surprises.

Horizontal wells are the optimal design for the Andrew field. They have allowed oil to be produced preferentially at high rates and low drawdowns, which is critical for reducing gas-coning rates. Drilling the horizontal wellbores from the structural crest out through the flanks, thereby crosscutting all the reservoir layers, has increased reservoir connectivity and has provided balanced pressure depletion and even production. Drilling long horizontal wellbores has reduced the drawdown pressures along each wellbore. The placement of the wellbore close to (19 m above) the oil-water contact and away from the gas-oil contact (one third:two thirds; 33:67 standoff design) has delayed gas coning.

Considerable time was spent before development discussing and modeling completion design and perforation strategy. The use of cemented and perforated liners has been a huge success; it has allowed a preferential perforation strategy in which production, low drawdown pres-

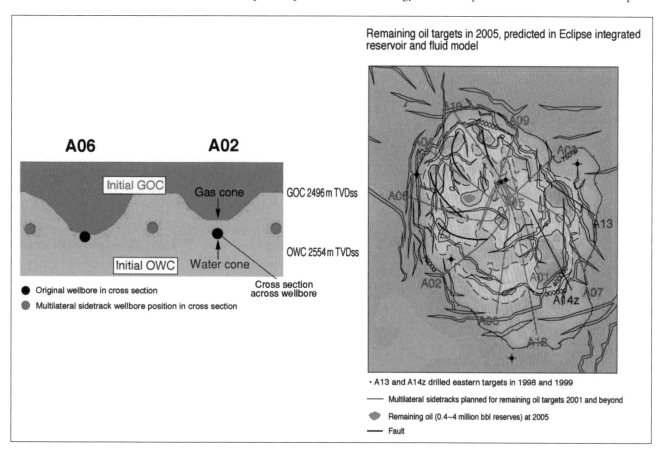

FIGURE 25. Andrew field, redeveloped with multilateral sidetracks.

sures, and the possibility of future interventions are optimized, while at the same time minimizing the risk of sand production and early gas and water coning.

Data collection during drilling and early production has allowed the reservoir management plan to be updated continually and has given the team a platform from which to manage surprises. The field GOR, however, has risen close to the facility limits during the last year, making data collection more difficult to justify because of the risk of damaging a wellbore and losing oil production during data collection.

Horizontal-well design has improved oil recovery. Oil reserves have increased by 38 million bbl, from 116 million bbl before development to 154 million bbl in 2000. Twenty million bbl can be attributed to the increase in Andrew STOOIP, and 18 million bbl to an increase in the oil-recovery factor.

The future objective is to maintain oil rate by increasing recovery factor and reducing water and gas production. The critical issue to delivering this is being able to more precisely image fluids (oil, water, and gas) in the reservoir and to deliver a program of options at an acceptable technical and operational risk. This has been enabled by acquiring and using 4-D seismic technology. From this understanding, several new oil targets, with higher certainty, sufficient size, and risk, have been mapped, and well-intervention options have been identified. This option set makes up Phase 3, which began in 2001 with infill well A15.

ACKNOWLEDGMENTS

This paper draws on ideas, data, and reports created over eight years presanction, through development drilling to the present by many members of the Andrew field team. Specifically, we would like to acknowledge Paul Bowden, Keith Everill, Bob Klein, John Petler, Anchala Ramasamay, and Simon Todd from the 1992–1996 field-development team. We would also like to acknowledge the field partners, Kerr-McGee Corp., Lasmo, MOEX, and Talisman, for their support and contribution to the understanding and development of the Andrew field.

REFERENCES CITED

Aadnoy, B. S., 1987, A complete elastic model for fluid-induced and in situ generated stresses with the presence of a borehole: Energy Sources, v. 9, p. 239–259.

Anderson, S. A., S. A. Hansen, and K. Fjeldgaard, 1988, SPE Paper 18349, presented at the SPE Horizontal Drilling and Completion Symposium, Copenhagen, Denmark.

Clark, J. D., and K. T. Pickering, 1996, Quantitative analysis of modern submarine channels, in J. D. Clark and K. T. Pickering, eds., Submarine channels processes and architecture: London, Valis Press, p. 75–95.

Fine, S., M. R. Yusas, and L. N. Jorgensen, 1993, Geological aspects of horizontal drilling in chalks from the Danish sector of the North Sea, in J. R. Parker, ed., Petroleum geology of north west Europe: Proceedings of the 4th Conference: Geological Society (London), p. 1483–1490.

Masie, I., O. Nygaad, and N. Morita, 1987, Gulfaks subsea wells: An operator's implementation of a new sand production prediction model, SPE Paper 16893, presented at the SPE Annual Conference, New Orleans, Louisiana, September 1987.

Smith, P. J., and Buckee, J. W., 1985, Calculating in-place and recoverable hydrocarbons: A comparison of alternative methods, SPE Paper 13776, presented at the SPE Hydrocarbons and Evaluation Symposium, Dallas, Texas, March 14–15, 1985.

Steele, R. P., R. M. Allen, G. J. Allinson, and A. J. Booth, 1993, Hyde: A proposed field development in the southern North Sea using horizontal wells, in J. R. Parker, ed., Petroleum geology of north west Europe: Proceedings of the 4th Conference: Geological Society (London), p. 1465–1472.

Tehrani, A. D. H., 1991, An overview of horizontal well targets recently drilled in Europe: SPE Paper 22390, presented at the Second Archie Conference, Houston, Texas, November 3–6, 1991.

6

Craig, P. A., T. J. Bourgeois, Z. A. Malik, and T. B. Stroud, 2003, Planning, evaluation, and performance of horizontal wells at Ram Powell field, deep-water Gulf of Mexico, *in* T. R. Carr, E. P. Mason, and C. T. Feazel, eds., Horizontal wells: Focus on the reservoir: AAPG Methods in Exploration No. 14, p. 95–109.

Planning, Evaluation, and Performance of Horizontal Wells at Ram Powell Field, Deep-water Gulf of Mexico

Peter A. Craig

Shell Exploration and Production Company
New Orleans, Louisiana, U.S.A.

Timothy J. Bourgeois

Shell Exploration and Production Company
New Orleans, Louisiana, U.S.A.

Zaheer A. Malik

Shell Exploration and Production Company
New Orleans, Louisiana, U.S.A.

Terrell B. Stroud

Shell Exploration and Production Company
New Orleans, Louisiana, U.S.A.

ABSTRACT

Ram Powell is one of the major tension-leg platform (TLP) developments in the eastern deep-water Gulf of Mexico. The three main turbidite reservoirs are producing from a total of five horizontal, open-hole, gravel-packed wells. The high rate and high ultimate recovery from these horizontal wells have reduced substantially the development well count (in one reservoir by 50%). Well A-3ST1 at one time held a Gulf of Mexico (GOM) record of highest rate for a single well—40,900 BOE per day.

Understanding reservoir architecture is key to execution of a trajectory plan. This knowledge is best obtained through use of pilot wells near the horizontal-well path. In using exploratory or appraisal wells as pilots, survey accuracy can be critical to success. Horizontal-well trajectories can be optimized in reservoirs (such as levees or sheet sandstones) in which lateral variations are well understood. At Ram Powell, horizontal wells in these depositional systems were drilled oblique to strike, so as to transect the entire section. In more laterally variable systems, such as channels, more well control or geosteering is required.

New applications of petrophysical tools that are unique to horizontal wells aided in drilling and reservoir analysis. Density images from a 121-mm (4¾-in.) OD logging-while-drilling (LWD) density tool were used to calculate the dip of the thin-bedded L sandstone. This tool was also used to evaluate the geometry of calcite-cemented zones in an unconsolidated sandstone. Quantitative fluorescence technique (QFT), a patented Texaco process, quantifies the fluorescence in cuttings for pay evaluation. Measurements were used to distinguish mudstone, thin-bedded sandstone, and massive sandstone.

INTRODUCTION

Ram Powell is one of the major tension-leg platform (TLP) developments in the deep-water Gulf of Mexico operated by Shell, in partnership with BP Amoco and ExxonMobil. It is located in the eastern Gulf of Mexico, 113 km (70 mi) east of the tip of the Mississippi delta, in Viosca Knoll Block 956. The TLP is situated on the modern-day slope in 980 m (3214 ft) of water (Figure 1).

Three primary reservoirs are under development from five horizontal, open-hole, gravel-packed wells. The primary development was conceived as a horizontal program because of the necessity of producing high rates from a thin oil rim (J sandstone) and from reservoirs with low permeability and thickness (L sandstone, N sandstone), and because of connectivity concerns (N sandstone). The development also consists of five conventional wells in four other reservoirs producing from multiple stacked pay zones, or in which the above considerations were not an issue.

GEOLOGIC OVERVIEW

Ram Powell is located in the eastern Gulf of Mexico (Figures 1 and 2) in the Viosca Knoll area on the modern-day slope. Unlike the central and western deep water, in which sedimentation occurs in predominantly ponded, salt-withdrawal minibasins (Prather et al., 1998), the area shown in Figure 2 is dominated by slope, or bypass, depositional processes. Based on paleontologic control, the primary reservoirs (J, L, M, and N sandstones) at Ram Powell were deposited in the Serravallian (middle Miocene) (Figure 3). The deposits represent deposition in slope and toe-of-slope accommodation space, or a lowstand systems tract, at approximately 12.0 Ma (Haq et al., 1989). Ram Powell reservoirs are the downdip equivalents of the slope-channel and channel-overbank systems at the Virgo (Elf) field (Figure 2).

The N sandstone was the first sandstone deposited as amalgamated channel deposits in a preexisting submarine slope canyon. The M sandstone was subsequently deposited as amalgamated channel and channel overbank in accommodation space created by differential subsidence of the N

sandstone. Avulsion of this submarine canyon to another feeder system to the west formed the L and J reservoirs. The J reservoir, with a more fanlike geometry, was deposited in accommodation space formed by differential subsidence of the L sandstone. The J is west of the L, onlapping the L to the east. The two share fluid contacts and pressure regimes. The M and N sandstones are amalgamated channel deposits, the L sandstone is a channel-levee complex, and the J sandstone is a hybrid deposit of amalgamated sheet and channel-levee (Bramlett et al., 1998).

The reservoirs dip south-southeast 2° to 4°. The sandstones pinch out updip to form a stratigraphic trap. The reservoirs form a syncline to the south and truncate against a salt diapir approximately 6.5 km (4 mi) to the south. Estimated aquifer volumes are approximately the same as the hydrocarbon volumes; production has dem-

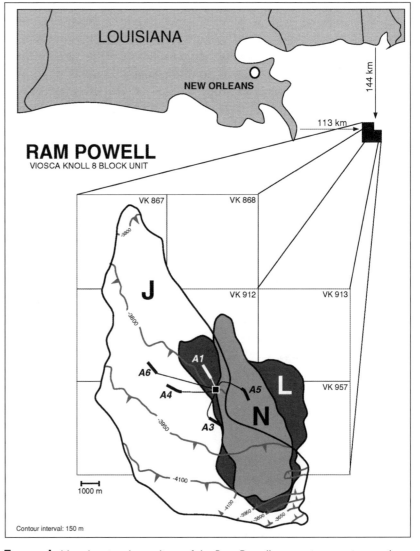

FIGURE 1. Map showing the outlines of the Ram Powell reservoirs superimposed over the J sandstone top structure map. Inset shows the location of the platform in the Gulf of Mexico.

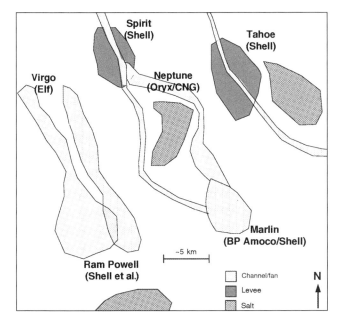

FIGURE 2. Middle Miocene paleogeographic map of the Viosca Knoll area. Commercial discoveries at the time of writing are named on the map.

onstrated that aquifer support is negligible at Ram Powell. Faulting is absent in the reservoirs except for a small fault that cuts the eastern edge of the J sandstone. This fault had no impact on development planning.

WELL PLANNING

The stratigraphy of the reservoir dictated the number of pilot wells needed. The necessity of precisely constraining the depth of the reservoir to locate a horizontal well in a thin oil rim (J sandstone) required the use of a single pilot. However, when the internal architecture of the reservoir was complex and uncertain, two pilots were needed (N sandstone) to maximize the quality of the horizontal well.

The objective was to place a pilot hole at the heel of the horizontal-well trajectory so that the 194 mm (7⅝-in.) casing shoe could be placed at the desired location in the sandstone. In the N sandstone, which had a complex internal architecture, a second pilot was used to constrain

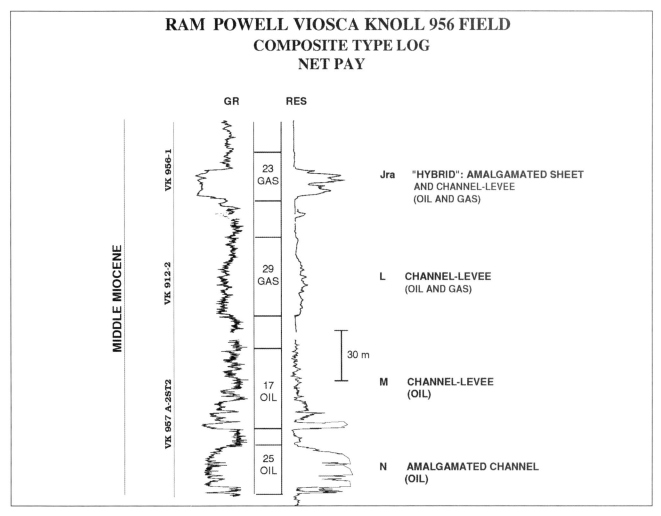

FIGURE 3. Composite type section of the J-N sandstones.

the geology at the toe, and the well was drilled between the two pilots. The horizontal sections were started at the base of the sandstone. In that way, if the top of the sandstone came in higher or lower than expected, minor course corrections could be made to position the heel and, therefore, the lateral section in the optimal location. This was particularly important when the landing point of the heel of the horizontal section did not exactly coincide with the pilot location.

In practice, the horizontal-well plans used existing penetrations (appraisal wells) as pilots, thereby reducing the total cost of the pilots by half (~U.S. $10 million). Four preexisting wells were used as horizontal pilots, compared to only three planned (and one unplanned) pilots.

The stratigraphy of the reservoir also guided the formulation of strategy for horizontal drilling. The trajectories of the wells were optimized to maximize the net-to-gross ratio (N/G) exposed to the wellbore, to improve the maximum rate of the well. This was successfully done in both the L and J sandstones. In layered systems, such as a sheet or levee, the wells were drilled obliquely to strike, allowing the wellbore to cut through the entire section. In the N sandstone, where multiple channel bodies are present and compartmentalization is a risk, the trajectory was designed to cut through multiple channel bodies.

DRILLING AND COMPLETION

The methodology for drilling and completion of the horizontal wells at Ram Powell is similar for all five horizontal wells; the main execution difference between each well was the selective use of pilot wells. Refer to Table 1 for a list of the horizontal wells.

Each well was drilled to a true vertical depth of 2750–3050 m (9000–10,000 ft), at which point a 299-mm (11¾-in.) casing string was set. (The objective for the horizontal wells varied from 3810 to 3900 m [12,500 to 12,800 ft] true vertical depth.) At this point, a pilot well (if needed) was drilled to the objective sandstone to obtain the necessary depth control or stratigraphic information. The well was then plugged back and sidetracked to land the casing shoe. The strategy was to drill the pilot hole to the desired horizontal location, then plan the horizontal trajectory to hit the exact location of the pilot well. In practice, this plan was not feasible in many cases, because unexpected results in the pilot would necessitate changing the location of the horizontal well. After the sidetrack was drilled, a 194-mm (7⅝-in.) casing shoe was set within the objective sandstone. Because of the high cost of sidetracking after running the 194-mm (7⅝-in.) casing, it was critical to be at the desired location in the sandstone.

After changing from oil-based mud to a synthetic drill-in fluid, a 762-m (2500-ft) horizontal section was drilled and evaluated using logging-while-drilling (LWD) triple combo and mud logging.

Following drilling, a wire-wrapped, prepacked gravel screen was run into the hole, then an external gravel pack was run (Lester et al., 1999). The tubing used was a string tapered from 114 mm (4½ in.) to 140 mm (5½ in.).

Note that because the first four horizontal wells were predrilled prior to the TLP installation, they were temporarily abandoned before being drilled as a horizontal well. The horizontal section in each well was a reentry drilled to a different target. As a result, the Minerals Management Service (MMS, a U.S. regulatory body for federal waters) required that the horizontal section drilled by the TLP be considered a sidetrack. Thus, the name of the well does not indicate the number of pilots (e.g., A-3ST1 had no pilot and A-4ST2 had only one).

EVALUATION

New applications of petrophysical tools that are unique to horizontal wells aided in the drilling of the wells and analysis of the reservoirs. Density imaging assisted in evaluating structural and stratigraphic informa-

TABLE 1. MAXIMUM RATES AND LENGTHS FOR EACH OF THE HORIZONTAL WELLS IN THE RAM POWELL FIELD. A-2ST2, A DEVIATED WELL, IS INCLUDED FOR COMPARISON PURPOSES.

Well name	Completion length (m)	Reservoir	Oil rate STB/D	Gas rate m³/sec	BEQ
A-1ST1BP2	688	LRA	9509	34	28,320
A-3ST1	623	JRA	25,312	28	40,908
A-4ST2	757	JRA	31,009	14	38,893
A-6ST2	670	JRA	17,058	7	21,054
A-5ST2	726	NRA	11,681	6	14,746
A-2ST2	*	MRB	6330	3	7857

tion in a thin-bedded formation. Image data from a 121-mm (4¾-in.) OD LWD density tool was used to calculate the dip of the thin-bedded L sandstone in well A-1ST1BP2. This dip data, coupled with visual interpretation of the images, aided the geologic interpretation of this formation. The LWD density tool also was used to evaluate calcite-cemented zones in an unconsolidated sandstone. The data showed that the shape of the cemented zone was nodular and not a flow boundary. Another tool, QFT, quantifies the fluorescence in cuttings, which enables the final interpreter to have greater confidence in pay evaluation and long-reach horizontal wells without LWD log confirmation. This data also aided in identifying massive versus laminated sandstone.

J SANDSTONE

J SANDSTONE OVERVIEW

The J sandstone is a hybrid deposit—an unconfined fan-lobe, sheetlike turbidite topped by a channel-levee (Figure 4). The sheet ranges in thickness from 15 to 18 m (50 to 60 ft). The levee ranges from more than 21 m (70 ft) near the channel in well A-4ST2 to less than 3 m (10 ft) in well A-3ST1. Permeability in the sheet sandstone ranges from 640 to 2680 millidarcys (md) in core; in contrast, the thin-bedded levee deposit is estimated to have a permeability of less than 100 md.

The J sandstone has a gas-oil contact at −3872 m (−12,705 ft) subsea and an oil-water contact at −3905 m (−12,812 ft) subsea. This 33-m (107-ft) oil rim is the target for the three horizontal J sandstone producers: A-3ST1, A-4ST2, and A-6ST2. The J sandstone is quite extensive; the oil rim covers 9.7 km² (2400 ac), and the gas cap has been mapped at more than 29.1 km⁻² (7200 ac). In-place volumes are approximately 80 million bbl and 17×10^9 m³ (600 bcf).

J SANDSTONE DEVELOPMENT STRATEGY

The J sandstone development plan in 1993 consisted of eight wells: six oil-rim producers and two gas-cap blowdown wells. Advances in horizontal drilling and completion technology created the ability to produce the oil rim more efficiently. An intensive modeling effort was undertaken to determine the number of horizontal wells needed and their placement in the oil column (Lerch et al., 1997). As the modeling showed, optimal placement of the wells was in the middle of the oil column. Also, because the J sandstone is a layered deposit, the wells were drilled oblique to strike in order to transect (and drain) the entire section (Figure 4).

The reservoir simulations also demonstrated that all of the J sandstone oil-rim reserves—nearly 50 million bbl—could be drained by only three wells. The conventional plan had a well spacing in the oil rim of 1.4 km² (340 ac); the horizontal well spacing is 3.2 km² (800 ac). This reduced the development cost of the J sandstone from approximately U.S. $100 million to U.S. $60 million, a substantial savings. The conventional wells in the 1993 plan were expected to produce 6000 BOPD, whereas the horizontal wells have delivered as much as 30,000 BOPD, a performance improvement factor (PIF) of 5.

J SANDSTONE WELL EXECUTION

As a hybrid deposit (sheet sandstone and levee), the J sandstone would seem to be an ideal candidate to forgo the use of pilot wells. However, because the objective was to optimally develop a thin oil rim, the placement of the wells needed to be precise.

The VK 956 A-3ST1 is a 623-m (2043-ft) horizontal well, the first to be drilled in the J sandstone (Figure 5). Amoco, in the VK 956-2 and in three sidetracks, had penetrated the oil rim, and these data were used to plan the location and constrain the depth of the A-3ST1 producer. First production began in December 1997. During the drilling, calcite-cemented zones approximately 3 m (10 ft) in length were encountered in the reservoir. The analysis of these zones is discussed in more detail in the evaluation section.

The VK 955 A-4ST2 is a 757-m (2484-ft) horizontal well. The A-4 was intended to be a pilot well at the heel of the horizontal section; however, the J sandstone was encountered deeper than expected, and the heel was then moved eastward, both along strike and slightly updip, to accommodate this. The initial horizontal trajectory, drilling straight toward the A-4 pilot, was modified slightly to optimize the wellbore's position in the reservoir. A continuous turn rate of 3°/30 m (3°/100 ft) was applied in the first half of the horizontal well to turn the path more nearly parallel to strike, thus increasing the amount of high-permeability sheet sandstone in the wellbore by 20%. The intent was to improve the rate potential of the well. The well was completed, and first production began in March 1998.

The VK 911 A-6ST2 is a 669-m (2196-ft) horizontal producer. The A-6ST1 pilot, like the A-4 pilot, was intended to be located in the heel of the horizontal section; however, the sandstone was 15 m (50 ft) deeper than expected. As a result, the A-6ST2 location was moved updip, 274 m (900 ft) away from the pilot. When drilling down the horizontal landing point, a stratigraphic surprise was encountered—the basal sheet sandstone was

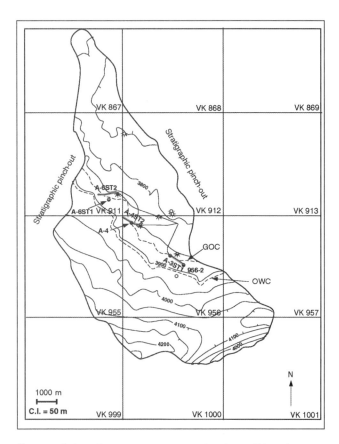

FIGURE 4. J sandstone structure map showing well locations discussed in the text. Lateral sections of horizontal wells are shaded gray.

much thinner than expected (Figure 6). Consideration was given to moving the location of the horizontal, but it was interpreted that the channel (associated with the levee) had eroded the sheet. With the understanding that this erosion was a phenomenon at this location, the decision was made to land the casing at this location for the horizontal well. By drilling the horizontal away from the channel, the expected 15–18 m- (50–60 ft-) thick sheet sandstone was penetrated, and the well has produced as expected.

J SANDSTONE PRODUCTION PERFORMANCE

The production performance of the three J sandstone horizontal oil-rim producers attests to the success of the development. Well VK 956 A-3ST1 held the Gulf of Mexico production record—40,900 BEQ per day (which was subsequently broken by Troika, Green Canyon 244 TA-3 well). Cumulative production from the J sandstone (as of December 31, 2001) was 29.0 million bbl and 5.83 $\times 10^9$ m^3 (205.9 bcf) of gas (Figure 7). Recent history matching of the two years of production has demonstrated the accuracy of the initial reservoir simulations and has confirmed that the oil rim is being drained by the three wells. Although they are located only 12–15 m (40–50 ft) from the oil-water contact, significant water production did not begin until a year and a half after first production.

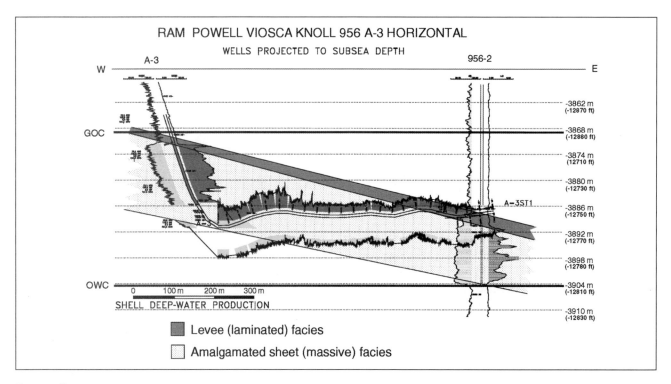

FIGURE 5. Cross section of well A-3ST1. The well design and relative placement in the oil rim are typical of the three J sandstone horizontal producers.

N SANDSTONE

N Sandstone Overview

The N sandstone is the basal sandstone of the J-N lowstand systems tract. It is an amalgamated channel deposit with complex internal architecture (Lerch et al., 1997). The N sandstone is an undersaturated oil reser-voir with an oil-water contact estimated to be at −4084 m (−13,400 ft), based on seismic evidence. Several other water levels, however, have been found updip of known, pay-to-base penetrations (Figure 8). This is indicative of reservoir compartmentalization. The reservoir perme-ability is estimated from pressure-transient analysis to be 300 md.

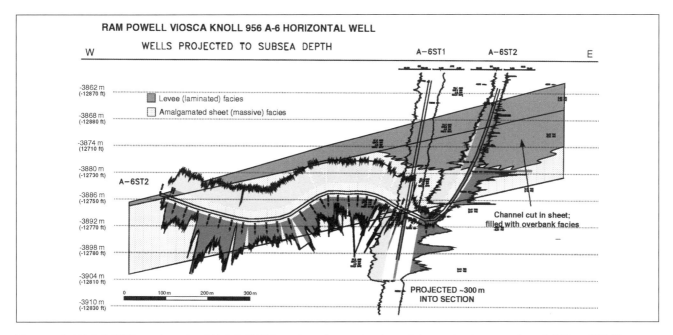

FIGURE 6. Well A-6ST2 cross section. Well A-6ST1 is projected onto the plane of the section by approximately 300 m (900 ft).

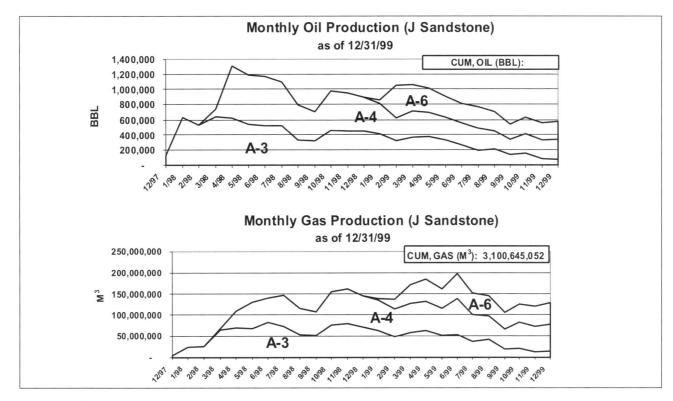

FIGURE 7. Graph showing monthly oil and gas production for the J sandstone, by well.

N Sandstone Development Strategy

Because of a high degree of internal complexity, the strategy in the N sandstone was to use two wells for stratigraphic control when drilling a horizontal well. The development plan for the N sandstone had envisioned two horizontal producers, one updip and one middip. (As many as four other wells were planned—a contingent downdip horizontal producer and two or three high-rate injectors—but those are not discussed here.)

Although no water levels had been penetrated during the initial stages of development planning, the risk of N sandstone compartmentalization had been acknowledged during development planning (Lerch et al, 1997; Bramlett et al., 1998). A location was chosen for a horizontal well in the central part of the reservoir, connecting a pilot well, the VK 956 A-2, to an appraisal well, the VK 956-3. (This is the pilot-to-pilot strategy discussed earlier.) The A-2 pilot was expected to encounter 12 true vertical net m (40 ft) of oil. Instead, it penetrated 18 true vertical net m (60 ft) of oil, and 24 true vertical net m (80 ft) of water (Figure 8). This location was deemed unsuitable for a producer, and the well was then sidetracked downdip to the thickest location penetrated in the appraisal program. It was completed conventionally because the thickness at that location did not require a horizontal well to achieve a high-rate producer.

An updip well was still needed, and a second attempt was made to drill a horizontal well in the northern part of the reservoir. The favored option was to drill a new pilot well (VK 911 A-5; Figures 8 and 9) to test an unproven portion of the reservoir and link it to another penetration, the VK 956 A-2ST1. Once again, the pilot encountered a surprise—a new water level, different from the one encountered in A-2. This contingency was planned for, so an alternate location was chosen for the horizontal. The VK 956 A-5ST2 was drilled between two existing penetrations that were used as pilots (A-2ST1 and 956-3). The horizontal well revealed in detail some of the internal complexities of this amalgamated channel reservoir (Figure 9).

N Sandstone Production Performance

The VK 955 A-5ST2 is a 725-m (2380-ft) horizontal producer in the N sandstone. Peak rate from this well is 11,681 BOPD. However, the relative placement of A-5ST2 in the N sandstone has hampered the productivity of the well. The horizontal well penetrates several shale sections (~90 m in length; Figure 6) and the borehole is near (<1 m) the top of the reservoir for approximately 30% of its length. Despite this, however, the A-5ST2 has a PIF of 2 over a conventional well. The A-2ST2, initially

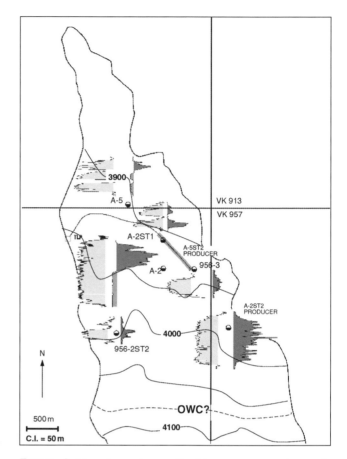

FIGURE 8. N sandstone log motif, with structure superimposed, showing the outline of the N sandstone and logs for each penetration. Note the water levels updip of the inferred oil-water contact (OWC).

completed in the N sandstone, is now producing from the overlying M sandstone. The M sandstone is an amalgamated channel reservoir, similar to the N sandstone. The two also share the same pressure regime and have similar PVT (pressure-volume-temperature) properties and permeability. The A-2ST2 M sandstone is a conventional completion well with a 66° deviation, with 12.5 true vertical net m (41 ft) of oil. The A-2ST1, the heel pilot for the A-5ST2, contains 9 true vertical net m (30 ft) of oil. The A-5ST2 produces at double the rate of the A-2ST2 M (maximum rate 6,330 BOPD; Table 1 and Figure 10).

L SANDSTONE

L Sandstone Overview

The L sandstone is an oil-rimmed gas reservoir deposited as a channel-levee complex. Only the eastern levee has been found to contain hydrocarbons. The eastern levee is a fining-upward, thin-bedded deposit with an average lamination thickness of approximately 2 cm (less than 1 in.). Core-plug permeability ranges from 10 to 1000 md. A permeability of 90 md is observed from

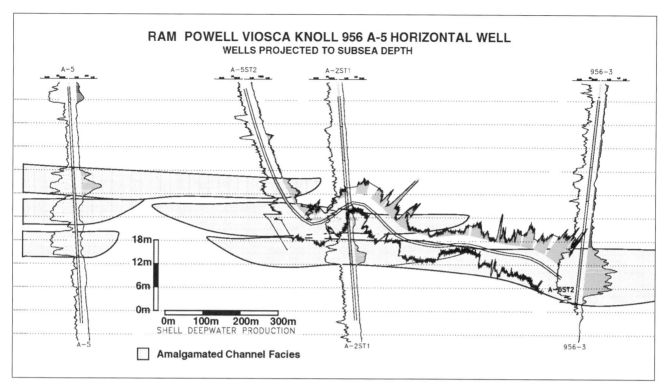

RAM POWELL VIOSCA KNOLL 956 A-5 HORIZONTAL WELL
WELLS PROJECTED TO SUBSEA DEPTH

SHELL DEEPWATER PRODUCTION

☐ Amalgamated Channel Facies

FIGURE 9. Interpreted cross section of A-5 pilot, A-2ST1, and 956-3 wells with the A-5ST2 horizontal well superimposed. Note the complex nature of the N sandstone. The hump at the beginning of well A-5ST2 results from mechanical difficulty and was not planned.

pressure-transient analysis. The gas cap of the L sandstone covers 18.2 km² (4500 ac) and contains an in-place volume of approximately 7.08×10^9 m³ (250 bcf) (Figure 11).

L SANDSTONE DEVELOPMENT

The L sandstone has a gas-oil contact at −3872 m (−12,705 ft) subsea and an oil-water contact at −3905 m (−12,812 ft) subsea (the same fluid levels as the J sandstone). The L sandstone oil rim was not developed; it was economically unfeasible because of the low permeability and low in-place volume. A horizontal-well concept was applied to develop the gas cap because of the benefit of increased rate.

The VK 912 A-1ST1BP2 was placed in the thickest penetration of the L sandstone, at the crest of the levee. The VK 912-2 L sandstone discovery well was used as a pilot for the horizontal well (VK 912-2 is shown in Figure 3). Because the levee is a layered system, the well was designed to transect the entire section to give drainage access to all the layers. The trajectory was also optimized to maximize the net to gross and the length of the wellbore exposed to the more permeable layers. The well was drilled at a high angle (82°) in the poorer-quality upper section, then the angle was built to near horizontal (89°) and drilled at the higher angle in the lower part of the section.

L SANDSTONE PRODUCTION PERFORMANCE

The production performance of the A-1ST1BP2 attests to the success of the horizontal well. First production was on September 6, 1997. From a kh of approximately 2750 md/m (9000 md/ft), a rate of 35.7 m³/sec (109 bcf/day) and 9700 BOPD was achieved. The well is currently producing 29.5 m³/sec (90 bcf/day), after more than two years of production. Total production as of the end of 2001 from the L sandstone was 2.89×10^9 m³ (102.1 bcf) and 6.5 million bbl (Figure 12).

PETROPHYSICAL TOOLS

DENSITY IMAGING USED TO DETERMINE FORMATION DIPS

An unexpected source of geologic information came from formation images generated by MWD/LWD density measurements. These images can be used to determine formation dip. Although formation imaging is not normally the domain of density logging, the special circumstances of horizontal-well drilling allowed images to be obtained in the highly laminated Ram Powell L sandstone (Figures 13 and 14). Typically, the axial resolution of a density tool is much greater than the thickness of individual sandstone laminations in wells drilled in a vertical

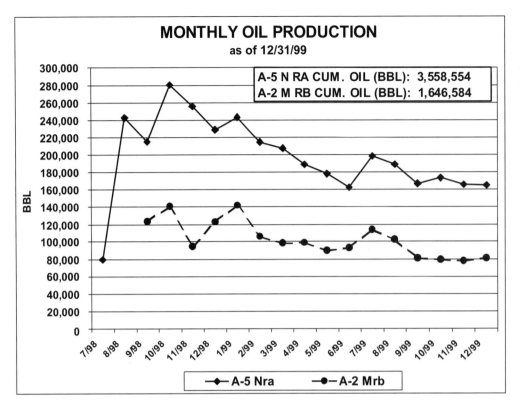

FIGURE 10. Graph comparing the A-5ST2 N sandstone horizontal-well performance with the A-2ST2 M sandstone deviated-well performance. Nra = sand, Reservoir A, Mrb = sand, Reservoir B.

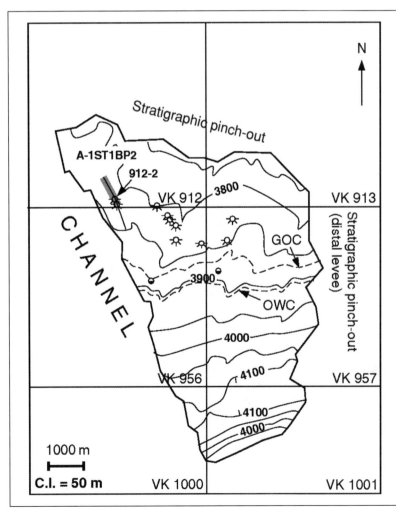

FIGURE 11. L sandstone structure map showing key well locations; labeled wells are discussed in the text. Lateral sections of horizontal wells are shaded gray.

sense. This makes the log measurement an average of the individual sandstones and shales within the axial resolution of the device. In a horizontal well drilled very near to the bedding plane, the axial resolution of the density tool can fall within the boundaries of the individual sandstone/shale laminations. A complicating factor is that the density measurement has a nonnegligible depth of investigation (~15 cm or 6 in.). This means that some averaging still is taking place, given that the bed thickness in most of this reservoir is less than 15 cm (6 in.). It is readily apparent, however, that the material closest to the tool face has the most effect on the measurement. To generate images, the contrast of the density measurements must be larger than 0.2 gm/cc (Bornneman et al., 1998). The density tool used in the Ram Powell horizontal wells was the Anadrill 121 mm (4¾-in.) OD device. This tool measures density in 16 sectors around the circumference of the borehole. Without the 16 sector measurements, imaging would not be possible.

In a 15-cm (6-in.) borehole, the estimated pixel size is ~3 cm (~1.2 in.) (Bornemann et al., 1998). This is, of course, much larger than that of the typical electromagnetic imaging device. Accordingly, dips computed from the density images are expected to be less accurate than those generated by electromagnetic imaging devices. At Ram Powell, the magnitude of dips computed from the density images compared closely to those generated from a dipmeter run in a very near offset well (Figure 15), al-

though the azimuth of the dip varied by ~25° (Bornemann et al., 1998). The larger depth of investigation of the density device has been postulated as the reason for this difference (Bornemann et al, 1998).

Density imaging aided the Ram Powell development team in an unexpected reservoir situation—the massive-sheet J sandstone. The A3 well was drilled horizontally oblique to strike from base to top of sandstone. In this wellbore, 21 events occurred, most of which were approximately 3 m (10 ft) in length, which indicated nonporous sandstone (Figure 16). These events were not seen in any of the exploration and appraisal well penetrations. The concern of the team was that these events could decrease well productivity. Questions about the size of these events were answered by the use of density images in conjunction with the LWD resistivity device (Figure 17). The intervals intersected the wellbore in a nontabular nature, as seen in the sinusoids drawn by the interpreter on the image log. This indicated a nodule shape, potentially. Analysis of the phase and attenuation responses indicated that the events extended outward perhaps 1.2–2.4 m (4–8 ft) from the well. This again suggested a nodular shape.

Petrographic evidence showed that these events were actually calcite cement completely filling the available pore space. The calcite was formed early in the depositional process, as suggested in thin sections by the lack of quartz grain deformation and suturing.

Knowledge of the shape of these events helped the

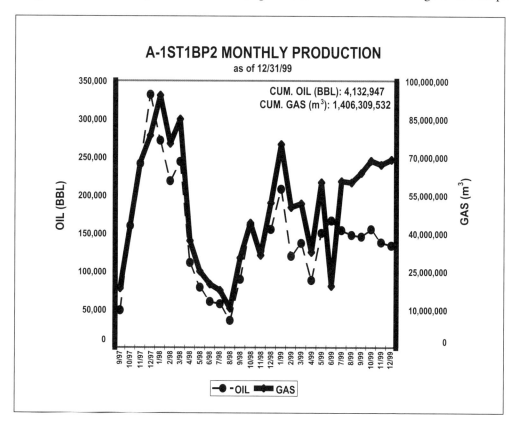

FIGURE 12. Graph showing monthly oil and gas production from well A-1ST1BP2. Beginning in April 1998, the well was both facilities- and pipeline-constrained. The decrease is not indicative of reservoir or well performance.

development team move forward with confidence that the well would produce with minimal impairment caused by the zones of calcite cementation. The drilling team also was aided by the image data, in that the bit type was changed to optimize bit life over rate of penetration. With knowledge that the events could not be avoided or predicted and that they would seriously degrade the life of milled tooth bits, a tougher but slower drilling PDC (polycrystalline diamond compact) bit was chosen to drill the final two-thirds of the well. It was estimated that at least one bit trip was saved as a result.

QUANTITATIVE FLUORESCENCE TECHNIQUE (QFT)

QFT (Delaune, 1992) provided petrophysical information not normally obtained in wells drilled in the Gulf of Mexico today. A significant majority of these wells are drilled with oil-based mud systems, either of diesel or syn-

thetic type. Oil-based mud systems render fluorescence measurements useless. Obviously, diesel-based systems innately would have very high values of fluorescence, but even nonfluorescing synthetic-based systems are typically reclaimed from other wells and contain enough contamination to render the experiment useless. Also, brand-new synthetic-based muds can be contaminated quickly in their first application if oil sandstones are encountered. At

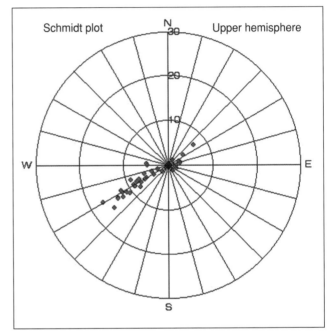

FIGURE 14. Stereonet of dips computed from density imaging. Reprinted with permission of Bornemann et al. (1998).

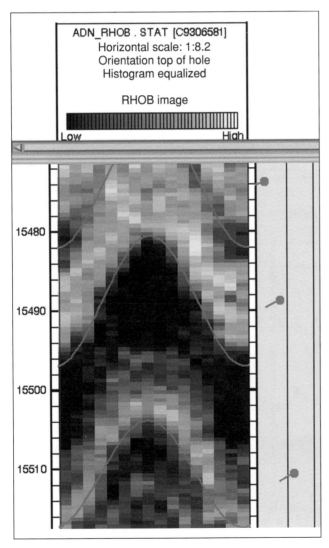

FIGURE 13. Density image of the Ram Powell L sandstone. Reprinted with permission of Bornemann et al. (1998).

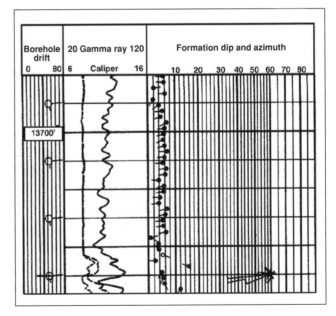

FIGURE 15. Dipmeter plot of the VK 912-2 L sandstone discovery well (pilot for A-1ST1BP2). Reprinted with permission of Bornemann et al. (1998).

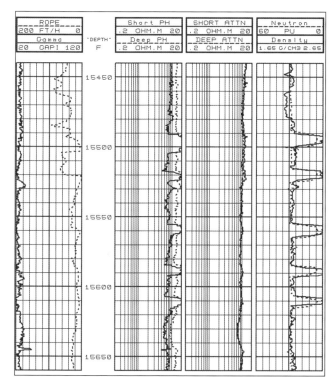

FIGURE 16. Well A-3ST1 log data showing nonporous events. Reprinted with permission of Bornemann et al. (1998).

Ram Powell, a water-based drill-in fluid was used to drill the horizontal wells.

QFT is a fluorescence-type measurement performed on drill cuttings. The cuttings are stripped of any native oils by the passage of a specific volume of nonfluorescing solvent through the cuttings. The effluent then is placed in a special device that measures the fluorescence in a repeatable, quantitative fashion. This allows for comparison of oil content from one portion of the well to another on a more consistent basis.

In Figure 18, we see the transition of the A3 well from massive-sheet sandstone to laminated sandstone near the toe of the well. Although laminated sandstone was not unexpected, it did add complexity to the determination of total depth. One of the criteria for total depth was to reach the upper-bounding shale. Encountering laminated sandstone at this point in the well brought up questions as to whether this event was sandstone or shale. QFT showed that the well was still in oil-bearing rock, which helped to confirm LWD measurements that the well was in laminated sandstone, not bounding shale. This gave the team confidence to drill on in search of the bounding shale.

In Figure 19, we see the transition of the A3 well from the laminated sandstone to the bounding shale. The QFT is at a background value, showing that the well is

out of oil-bearing rock. At this point in the drilling of the well, the drilling rate was slow enough that cuttings were being analyzed prior to LWD measurements reaching the same place in the well. In other words, the mud-log information was leading the LWD information. Usually, lag time associated with cuttings coming from bottoms up causes LWD readings to lead the mud-log readings. Because the situation was reversed in this well, the development team concluded from QFT that the well was in the bounding shale and decided to terminate drilling prior to the LWD measurements reaching the shale. The team expected that the shales would have to be logged with LWD to confirm rock type. The drill bit was drilling very slowly and nearing the end of its life in terms of rotating hours. The risk of losing a cone was high with that type of bit, so the use of QFT measurement saved significant time and risk at that decision point in the well.

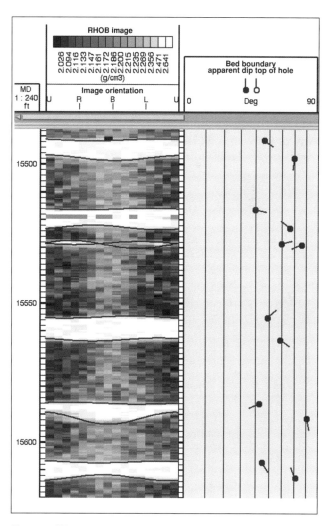

FIGURE 17. Well A-3ST1 log density images of nonporous events. Note the scattered dip pattern. Reprinted with permission of Bornemann et al. (1998).

CONCLUSIONS

The use of horizontal wells at Ram Powell has significantly improved the field's profitability by reducing the number of wells drilled and increasing production rates. Performance improvement factors from 2 (N sandstone) to 5 (J sandstone) have been observed. The development

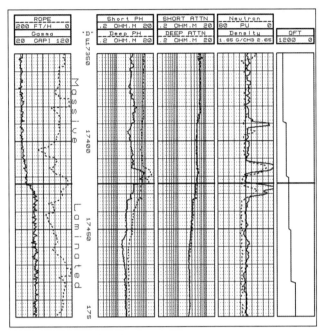

FIGURE 18. Well A-3ST1 log data including QFT showing the transition from massive sandstone to laminated sandstone.

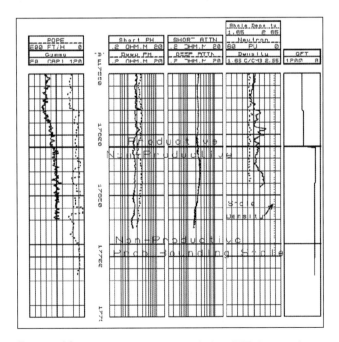

FIGURE 19. Well A-3ST1 log data including QFT showing the transition from laminated sandstone to bounding shale.

cost of the J sandstone oil rim has been reduced by 40% from the use of horizontal wells.

The stratigraphy of the reservoirs dictated the horizontal-well planning strategy. Single pilot wells were used in reservoirs with good lateral continuity (amalgamated sheet and levee deposits). Understanding the net-to-gross distribution in the reservoir enabled optimization of the trajectory, such as in the A-1ST1BP2 (L sandstone) and the A-4ST2 (J sandstone). In reservoirs with a high degree of lateral variability, such as the channelized N sandstone, two pilot wells were used. Where possible, exploratory and appraisal penetrations were used to cut costs.

The use of new evaluation techniques also added value to the project. Density imaging facilitated the determination of bedding orientation in the thin-bedded L sandstone. Density imaging was also used to understand the nature of nodular calcite cement and to establish that the cementation did not create flow barriers in the reservoir. QFT analysis also was used to make lithologic predictions independent of LWD.

ACKNOWLEDGMENTS

The authors gratefully acknowledge the assistance of Tina Johnson and Cyndi Jorgensen in preparation of this manuscript. Also, for advice and technical contribution in the Ram Powell development, we thank Ken Bramlett, Jem Scales, Jim Crump, Chris Lerch, Kaz Javanmardi, Scott Lester, and Gary Lanier. We would also like to thank Eric Mason for giving us the opportunity to share the Ram Powell story. Our thanks also to the reviewers of this paper, Tim Carr, Erik Mason, and Ray Sorenson, for their suggestions.

REFERENCES CITED

Bornemann, E., T. Bourgeois, K. Bramlett, K. Hodenfield, and D. Maggs, 1998, The application and accuracy of geological information from a logging-while-drilling density tool: Transactions of the Society of Professional Well Log Analysts 39th Annual Logging Symposium, Houston, Texas, Paper L, 14 p.

Bramlett, K. W., T. Bourgeois, P. Craig, Z. Malik, J. Scales, and T. Stroud, 1998, Ram Powell 1998: Significant challenges met, significant challenges ahead: Distinguished Lecture for the New Orleans Geological Society and the Houston Geological Society.

Delaune, P. L., 1992, Surface techniques to measure oil concentration while drilling: Transactions of the Society of Professional Well Log Analysts 33rd Annual Logging Symposium, Oklahoma City, Oklahoma, Paper KK.

Haq, B. U., Hardenbol, J., and Vail, P.R., 1989, Mesozoic and Cenozoic chronostratigraphy and cycles of sea-level change, *in* C. K. Wilgus, B. K. Hastings, H. Posamentier, J. Van Wagoner, C. A. Ross, and C. G. St. C. Kendall, eds., Sea-level change: An integrated approach: Society for Sedimentary Geology (SEPM) Publication 42, p. 71–108.

Lerch, C., K. Bramlett, B. Butler, J. Scales, T. Stroud, and C. Glandt, 1997, Ram-Powell partners see big picture with integrated modeling: The American Oil and Gas Reporter, v. 40, p. 50–67.

Lester, G. S., G. H. Lanier, K. Javanmardi, T. Bernardi, and A.

S. Halal, 1999, Ram/Powell deep-water tension-leg platform: Horizontal well design and operational experience: Offshore Technology Conference, Houston, Texas, Paper #11028.

Prather, B. E., G. S. Steffens, J. R. Booth, and P. A. Craig, 1998, Classification, lithologic calibration, and stratigraphic succession of seismic facies of intraslope basins, deep-water Gulf of Mexico: AAPG Bulletin, v. 82, no. 5A, p. 701–728.

Fluvial
and
Eolian
Reservoirs

7

Tye, R. S., B. A. Watson, P. L. McGuire, and M. M. Maguire, 2003, Unique horizontal-well designs boost primary and EOR production, Prudhoe Bay field, Alaska, *in* T. R. Carr, E. P. Mason, and C. T. Feazel, eds., Horizontal wells: Focus on the reservoir: AAPG Methods in Exploration No. 14, p. 113–125.

Unique Horizontal-well Designs Boost Primary and EOR Production, Prudhoe Bay Field, Alaska

R. S. Tye[1]

ARCO, Plano, Texas, U.S.A.

B. A. Watson[2]

ARCO Alaska, Inc., Anchorage, Alaska, U.S.A.

P. L. McGuire[3]

ARCO Alaska, Inc., Anchorage, Alaska, U.S.A.

M. M. Maguire[4]

ARCO Alaska, Inc., Anchorage, Alaska, U.S.A.

ABSTRACT

Using horizontal wells and an unprecedented enhanced oil-recovery (EOR) process, ARCO Alaska, Inc., capitalized on structural, stratigraphic, sedimentologic, and fluid-distribution complexities in the Triassic Ivishak Sandstone of Prudhoe Bay field to recover untapped or bypassed reserves. Millions of barrels of reserves reside in a basal, 120 ft- (37 m-) thick succession of en echelon, offlapping, deltaic wedges. Distributary-mouth bar and distributary-channel sandstones are targeted. Onlapping, retrogradational mudstones form laterally extensive flow barriers that impede the gravity-drainage process.

Parts of the Ivishak reservoir are structurally isolated in fault blocks covering 100 or fewer acres. Reserves are sacrificed with conventional wells because of small drainage areas and rapid gas coning. However, a horizontal well with a 90° bend, or fishhook shape, maximizes exposure to productive sandstones. Improvements in horizontal-well planning and completion practices learned and implemented during three phases of development drilling between 1992 and 1996, combined with new drilling and geosteering technology, increased initial-oil rates by a factor of two.

Horizontal wells in a miscible-injectant stimulation treatment (MIST) program recover millions of barrels of EOR oil. In a lateral MIST project, a horizontal well is drilled from an existing well in a watered-out pattern. Maximum-sweep efficiency is achieved by keeping the wellbore near the reservoir base as it arcs between an injector and outlying producers. Miscible injectant is introduced at the well tip, forming a gas bulb that pushes oil to the pro-

[1]Present affiliation: ConocoPhillips, Bartlesville, Oklahoma, U.S.A.
[2]Present affiliation: Petrochemical Resources of Alaska, Anchorage, Alaska, U.S.A.
[3]Present affiliation: BP Alaska, Inc., Anchorage, Alaska, U.S.A.
[4]Present affiliation: ConocoPhillips Alaska, Anchorage, Alaska, U.S.A.

ducers. After adequate injection, the perforations are squeezed, and the well is reperforated nearer its heel. By sequentially injecting gas, squeezing perforations, and reperforating in a more proximal position, postwaterflood oil is stripped from the reservoir. To date, incremental EOR from three horizontal wells exceeds 1.3 MMSTB. Simulations predict incremental EOR of greater than a million barrels per horizontal well.

INTRODUCTION

Prudhoe Bay field lies on the Alaska coastal plain between Naval Petroleum Reserve No. 4 (NPRA) and the Arctic National Wildlife Refuge (ANWR; Figure 1). Present-day production from Permian-Triassic sandstones and conglomerates of the Ivishak Formation (Figure 2) is approximately 550 MSTBD.

In August 1995, Prudhoe Bay field produced its nine billionth barrel of oil and has since passed the 10-billion-barrel mark. Original predictions for recoverable reserves

at Prudhoe approximated 9.6 billion barrels (Morgridge and Smith, 1972). Today's expectations exceed 13 billion barrels. Many advances in reservoir development, including prudent waterflood management and the largest miscible-injectant (MI) recovery project in the world, account for the increased recoverable reserves. Furthermore, by using coiled-tubing drilling, advanced directional-drilling technology, and refined depositional and stratigraphic interpretations of the reservoir, previously uneconomic targets are now attractive and attainable.

This paper describes two examples of how the align-

FIGURE 1. Location of Prudhoe Bay field, Alaska. Inset maps highlight the ARCO-operated Eastern Operating Area (EOA). ARCO PBU 15-45 is located in the north-central part of the field. Details of Drill Sites 3 and 9 show where lateral MIST projects are under way. Locations for cross sections in Figures 3, 8, and 12 are shown.

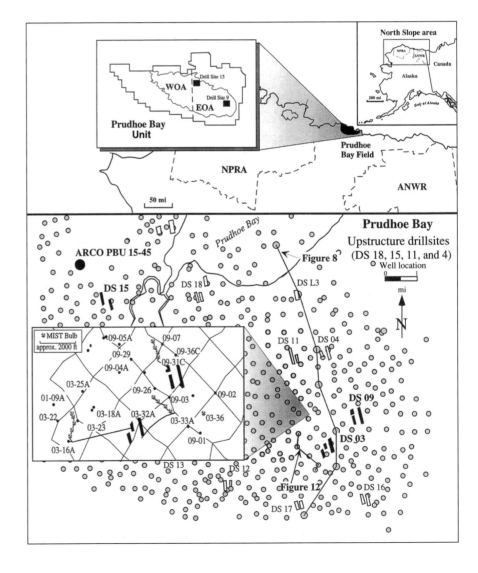

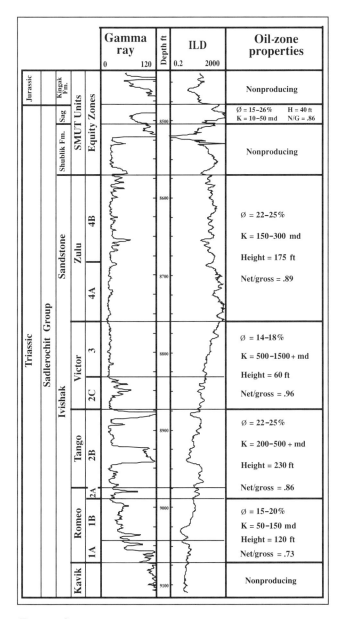

FIGURE 2. Prudhoe Bay type log depicting formal stratigraphic nomenclature and informal, operationally based reservoir-layering schemes (SMUT Units and Equity Zones). Reservoir properties are generalized, full-field values. The Romeo (Zones 1A and 1B) and Victor (Zones 2C and 3) Intervals are the reservoir zones discussed in this study.

ment of reservoir geology, drilling and completion capabilities, EOR (enhanced-oil recovery) mechanisms, and business demands created economic successes. The first example demonstrates how customizing well design and completion type optimizes offtake from thin, structurally isolated targets located immediately below the gas cap. The second example illustrates an unconventional EOR process in which horizontal wells introduce miscible injectant at the base of a thick (46 m [150 ft]) highly permeable and watered-out zone to recover millions of barrels of incremental oil.

GEOLOGIC OVERVIEW

Several publications relate the exploration process leading to the discovery of Prudhoe Bay field and its geologic characteristics (Morgridge and Smith, 1972; Jones and Speers, 1976; Jamison et al., 1980). Briefly, Prudhoe Bay field lies along the crest of the Barrow Arch, an east-west-trending, Cretaceous anticline formed by the northward thrusting of the Brooks Range (Figure 3). The southern flank of the Barrow Arch dips gently (2°), whereas the northern flank is much steeper. Sandstones and conglomerates of the Ivishak Formation record the north-to-south advancement of shallow-marine and fluvial depositional systems. At Prudhoe Bay, the Ivishak Formation is unconformably overlain by shallow-marine mudstones, limestones, and fine-grained sandstones of the Shublik and Sag River Formations. The Jurassic Kingak Shale forms the seal.

Prudhoe Bay field is 52 km long by 19 km wide (32 mi by 12 mi). Structurally, the field occurs between 2500 and 2800 m subsea (8200 and 9200 ft), and original oil- and gas-column thicknesses were 142 m (465 ft) and 122 m (400 ft), respectively. Numerous west-to-east- and northwest-to-southeast-trending faults segregate the reservoir into isolated blocks and, in places, enhance fluid migration (Figure 4).

Geologic investigations undertaken since field discovery are almost universal in their acceptance of a fluviodeltaic origin for the Ivishak Formation (Detterman, 1970; Morgridge and Smith, 1972; Eckelmann et al., 1975; Jones and Spears, 1976; Wadman et al., 1979; Jamison et al., 1980; Melvin and Knight, 1984; Lawton et al., 1987; McMillen and Colvin, 1987; Atkinson et al., 1988; Atkinson et al., 1990; Crowder, 1990; Begg et al., 1992; Tye et al., 1999). Fluvial strata have been described as braided-river deposits on a coastal plain (Eckelmann et al., 1976; Jones and Spears, 1976; Wadman et al., 1979; Melvin and Knight, 1984; Lawton et al., 1987; Atkinson et al., 1990) or on a large alluvial fan (McGowen and Bloch, 1985; McGowen et al., 1987). Petrophysically defined layers (SMUT Units; Figure 2) constitute a gross reservoir-layering scheme for this interval, which is approximately 180 m (600 ft) thick (Wadman et al., 1979; Melvin and Knight, 1984).

Four reservoir-depletion mechanisms (gravity drainage/gas-cap expansion, waterflooding, miscible-gas flooding, and gas cycling) operate simultaneously at Prudhoe Bay (Figure 5; Szabo and Meyers, 1993). Each process is managed to work in concert with the existing fluid distribution (expanding gas cap, injected water, and oil), regional geology (2° southward structural dip; intersecting east-west and northwest-southeast fault trends),

FIGURE 3. Regional south-to-north cross section from the Brooks Range to the Beaufort Sea illustrates the stratigraphy and major structural features of the North Slope. Prudhoe Bay field formed where the Ivishak Formation is draped over the Barrow Arch, truncated by the Lower Cretaceous unconformity, and overlain by marine shales. Location is shown in Figure 1.

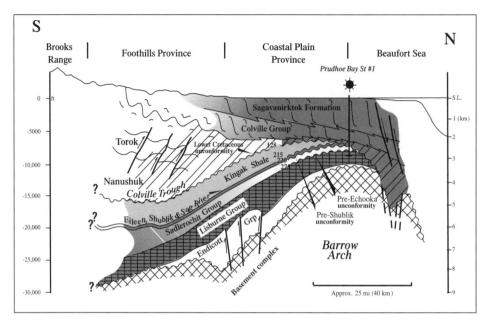

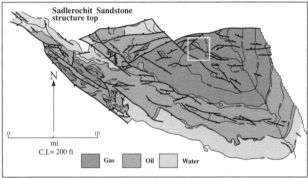

FIGURE 4. Generalized structure map on the top of the Sadlerochit Group. Major northwest-southeast- and west-east-trending faults and approximate fieldwide fluid distributions are shown. The white square shows the location of Drill Site 15.

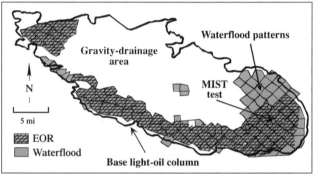

FIGURE 5. Map of recovery processes active in Prudhoe Bay field. Gravity drainage is the dominant recovery mechanism in the north-central part of Prudhoe, whereas the eastern, southern, and western field margins are under active waterflood and miscible-injectant recovery.

and field-scale geology. Interbedded conglomerates, sandstones, and mudstones create a spatially variable rock volume containing permeabilities ranging from a few millidarcys to tens of darcys. Tailoring horizontal wells and their completions to suit specific engineering and geologic conditions significantly increases primary and tertiary recovery from fine-grained sandstones and conglomerates in Prudhoe Bay field.

DEVELOPMENT CHALLENGES

THE ROMEO INTERVAL IN THE GRAVITY-DRAINAGE AREA

Along the northern field boundary, remaining recoverable reserves (>1 billion bbl) reside in medium- to fine-grained deltaic sandstones in the Romeo interval (Figures 2 and 5). Progradation and abandonment of fluvially

dominated deltaic depocenters produced a vertical sequence roughly 37 m (120 ft) thick comprising multiple stacked, upward-coarsening parasequences (average thickness 20–30 ft [6–9 m]) of prodelta, delta-front, distributary-mouth bar, distributary-channel, and fluvial-facies associations (Figure 6). The geomorphology and depositional environments of the Colville Delta, Alaska, exemplify the interpreted facies-association distribution in a single Romeo delta lobe (Figure 7).

Tabular distributary-channel and distributary-mouth bar sandstones, whose lengths and widths are on the order of thousands of feet, constitute the productive zones (Tye et al., 1999; Tye and Hickey, 2001). Distributary-channel and distributary-mouth bar (approximately 3 m- [10 ft-] thick) sandstones with mean permeabilities of 300 and 150 md, respectively, are interbedded with laterally continuous shales that strongly inhibit vertical communica-

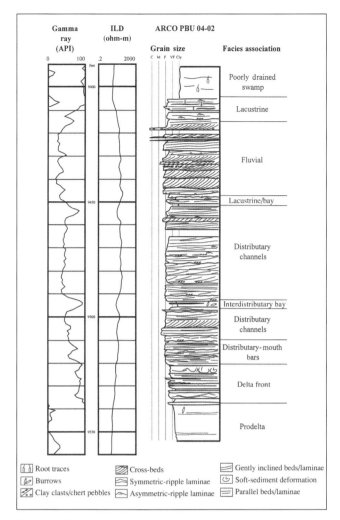

Gamma ray (API)

ILD (ohm-m)

ARCO PBU 04-02

Grain size

Facies association

Poorly drained swamp

Lacustrine

Fluvial

Lacustrine/bay

Distributary channels

Interdistributary bay

Distributary channels

Distributary-mouth bars

Delta front

Prodelta

Root traces
Burrows
Clay clasts/chert pebbles
Cross-beds
Symmetric-ripple laminae
Asymmetric-ripple laminae
Gently inclined beds/laminae
Soft-sediment deformation
Parallel beds/laminae

FIGURE 6. Typical Romeo vertical section tied to wireline logs. Basal deltaic facies associations interpreted as prodelta, delta front, distributary-mouth bar, and distributary channel, form an upward-coarsening stratigraphic succession. In Drill Sites 4, 11, 15, and 18, distributary-mouth bar sandstones are the preferred reservoir target. Location is given in Figure 1.

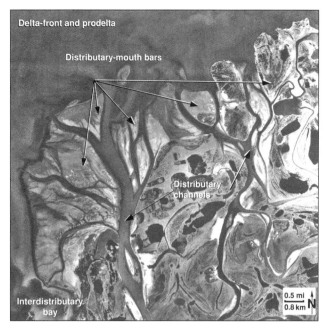

FIGURE 7. Modern depositional example of part of the fluvially dominated Colville Delta, Alaska. This style of deltaic deposition and the depositional environments present are thought to be similar to those in existence during deposition of Romeo delta lobes in Prudhoe Bay field.

Moreover, the proximity of a vertical well to some faults can result in excessive lost circulation and increased costs while drilling. The vertical-well orientation also can prevent effective isolation of the productive interval from the gas cap and underruns.

Customized Wells Drain Isolated, Thin Light-oil Column Targets

Three development programs, during which 30 high-angle to horizontal wells were drilled in Drill Site 15, occurred between 1992 and 1996. As development progressed, lessons learned from preceding successes and failures were incorporated into future development plans. As a result, by 1996, initial oil rates were greater than twice the 1992 values (Table 1). General lessons learned from drilling horizontal wells in Drill Site 15 are: (1) design wellbores to optimize exposure to the reservoir; (2) complete with cemented and perforated liners; (3) use shales as vertical barriers to coning gas but, where possible, crosscut shales to access discontinuous sandstones; and (4) selectively use pilot holes to preview target stratigraphy. Advantages of using horizontal wells include increased Romeo interval productivity, larger drainage areas attained, and an increase in the standoff between perforations and the gas-oil contact.

One example in which implementation of these development guidelines resulted in success is well 15-45 (Figure 1), which initially was proposed as a straight hori-

tion (Figure 8a; Tye et al., 1999). Moreover, east-west- and northwest-southeast-trending faults structurally compartmentalize the reservoir (Figure 4). Further complicating reservoir development is the ever-expanding gas cap that invades light-oil targets causing prohibitively high gas production (Figure 8b).

The development challenge in this part of Prudhoe Bay field is to efficiently drain stratigraphically and structurally isolated targets in fault blocks covering 100 or fewer acres and to avoid drilling wells that have a high gas-oil ratio. Traditionally, conventional wells targeted these fault blocks that were deemed too risky for a horizontal well. However, on average, vertical wells drain 70 acres, and even a horizontal well with a gentle bend has a small drainage area (200 m [700 ft] of perforations). Thus, a large amount of reserves is sacrificed (Figure 9).

a) Reservoir stratigraphy

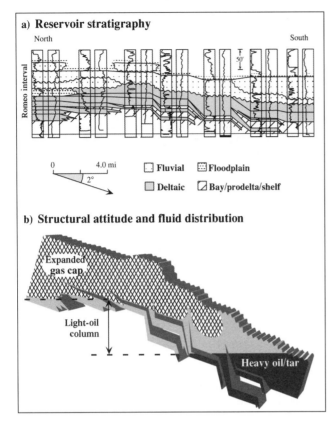

b) Structural attitude and fluid distribution

FIGURE 8. (a) Stratigraphically dip-oriented cross section through the Romeo interval. Discrete deltaic tongues are shingled from north to south and pinch out into marine shale. Distributary-mouth bar and distributary-channel sandstones as thin as 3 m (10 ft) constitute viable targets. Intervening shales create vertical and lateral permeability barriers. (b) Hypothetical fluid distribution in the Romeo interval. Gentle southerly structural dip (2°) augments the gravity-drainage process; however, the gas cap rapidly invades the light-oil column. A tar mat forms the reservoir base. See Figure 1 for cross-section location.

zontal well with a 210-m (688-ft) perforated interval within a single fault block (Figure 10). Subsequently, it was determined that much of the target polygon would be undrained by the originally proposed well. Therefore, to more efficiently drain the fault block, well 15-45 was redesigned, incorporating new directional drilling technology to create a hook-shaped horizontal section 555 m (1820 ft) long.

To avoid leaving reserves, well 15-45 penetrated producible base, then climbed stratigraphically through a distributary-mouth bar sandstone and a distributary-channel sandstone (Figure 10). The well was kept horizontal during a 90° westward turn. Ultimately, a 350-m (1150-ft) horizontal section of cemented and perforated liner (4.5-in. diameter), of which 268 m (880 ft) was perforated, maximized the well's exposure to productive sandstones.

To prevent gas coning along faults, this well cut faults

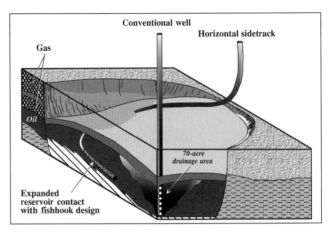

FIGURE 9. Cartoon demonstrating how a curved horizontal well more effectively drains a structural reservoir compartment in comparison to a conventional well.

high in the stratigraphic section and was logged to identify the gas-oil contact in the targeted fault block. In addition, a 60-m (200-ft) buffer was maintained between the well and adjacent faults in the production interval. Moreover, once in the targeted sandstone, the wellbore was kept below a shale interpreted to be laterally continuous in order to impede gas influx. First stable production was 3750 BOPD and 2600 GOR. Well 15-45 production was 1.2 million bbl as of April 1997.

ENHANCED OIL RECOVERY IN A MATURE WATERFLOOD AREA

In a geographically and stratigraphically different part of the field, located in a mature waterflood-recovery stage (Figures 1 and 5), horizontal wells coupled with a miscible-injectant stimulation treatment (MIST; McGuire et al., 1999) process are recovering millions of barrels of incremental EOR oil. This process capitalizes on the existing well-pattern design, coiled-tubing drilling, and geologically controlled flow characteristics of the reservoir. Two lithofacies, mudstone and open-framework conglomerate, dominate fluid flow in this part of the reservoir (Stalkup and Crane, 1991; McGuire et al., 1994).

The Victor interval (Figure 2) is typically a 46-m (150-ft-) thick, high net-to-gross (0.96) succession of sandstones, pebbly sandstones, and conglomerates with few extensive shales or other vertical permeability barriers (Figures 11 and 12). Where present, mudstones create vertical permeability barriers or baffles, but their areal coverage is patchy. Embedded in the coarsest-grained strata are thin beds of open-framework conglomerate possessing extremely high (approximately 20-d) permeabilities. The areal and vertical distribution of open-framework conglomerates is variable. Vertical-permeability segregation is most evident near the top and base of the Victor interval.

TABLE 1. SUMMARY OF WELL COMPLETIONS, PROFILES, AND PERFORMANCE FOR THREE DRILL SITE 15 DEVELOPMENT PLANS.

Development year	1992	1994	1996
Well completions	Slotted liners	Slotted liners	Cemented and perforated liners
Well profile	Horizontal: heel in lowest producible sandstone (distributary-mouth bar), toe inverted into distributary-channel sandstone	Horizontal: wellbore kept in sandstone of similar permeability	Horizontal: short-radius build, tight turns, and long completions (average = 400 m [1327 ft]) in as many distributary-mouth bar sandstones as possible
Well performance	High initial rates; steadily increasing GORs	Distributary-channel completions = high initial rates; steadily increasing GORs; distributary-mouth bar completions = low initial rates	High initial rates; wells put on cycle/shut-in status as GORs rise
Average initial rate (BOPD)	2200	2375	5450

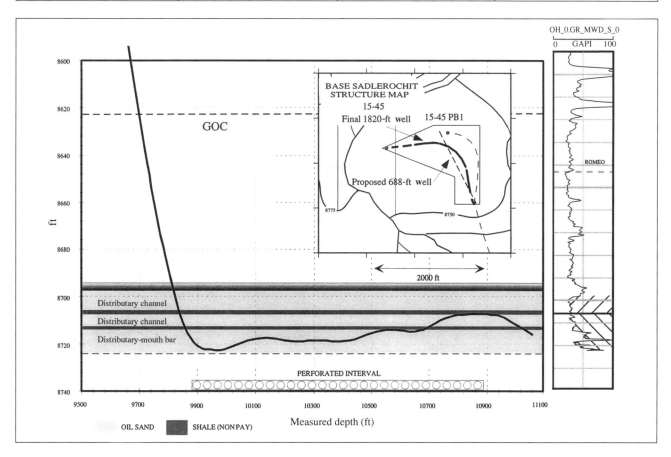

FIGURE 10. Wireline logs, structure map, and well profile for the ARCO PBU 15-45 well. Two distributary-mouth bars and one distributary-channel sandstone were perforated. Inset shows the base Sadlerochit structure map over the Drill Site 15-45 infill location and plan view of the target polygon, the proposed original well trajectory, a plugged-back first attempt, and the final path for the ARCO PBU 15-45 well.

Depositional Interpretation

In a generalized interpretation, Victor conglomerates, sandstones, and mudstones were deposited in nonmarine-channel (point, longitudinal, transverse, or lateral bars), lacustrine-delta, crevasse-splay, and floodplain facies associations. Discrete channels having thicknesses of 2 to 6 m (6 to 20 ft) are identifiable in cores; however, channels are lumped into channel belts (a composite of many channels; Bridge and Leeder, 1979) to facilitate well-to-well correlation (Figure 12). Coarse-grained sediments that accumulated outside of channels (lacustrine deltas, levees, crevasse splays) are overbank sandstones. Fine-grained lithologies (lacustrine and soil) are grouped as floodplain deposits. Abandoned-channel and bar-drape mudstones occur locally. Figure 13 illustrates sediment distribution over part of a modern braided channel belt that may resemble Victor nonmarine depositional systems.

Where present, mudstones can form vertical-permeability barriers and can locally impede horizontal permeability (Stalkup and Crane, 1991). Depositional settings determine the thickness and lateral extent of mudstones, as follows: (1) amalgamated floodplain soils; (2) floodplain soils; (3) abandoned channels; and (4) bar drapes. Amalgamated floodplain soils consist of multiple paleosol horizons and form the most laterally extensive flow barriers, particularly near the base and top of the Victor interval (Figure 12). Floodplain soils are thinner and more restricted than the amalgamated floodplain soil horizons; however, they may correlate across several well spacings. Abandoned-channel and bar-drape mudstones may be genetically related, in that both represent low-energy deposition within a channel. These deposits are generally less than one well spacing in extent and have variable thicknesses (Figure 12).

Although not recognized as open-framework conglomerates, previous work (Atkinson et al., 1990) documented the presence of high-permeability streaks or thief zones in the Victor interval. Deposition of open-framework conglomerates forms thin beds (approximately 30 cm [1.0 ft]) of granule- to pebble-size grains through the aggradation and downstream accretion of river gravels on bars. Because pores are devoid of matrix or cement (Figure 14), they are well connected, and open-framework conglomerates have permeabilities of multiple darcys. Beds of open-framework conglomerate are poorly preserved in Prudhoe Bay cores; therefore, their occurrence and reservoir properties are unknown. However, flow simulations indicate that these thin layers can have permeabilities of 20 d or greater (McGuire et al., 1994).

Open-framework conglomerates sampled in modern depositional environments have permeabilities of 5–7 d.

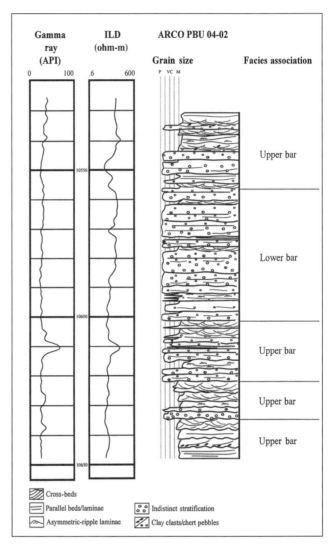

FIGURE 11. Typical gamma-ray and induction logs and lithologic section for the Victor interval. Reserves targeted by MIST projects reside in this 46-m- (150-ft-) thick section dominated by conglomerate. Geologic issues affecting production are lithofacies and permeability distribution, shale stratigraphy, and the 3-D character of high-permeability (thief) zones in the conglomerates.

They occur commonly in cross-bedded and horizontal strata (Smith, 1974). Individual open-framework beds can extend laterally from tens to thousands of feet (as much as 1000 m).

The MIST (Misciple Injectant Stimulation Treatment) Concept

Conventional water-alternating-gas (WAG) floods are dominated strongly by gravity in the Victor interval, with rapid vertical segregation of the miscible injectant. Moreover, horizontal flow in the reservoir is dominated by open-framework conglomerate beds that cause flood-front fingering and rapid breakthrough. The miscible injectant sweeps oil near the injection wellbore, but because

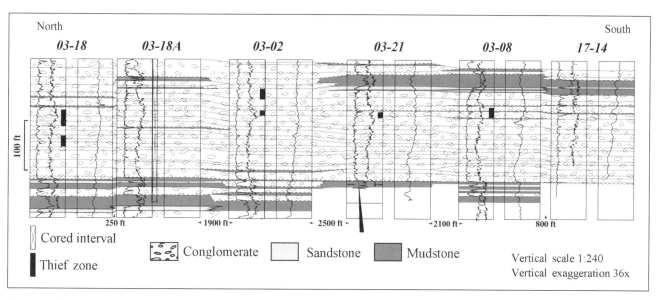

North   South

03-18   *03-18A*   *03-02*   *03-21*   *03-08*   *17-14*

100 ft

250 ft   ◄ 1900 ft ►   ◄ 2500 ft ►   ◄ 2100 ft ►   800 ft

Cored interval

Thief zone   Conglomerate   Sandstone   Mudstone

Vertical scale 1:240
Vertical exaggeration 36x

FIGURE 12. North-to-south stratigraphic cross section through the Victor interval. Fluvial channel belts are juxtaposed vertically and laterally, forming a high net-to-gross reservoir zone. Channel-belt bases are represented by the wavy lines. Note the paucity of mudstone beds in the central part of the section. Possible thief zones are determined by production data. See Figure 1 for cross-section location.

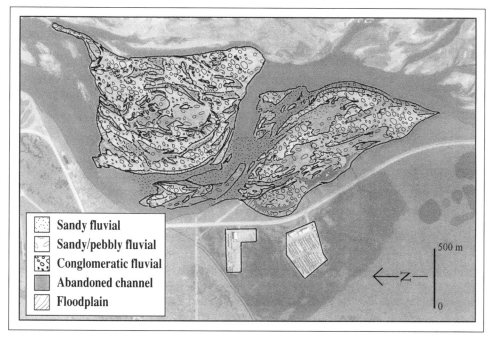

Sandy fluvial
Sandy/pebbly fluvial
Conglomeratic fluvial
Abandoned channel
Floodplain

500 m

◄—z—

0

FIGURE 13. Braid bars and channels in part of the channel belt of the Sagavanirktok River, Alaska. Lithofacies similar to those described in the Victor interval are mapped on the bar surfaces. Shallow trenches in conglomeratic regions typically reveal one or two open-framework conglomerate beds. Within channel belts, mud deposits are restricted areally, whereas floodplains, where preserved, can form extensive mudstone deposits.

of rapid vertical solvent migration, large areas of the reservoir are uncontacted (Figure 15a).

The actual miscible-injectant sweep efficiency in a conventional Victor well pattern was determined by coring a sidetrack well 91.5 m (300 ft) from the injector (McGuire et al., 1994). A history-matched, fully compositional reservoir simulation of WAG in the Victor interval showed a very limited area in which EOR oil was actually mobilized. Although the entire interval was open to injection, the bottom 30 m (100 ft) was not contacted by solvent. This unaffected area is the target for MIST, and it is particularly advantageous to utilize the reservoir's strong

vertical-drive process in conjunction with horizontal wellbores.

In the lateral MIST process, a coiled-tubing horizontal-sidetrack well from either a production or injection well is drilled along the base of the reservoir. To mobilize EOR oil in unswept areas, miscible injectant is injected sequentially, in a toe-to-heel fashion, into several intervals along the well's horizontal reach (Figure 15b). In a best-case scenario, the wellbore lies beneath a discontinuous mudstone that retards upward miscible-injectant migration and increases lateral sweep. Each slug of miscible injectant is about 3 bscf, and it is injected as quickly as pos-

FIGURE 14. Axial CT-scan slice through a cored open-framework conglomerate. The white space corresponds to empty pore space in the conglomerate; black areas are chert pebbles that average 1.5 cm long. Core-measured permeabilities for four open-framework conglomerate samples average 5300 md; however, simulations indicate that these thin layers can have 20-d permeabilities or greater. Sample is approximately 4.0 in. wide.

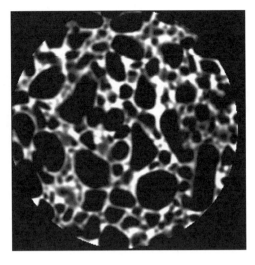

sible (typically about 30 MMSCFD). Solvent injection is followed by a short period of chase-water injection. After injection into each interval is completed, the perforations are squeezed and the well is perforated closer to its heel. Upon completion of solvent injection (several miscible-injectant bulbs), the horizontal well is converted to normal production or injection service.

A Horizontal MIST Field Test

Well 9-31C was the first of three wells sidetracked to test the lateral MIST concept. It was a long (560-m [1850-ft]) coiled-tub-

FIGURE 15. (a) Schematic diagram illustrating Victor interval permeability and remaining oil left by an irregular waterflood and WAG sweep. (b) The MIST process is shown during the third phase of MI injection. Note that the horizontal well has undergone sequential MI injection, perforation squeezes, and reperforations. Spent MI bulbs efficiently swept the reservoir to recover incremental EOR reserves.

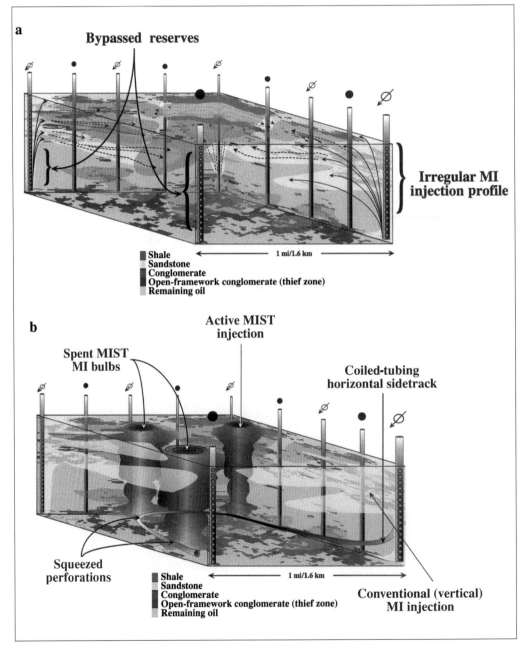

ing sidetrack with 460 m (1500 ft) of horizontal section along the base of the Victor interval. Four separate Victor injection intervals were planned (Figure 1). Major concerns to be addressed by this well were:

1) the feasibility of using coiled tubing to build angle at 30–35° per 30 m (100 ft) and to drill a horizontal section in the conglomerates (a drilling challenge even in vertical wells)
2) operational feasibility of solvent injection and multiple plugbacks in a long, small-diameter coil-drilled horizontal completion
3) miscible-injectant sweep efficiency and recovery potential of the lateral MIST process

A comprehensive surveillance plan evaluated the 9-31C performance. Weekly well tests were performed on all offset producers to monitor oil rate and water-cut changes. Oil and gas samples were analyzed to measure returned miscible injectant, API gravity, and changes in black-oil composition.

Solvent injection into the first bulb began on June 10, 1996. A total of 2.1 bscf was injected into the first bulb and 2.1 bscf was injected into the second bulb. A streamline map indicates that most of the injection into the first bulb was flowing through a small elliptical area between wells 9-31C and 9-07 (Figure 16). Major production response occurred in well 9-07 fewer than 20 days after in-

jection began. Oil rates increased from pre-MIST levels of about 700 STBD to 1500 STBD, whereas gas rates increased from 8 MMSCFD to 15 MMSCFD. Oil rates stayed between 1000 and 2000 STBD until April 1997, when the well was shut in because of excessive gas production.

To determine if more distant pattern wells could be affected, a very large slug (7.8 bscf) was put into the third injection interval. Producers 9-04, 9-05, and 9-29 all responded favorably. Cumulative MIST oil through September 1998 was roughly 1100 MSTB, with most of the response from wells 9-07 and 9-35A. Simulations indicate potential recovery of more than a million barrels of incremental EOR per horizontal well.

CONCLUSIONS

In Prudhoe Bay field, Alaska, horizontal wells recover millions of barrels of untapped oil from heterogeneous deltaic reservoirs and are paramount in an EOR process that strips bypassed reserves from previously waterflooded conglomeratic reservoir intervals in the Triassic Ivishak Sandstone. Along the northern field boundary, oil-filled sandstones in stratigraphically offlapping, fluvially dominated deltaic wedges are produced by gravity drainage. Primary targets are fine-grained distributary-mouth bar and distributary-channel sandstones within en echelon, offlapping, fluvially dominated deltaic wedges. Mudstones deposited after delta-lobe abandonment form laterally extensive flow barriers and baffles, which impede the gravity-drainage depletion process. Because of expansion of the large (47-tcf) gas cap, the risk of drilling an uneconomic, high GOR well is high.

Locally continuous and numerous, east-west-trending faults structurally isolate the deltaic sandstones in fault blocks covering 100 or fewer acres, too risky for horizontal wells. The limited drainage areas of conventional wells results in reserves being sacrificed where wells inefficiently recover oil confined in small, isolated fault blocks. Therefore, in an attempt to maximize reserve recovery, a hook-shaped, horizontal well was drilled to drain a single fault block. Once the well penetrated producible base, it climbed stratigraphically into the reservoir sandstones, where it was kept horizontal during a 90° westward turn. To retard gas coning, the wellbore stayed below a shale and maintained a minimum buffer of 60 m (200 ft) from adjacent mapped faults. First stable production from 288 m (880 ft) of cemented and perforated liner was 3750 BOPD and 2600 GOR.

Where the reservoir contains a thick, watered-out conglomeratic section, horizontal wells coupled with a

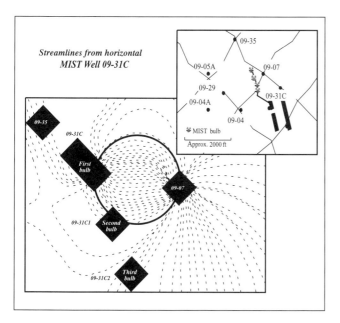

FIGURE 16. Streamline map of the 09-31C MIST project. Injected into the first and second bulbs was 2.1 bscf of solvent. Most of the injection from the first bulb swept the area between 09-31C and 09-07. Injected into the third bulb was 7.8 bscf. Surrounding producers all responded favorably.

MIST process are recovering millions of barrels of EOR oil. This process exploits both the existing well-pattern design and the geologically controlled flow characteristics of nonmarine conglomerates, conglomeratic sandstones, and sandstones. Because of the high Kv/Kh of the conglomerates and few shale barriers to vertical flow, gravity segregation of the solvent is high. Moreover, embedded in the coarsest-grained strata are thin beds of open-framework conglomerate possessing extremely high (approximately 20-d) permeabilities. These geologic traits cause the solvent to inefficiently sweep the reservoir.

A lateral MIST project works in concert with the reservoir geology to maximize oil recovery. A horizontal well is drilled from a production or injection well in an inverted nine-spot pattern. Maximum-sweep efficiency is achieved by keeping the wellbore at the base of the conglomerate section as its route forms an arc between the central injector and the outlying producers. In a best-case scenario, the wellbore lies beneath a discontinuous mudstone, thus retarding upward miscible-injectant migration and increasing lateral sweep. The well is perforated at its tip, and miscible injectant is introduced to form a gas bulb to push oil to the producers. After adequately sweeping this reservoir volume, the original perforations are squeezed, and the well is reperforated closer to the heel. By sequentially injecting gas, squeezing perforations, and reperforating in a more proximal position, incremental reserves are stripped from the rocks. Incremental EOR from three horizontal MIST wells is more than 1.3 MMSTB. Simulations predict incremental EOR of greater than a million barrels per horizontal well.

ACKNOWLEDGMENTS

We thank the Prudhoe Bay Working Interest owners for their support and permission to publish this paper. Many individuals aided our work by sharing their experience and knowledge. Among those to whom we are especially grateful are J. A. Lorsong, Tim Verseput, Eric West, Beverly Burns, and Robert Morse. Lillian Tiulana and Katherine Hale provided data and graphic support. D. Przyowjski and M. McCracken cheerfully helped in the core lab. The enthusiastic and cheerful encouragement of Tim Carr and Erik Mason made this experience rewarding and pleasurable. Ray Sorenson, Erick Mason, and an anonymous reviewer are thanked for their comments and suggested improvements.

REFERENCES CITED

Atkinson, C. D., P. N. Trumbly, and M. C. Kremer, 1988, Sedimentology and depositional environments of the Ivishak Sandstone, Prudhoe Bay field, North Slope, Alaska, *in* A. J. Lomando and P. M. Harris, eds., Giant Oil and Gas Fields: Society for Sedimentary Geology (SEPM) Core Workshop No. 12, p. 561–613.

Atkinson, C. D., J. H. McGowen, S. Bloch, L. L. Lundell, and P. N. Trumbly, 1990, Braidplain and deltaic reservoir, Prudhoe Bay, Alaska, *in* J. H. Barwis, J. G. McPherson, and R. J. Studlick, eds., Sandstone Petroleum Reservoirs: New York, Springer-Verlag, p. 7–29.

Begg, S. H., E. R. Gustason, and M. W. Deacon, 1992, Characterization of a fluvial-dominated delta: Zone 1 of the Prudhoe Bay Field, Paper No. 24698: Society of Petroleum Enginers 67th Annual Technical Conference, Richardson, Texas, p. 351–364.

Bridge, J. S., and M. R. Leeder, 1979, A simulation model of alluvial stratigraphy: Sedimentology, v. 26, p. 617–644.

Crowder, R. K., 1990, Permian and Triassic sedimentation in the northeastern Brooks Range, Alaska: Deposition of the Sadlerochit Group: AAPG Bulletin, v. 74, no. 9, p. 1351–1370.

Detterman, R. L., 1970, Sedimentary history of the Sadlerochit and Shublik Formations in northeastern Alaska, *in* W. L. Adkison and M. M. Brosge, eds., Proceedings of the geological seminar on the North Slope of Alaska: AAPG Pacific Section, p. o1–o13.

Eckelmann, W. R., R. J. Dewitt, and W. L. Fisher, 1975, Prediction of fluvial-deltaic reservoir geometry, Prudhoe Bay Field, Alaska: Proceedings of the 9th World Petroleum Congress: John Wiley and Sons, International, v. 2, p. 223–227.

Jamison, H. C., L. D. Brockett, and R. A. McIntosh, 1980, Prudhoe Bay—A ten year perspective, *in* M. T. Halbouty, ed., Giant oil and gas fields of the decade: AAPG Memoir 30, p. 289–310.

Jones, H. P., and R. G. Speers, 1976, Permo-Triassic reservoirs of Prudhoe Bay field, North Slope Alaska: AAPG Memoir 24, p. 23–50.

Lawton, T. F., G. W. Geehan, and B. J. Voorhees, 1987, Lithofacies and depositional environments of the Ivishak Formation, Prudhoe Bay Field, *in* I. Tailleur and P. Weimer, eds., Alaskan North Slope Geology: Society for Sedimentary Geology (SEPM) Pacific Section, v. 50, p. 61–76.

McGowen, J. H., and S. Bloch, 1985, Depositional facies, diagenesis, and reservoir quality of the Ivishak Sandstone (Sadlerochit Group), Prudhoe Bay Field: AAPG Bulletin, v. 69, p. 286.

McGowen, J. H., S. Bloch, and D. Hite, 1987, Depositional facies, diagenesis, and reservoir quality of Ivishak Sandstone (Sadlerochit Group), Prudhoe Bay Field (abs.), *in* I. Tailleur and P. Weimer, eds., Alaskan North Slope Geology: Society for Sedimentry Geology (SEPM) Pacific Section, p. 84.

McGuire, P. L., A. P. Spence, F. I. Stalkup, and M. W. Cooley, 1994, Core acquisition and analysis for optimization of the Prudhoe Bay miscible gas project, Paper No. 27759: 9th Society of Petroleum Engineers/Department of Energy Symposium on Improved Oil Recovery, Tulsa, Oklahoma, p. 253–265.

McGuire, P. L., R. S. Redman, W. L. Mathews, and S. R. Carhart, 1999, Unconventional miscible EOR Experience at Prudhoe Bay: Journal of Petroleum Technology, v. 51, no. 1, p. 38–41.

McMillen, K. J., and M. D. Colvin, 1987, Facies correlation and basin analysis of the Ivishak Formation, Arctic National Wildlife Refuge, Alaska, in I. Tailleur and P. Weimer, eds., Alaskan North Slope Geology: Society for Sedimentary Geology (SEPM) Pacific Section, v. 50, p. 381–390.

Melvin, J., and A. S. Knight, 1984, Lithofacies, diagenesis and porosity of the Ivishak Formation, Prudhoe Bay Area, Alaska, in D. A. McDonald and R. C. Surdam, eds., Clastic diagenesis: AAPG Memoir 37, p. 347–365.

Morgridge, D. L. and W. B. Smith Jr., 1972, Geology and discovery of Prudhoe Bay Field, eastern Arctic Slope, Alaska, in R. E. King, ed., Stratigraphic oil and gas fields: AAPG Memoir 16, p. 489–501.

Smith, N. D., 1974, Sedimentology and bar formation in the upper Kicking Horse River: A braided meltwater stream: Journal of Geology, v. 82, p. 205–223.

Stalkup, F. I., and S. D. Crane, 1991, Reservoir description detail required to predict solvent and water saturations at an observation well, Paper No. 22897: 66th Annual Society of Petroleum Engineers Technical Conference and Exhibition, Dallas, Texas, p. 151–164.

Szabo, D. J., and K. O. Meyers, 1993, Prudhoe Bay: Development history and future potential, Paper No. 26053: Society of Petroleum Engineers Western Regional Meeting, Anchorage, Alaska, p. 1–9.

Tye, R. S., and J. J. Hickey, Permeability characterization of distributary-mouth bar sandstones in Prudhoe Bay field, Alaska: How horizontal cores reduce risk in developing deltaic reservoirs: AAPG Bulletin, v. 85, no. 3, p. 459–475.

Tye, R. S., J. P. Bhattacharya, J. A. Lorsong, S. T. Sindelar, D. G. Knock, D. D. Puls, and R. A. Levinson, 1999, Geology and stratigraphy of fluvio-deltaic deposits in the Ivishak Formation: Applications for development of Prudhoe Bay Field, Alaska: AAPG Bulletin, v. 84, no. 8, p. 1205–1228.

Wadman, D. H., D. E. Lamprecht, and I. Mrosovsky, 1979, Joint geologic/engineering analysis of the Sadlerochit Reservoir, Prudhoe Bay Field: Journal of Petroleum Technology, v. 31, p. 933–940.

Hamilton D. S., R. Barba, M. H. Holtz, J. Yeh, M. Rodriguez, M. Sánchez, P. Calderon, and J. Castillo, 2003, Horizontal-well drilling in the heavy-oil belt, eastern Venezuela Basin: A postmortem of drilling experiences, *in* T. R. Carr, E. P. Mason, and C. T. Feazel, eds., Horizontal wells: Focus on the reservoir: AAPG Methods in Exploration No. 14, p. 127–141.

8

Horizontal-well Drilling in the Heavy-oil Belt, Eastern Venezuela Basin: A Postmortem of Drilling Experiences

Douglas S. Hamilton[1]

Bureau of Economic Geology
Austin, Texas, U.S.A.

Robert Barba[2]

Bureau of Economic Geology
Austin, Texas, U.S.A.

M. H. Holtz

Bureau of Economic Geology
Austin, Texas, U.S.A.

Joseph Yeh

Bureau of Economic Geology
Austin, Texas, U.S.A.

M. Rodriguez

Petróleos de Venezuela
Caracas, Venezuela, S.A.

M. Sánchez

Petróleos de Venezuela
Caracas, Venezuela, S.A.

P. Calderon

Petróleos de Venezuela
Caracas, Venezuela, S.A.

J. Castillo

Petróleos de Venezuela
Caracas, Venezuela, S.A.

ABSTRACT

The Bureau of Economic Geology and Corpoven S.A. (now Petróleos de Venezuela) jointly undertook a detailed reservoir characterization study of the Merecure and Oficina Formations in the Arecuna field of the Faja region, Eastern Venezuela Basin. The primary objective of the study was to delineate the volumes and residency of remaining oil saturation and to develop appropriate advanced recovery strategies for maximizing recovery efficiency of the heavy oil in the Faja region. Initial completions in the field were all from vertical wells, and production rates were subeconomic. The field-development strategy was to accelerate production rates by taking advantage of increased drainage efficiencies of horizontal wells. A horizontal-well drilling program, however, requires a very accurate reservoir model, particularly in the Faja, where the reservoirs were highly compartmented because of complex faulting and lateral and vertical facies heterogeneity of the fluvial reservoirs. Moreover, hydrocarbon- and water-bearing reservoirs are strongly interlayered in the field, requiring the trajectory of the horizontal wells to be targeted accurately.

A 46-well drilling program, implemented after the reservoir characterization, validated the geologic model (in some cases confirming geologic predictions as much as 1050 m [3500 ft] from the nearest well data). Although the overall drilling program was highly successful, eight wells experienced water-production problems and were deemed uneconomic. A postmortem analysis of the unsuccessful wells indicates that most of the water problems were mechanical in nature, but coning of the aquifer and uncertainty in defining pay-resistivity cutoffs (as a result of variable water resistivities) also contributed to failure of some wells.

[1]Present affiliation: Hamilton Geosciences, Austin, Texas, U.S.A.
[2]Present affiliation: Integrated Energy Services, Austin, Texas, U.S.A.

INTRODUCTION

Arecuna field is in the prolific Faja heavy-oil-producing region of eastern Venezuela (Audemard et al., 1985; Burkill and Rondon, 1991), which is in its earliest stages of primary recovery. Defined simply by the limits of the 100-km² Arecuna 3-D seismic survey (Figure 1), Arecuna field contains an estimated 2.8 billion barrels of original oil in place, but to year-end 1996, it had produced only 2.2 million barrels of 7.5–16° API oil from 35 producing wells. Initial completions were all from vertical wells, and typical flow rates, which were on the order of 200 barrels of oil per day (BOPD), were subeconomic. Moreover, some of the wells were already producing at 20–40% water cut.

Experience in producing areas adjacent to Arecuna showed that flow rates could be increased to more than 1000 BOPD by producing from horizontal wells, and that horizontal-well costs were only 1.5 times those of vertical wells. The cost-effectiveness of the horizontal-well completions prompted the operating company, Corpoven S.A. (then an affiliate company of Petróleos de Venezuela), to initiate a horizontal-well drilling program to maximize recovery efficiency in the Arecuna field. Design of the program was undertaken jointly between our research group at the Bureau of Economic Geology and the Corpoven S.A. staff. An integrated team of geologists, geophysicists, petrophysicists, and engineers was assembled to investigate the geologic and fluid characteristics of the area to allow optimal horizontal-well drilling design.

Reservoir architecture, oil mobility, and water production were identified as issues to address in the investigation. A lack of oil mobility, for example, is the main reason for economic failure of the vertical wells, and excessive water production is well known in the neighboring blocks, where 28 million barrels of cumulative oil production was accompanied by 15 million barrels of water production. Additional concerns were readily identified as the project proceeded, however, and important problems to solve included understanding reservoir continuity, structural complexity, interbedding of oil- and water-saturated sandstones, fractional flow behavior of the oil and water, and variability in water resistivity.

Forty-six horizontal wells were drilled after reservoir characterization, and 38 wells produced successfully at rates in excess of the 1000 BOPD target. Eight of the wells, however, experienced water-production problems and were uneconomic. A postmortem of drilling results indicates that most of the water problems were mechanical in nature, but complexity of the fluid characteristics, specifically coning of the aquifer because of water mobility and water resistivity that complicated petrophysical determination of pay-resistivity cutoffs, also contributed to the failure of some wells. The objectives of this paper are to describe the complexity of the geologic architecture and fluid characteristics of the Arecuna field and to discuss the impact of these complexities on horizontal-well design and subsequent drilling.

FIGURE 1. Location of the Arecuna field study area in the heavy-oil belt (modified from Erlich and Barrett, 1992).

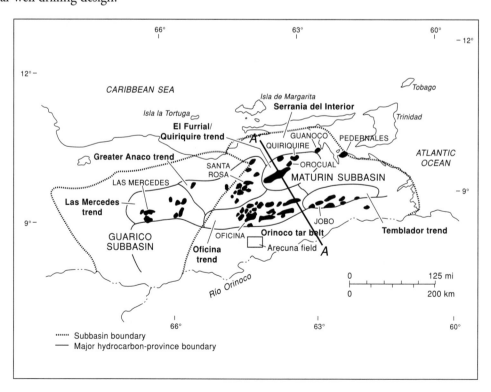

GEOLOGIC ARCHITECTURE OF THE ARECUNA FIELD

STRATIGRAPHIC FRAMEWORK

The main oil-bearing zones in the Arecuna field occur in the upper part of the Merecure Formation and throughout the Oficina Formation. The productive reservoir section is as much as 450 m (1500 ft) thick, and the stratigraphic analysis indicates that these productive intervals can be subdivided into 11 major stratigraphic units, three in the Merecure Formation and eight in the Oficina Formation (Figure 2). The units were identified by systematic vertical changes in bedding architecture (readily apparent on well-log traces) occurring above successive, laterally persistent high gamma-ray (or minimum spontaneous potential [SP]) shale markers that are interpreted as surfaces of maximum marine or lacustrine flooding. Widespread lignite beds also were used to divide the stratigraphic section (Figure 3). Because peat accumulation and preservation as lignite can occur only in the absence of significant clastic deposition (Hamilton and Tadros, 1994), lignite beds represent substantial periods of shutoff in sediment supply and approximate time lines separating successive depositional systems.

Architecture of the stratigraphic framework defined from well control mirrors that of the 3-D seismic data (Figure 3). Key stratigraphic surfaces readily identified on the well-log correlations (FS 40, 43, 45, 50, 60, 70, 90, and 100) were also readily identified on the seismic data (Figure 4). Both on the well logs and on the seismic data set, these sequences were characterized by uniform thicknesses and conformable seismic surfaces (except where postdepositional faulting removed section). Stratigraphic subdivision from well control was difficult between FS 60 and 70, which is consistent with the downlapping and truncated seismic geometries evident on the 3-D data log in this stratigraphic interval.

The stratigraphic nomenclature was defined arbitrarily in this study to avoid confusion with existing Corpoven sandstone nomenclature. In general, however, the existing nomenclature and the flooding-surface-defined stratigraphy correspond, as shown in Table 1.

DEPOSITIONAL SYSTEMS

Once the stratigraphic framework was established, gross sandstone and log-facies maps (Galloway and Hobday, 1983) were constructed for the major stratigraphic units, to interpret the depositional setting of the Arecuna reservoirs and to allow prediction of reservoir thickness and continuity along the proposed horizontal-well trajectories in the interwell areas. The Arecuna reservoirs are interpreted as deposits of bed-load and mixed-load fluvial systems. Depositional systems of genetic stratigraphic units Merecure A and Oficina A and B illustrate the range of reservoir styles at Arecuna field.

Merecure Unit A

Merecure Unit A, defined between flooding surfaces 90 and 100 and locally referred to as the U2 and U3 sandstone (Figure 2), is interpreted as the deposit of a large-scale, braid-plain system. The unit varies from 33 to 51 m (110 to 170 ft) thick, and gross sandstone varies from 6 to 36 m (20 to 120 ft) thick. Although gross-sandstone trends are generally very broad, several digitate, north-oriented sandstone-rich axes are apparent (Figure 5). Small pods where gross sandstone is thin occur only locally in the broad, sandstone-rich trends. These sandstone trends suggest weakly confined river flow across a broad alluvial

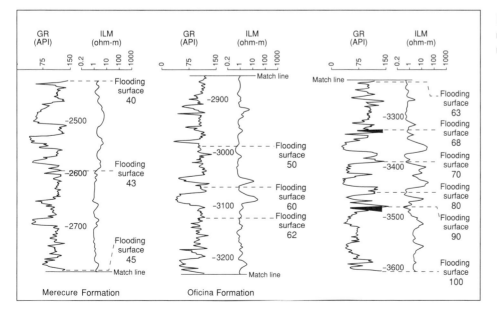

FIGURE 2. Reference log of the main producing interval at Arecuna field. Depth is in feet.

FIGURE 3. Simplified stratigraphic cross section illustrating the flooding surface and lignite occurrence that defines the key stratigraphic subdivisions of Arecuna field.

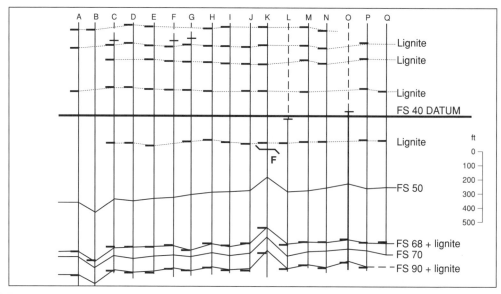

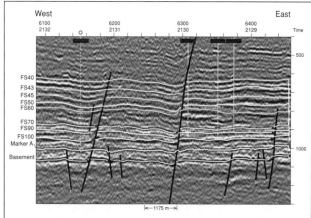

FIGURE 4. Seismic cross section illustrating the key seismic horizons of Arecuna field.

TABLE 1. COMPARISON OF THE STRATIGRAPHIC NOMENCLATURE OF THIS STUDY WITH EXISTING CORPOVEN SANDSTONE NOMENCLATURE.

Genetic stratigraphy	Corpoven nomenclature
FS 90–100	U2 and U3
FS 80–90	U2
FS 70–80	UI
FS 68–70	T
FS 63–68	S1–S5
FS 62–63	R4
FS 60–62	R1–R3
FS 50–60	Ro, P1–P3
FS 45–50	L3–L4, M, N, and O
FS 43–45	J2–J3, K and L
FS 40–43	F8, G, H, I, and J1

plain. The digitate sandstone-rich trends define axial channel belts, and the local pods, where net sandstone thins, represent areas of lacustrine ponds that developed intermittently during coarse, clastic-sediment bypass. Facies analysis, based on gamma-ray and SP log patterns, indicates that the unit is dominated by aggradational, blocky log patterns, and these are attributed to amalgamated sandstone channel fills that accumulated in the axial channel complexes (Figure 6). Interbedded mudstones displaying thin, spiky, and serrate gamma-ray and SP log motifs accumulated in the interaxial areas where intermittent coarse-clastic bypass led to local lacustrine inundation.

The sandstone-dominant lithology, broad sandstone distribution (Figure 5), and widespread blocky bedding architecture (Figure 6) provide evidence of river flow that was weakly confined across a broad alluvial plain, although there are well-developed interaxial belts that are sandstone poor. Unconfined bed-load fluvial systems of the Canterbury Plain, New Zealand (Figure 7; Leckie,

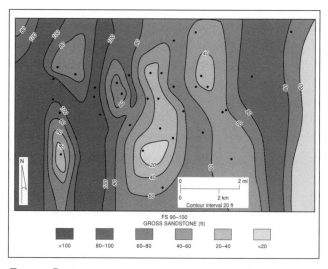

FIGURE 5. Gross sandstone, Merecure Unit A (FS 90–100).

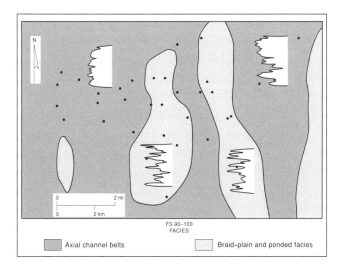

FIGURE 6. Log facies, Merecure Unit A (FS 90–100).

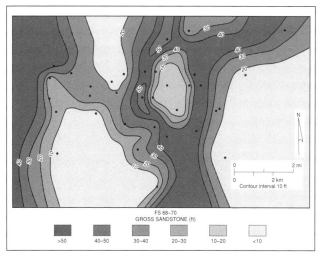

FIGURE 8. Gross sandstone, Oficina Unit A (FS 68–70).

FIGURE 7. The bed-load fluvial system of the Canterbury Plain, New Zealand, is a modern analog for Merecure Unit A at Arecuna field.

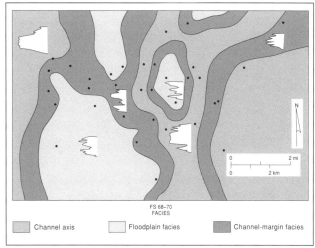

FIGURE 9. Log facies, Oficina Unit A (FS 68–70).

1994), perhaps provide the closest modern analog to Merecure Unit A, although these modern systems principally carry boulders, gravel, and very coarse sandstone.

Oficina Unit A

Oficina Unit A is defined between flooding surfaces 68 and 70 (locally referred to as the T sandstone) and is interpreted as an example of a low-moderate sinuosity, mixed-load fluvial system. The unit varies from 11 to 28 m (37 to 92 ft) thick, and gross sandstone varies from 2 to 17 m (7 to 57 ft) thick. Discrete sandstone-rich axes that trend north-northeast with low to moderate sinuosity are interpreted as channel-point bar complexes (Figure 8). These sandstone-rich axes are confined between broad sandstone-poor areas that are attributed to floodplain deposition. Log-facies characters of the sandstone-rich channel belts are blocky and blocky-upward fining, suggesting predominantly aggradational bedding with minor lateral accretion processes (Figure 9). In contrast, log characters of the sandstone-poor areas are serrate and upward coarsening, consistent with suspension settling of floodplain muds and sporadic crevasse-splay deposition.

The approximately equal proportions of sandstone and shale, discrete sinuous sandstone-rich belts confined by mud-rich areas, and log-facies architecture are all consistent with deposition from a mixed-load fluvial system. The Tanana River near Fairbanks, Alaska, is a modern example of a low-to-moderate sinuosity, mixed-load fluvial system (Figure 10) and thus provides an appropriate analog for Oficina Unit A.

Oficina Unit B

Oficina Unit B, which is defined between flooding surfaces 63 and 68 and includes the S1–5 sandstones (PDVSA nomenclature), is interpreted as a sandy, braided fluvial system. The unit is 27 to 41 m (81 to 137 ft) thick, and gross sandstone varies from 5 to 29 m (16 to 95 ft).

The sandstone is stored primarily in a well-defined north-northwest-trending axis that is approximately 6 km wide and splits into a second channel axis in the north part of the field study (Figure 11). Sand-body geometry in the axial trend is low sinuosity, and the bedding architecture is aggradational, as indicated by the uniformly blocky gamma-ray and SP log patterns (Figure 12). The sandstone-dominant, blocky bedding architecture is attributed to deposition from migrating transverse and longitudinal bar forms that are typical of sandy, braided river systems. The well-defined mud-rich areas of this interval are characterized by uniformly serrate log packages and are interpreted as interchannel floodplain facies.

The external sand-body distribution and internal bedding architecture of the Oficina Unit B resemble that of a sandy, braided fluvial system that is confined by stable (vegetated?) floodplains. The modern sandy, braided William River in Canada (Figure 13; Miall, 1992) is regarded as a possible analog.

STRUCTURAL COMPLEXITY

Because of its proximity to the craton or, conversely, its considerable distance from the thrust-and-fold belt, structure in the Arecuna area was historically thought of as simple and resulting from regional extension. Acquisition of the 3-D seismic data revealed a much more complex structural picture, however, and the 3-D seismic data proved essential to resolving the structural complexity that allowed accurate horizontal-well design. Nine key seismic-reflection events were identified (Figure 4), tied by synthetic seismograms to wellbore-defined stratigraphic horizons, and mapped throughout the data volume. Nine wells, all having sonic and density logs and some having check-shot surveys, were used to generate the synthetic tie between seismic and well control. Depth-structure

FIGURE 10. Tanana River, Fairbanks, Alaska, is a modern analog for FS 68–70. Photo courtesy of Roger Tyler.

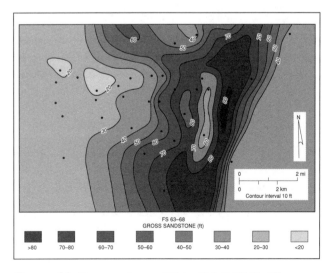

FIGURE 11. Gross sandstone, Oficina Unit B (FS 63–68).

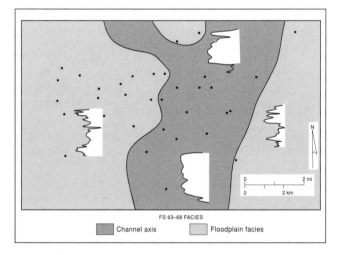

FIGURE 12. Log facies, Oficina Unit B (FS 63–68).

FIGURE 13. The William River, Canada, is a possible modern analog for the sandy, braided fluvial system of the Oficina Unit B at Arecuna field.

maps were generated for each of the major stratigraphic units, providing the fundamental control surfaces for designing horizontal-well trajectories.

The depth map on FS 70 illustrates the structural trends and faulting pattern of the field area (Figure 14). The map reveals numerous fault-bounded blocks that display variable degrees of dip and orientation, although structural strike is generally east-west. The variability in structural orientations resulted from multiple episodes of faulting, with the largest faults oriented east-west and generally south-dipping. Throw on the main fault is variable, but generally 45–60 m (150–200 ft). This main fault provides the trap for the hydrocarbons in the northern half of the field area. The trapping mechanism for oil in the southeastern part of the field is another fault located south of the study area. The main fault in the center of the field is accompanied by a variety of parallel and antithetic faults, with fault throws typically on the order of 15 m (50 ft). A second set of smaller-scale faults is oriented north-northwest–south-southeast and dips mostly southwest. Although highly variable, fault throw is generally 3–15 m (10–50 ft), but locally can be as much as 30 m (100 ft).

The faulting pattern suggests that there were two main episodes of extension. The north-northwest–south-southeast-oriented fault set appears to be the oldest and was caused by regional extension of the basin (including basement). Bending at the basin rim probably caused the younger east-west fault set, but regional right-lateral strike-slip motion caused by collision of the South American and Caribbean Plates is superimposed on the younger fault set. The large structural low on the south side of the main east-west fault provides evidence of right-lateral shear caused by extensional forces at the fault bend (Figure 14). Several of the small normal faults appear to have been displaced by left-lateral shear, but this apparent displacement is misleading because the dip-slip movement is much greater than the strike-slip motion.

ENGINEERING ANALYSIS

Arecuna field is in the early stages of primary recovery. The drive mechanism is a combination of solution gas drive and aquifer drive. Because the field is in the earliest stages of primary production, fluid-flow trends were established primarily by using the initial fluid levels, oil-water contacts, lowest known oil, and highest known water. The initial hydrocarbon fluid properties also were investigated because they are fundamental to the performance of this heavy-oil reservoir.

A key finding of the research was the identification of two types of oil in the study area—foamy oil of moderate

viscosity and dead oil of high viscosity. The foamy oil, because of its lower viscosity, has a far superior production performance. The stratigraphic distribution of these two oil types is complex, and the data set of only nine PVT samples was inadequate to resolve this problem definitively. It appeared, however, based on available data, that reservoir pressure was one controlling factor. An abrupt change occurs in methane and heptane-plus content of the oil at reservoir pressures between 1530 and 1544 psi, with the foamy oils (higher methane and lower heptane-plus content) occurring above these reservoir pressures (Figure 15). Samples taken at lower reservoir pressure displayed lower methane and higher heptane-plus percentages. The average hydrocarbon characteristics of the two oil types are summarized in Table 2. The dead oil is char-

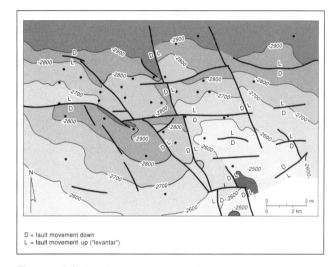

D = fault movement down
L = fault movement up ("levantar")

FIGURE 14. Depth-structure map in feet subsea on FS 70 (top of U1 sandstone).

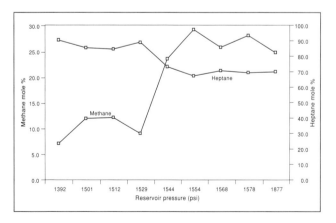

FIGURE 15. Mole % of light hydrocarbons (methane) and heavy hydrocarbons (heptane-plus) in an oil sample is related to reservoir pressure. At lower reservoir pressures, oil has a heavy character with relatively low methane and high heptane content. At high reservoir pressure, oil displays a relatively higher methane content and a relatively lower heptane content.

acterized by lower methane content, lower gas-oil ratio (GOR), lower bubble-point pressure (Pb), and lower formation volume factor (FVF), whereas the foamy oil is characterized by higher methane content, higher GOR, higher bubble-point pressure, and higher FVF.

Another important finding of the research was the extreme sensitivity of oil production to water saturation. The fractional-flow behavior of the oil and water indicates that the fluid-rock system is very sensitive to water saturation, and that a change in water saturation of only a few percent can lead to excessive water production (Figure 16). Fractional flow is controlled mainly by the relative permeability of the water and oil viscosity. A representative relative-permeability curve illustrates the steep decline of oil-relative permeability with increasing water saturation in the Arecuna field (Figure 17). Applying this curve and the average dead-oil viscosity to the Buckley-Leverett horizontal fraction flow demonstrates that a small change in water saturation at a critical range (26% to 30%) changes production from 100% oil to 100% water (Figure 16). The relative permeability characteristics of four samples from the Arecuna field area are summarized in Table 3.

Average irreducible water saturation is 0.24, and the average residual oil saturation is a high value of 0.40. This results in an average pore-volume, initial-oil saturation of 0.76 with just under half of the oil being movable to water. The relative-permeability character indicates that aquifer drive will result in significant residual oil saturation.

A further finding of the engineering investigations was the potential for two-phase flow in vertical-well completions. Investigation of production performance of the Arecuna wells is limited by the short production time; therefore, production of some neighboring wells, including horizontal-well completions, was included in the analysis. Initial oil production rates are affected strongly by the producing gas-oil ratio (GOR). As the oil rate decreases, the producing GOR increases (Figure 18). At high oil-production rates, the GOR is equal to the solution GOR, but at lower oil rates, the GOR is much higher than the solution GOR. These production trends suggest that two-phase flow (oil and gas) is occurring at lower production rates. More important, the high oil rates and solution gas GOR production are occurring in the horizontal wells (from the neighboring areas), whereas the lower oil rates and GORs higher than solution GOR are occurring in vertical wells. This production characteristic is interpreted to be a result of near-wellbore pressure drawdown. The horizontal wells have low drawdown pressures that maintain one-phase flow in the reservoir, but the vertical wells have high drawdown pressures that induce two-phase flow in the reservoir. Horizontal wells

TABLE 2. AVERAGE FLUID CHARACTERISTICS OF DEAD OIL AND FOAMY OIL.

Character type	Dead oil	Foamy oil
Viscosity (cp)	1866	398
Formation volume factor	1.04	1.08
Initial GOR (SCF/STB)*	30	114
Bubble point (psi)	395	1412
Methane content (%)	10	26
Heptane content (%)	88	70

*SCF/STB = standard cubic feet/stock tank barrels.

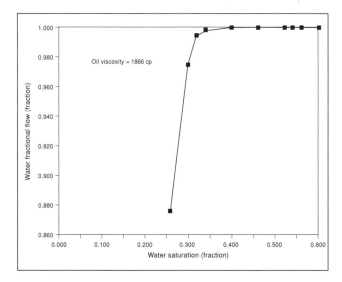

FIGURE 16. Buckley-Leverett horizontal fraction flow for the sample 7H relative-permeability curve and dead oil demonstrates that water cut is highly sensitive to small increases in water saturation.

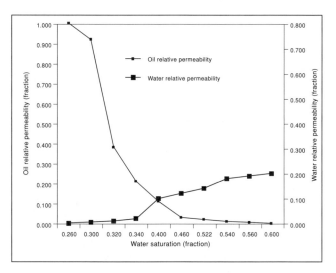

FIGURE 17. The relative permeability measured for sample 7H is representative of four samples measured in the Hamaca field. It displays a steep decline in oil relative permeability with a small increase in water saturation above irreducible.

TABLE 3. SUMMARY OF RELATIVE-PERMEABILITY CHARACTERISTICS FOR THE FOUR SAMPLES IN THE ARECUNA FIELD AREA.

Sample	Absolute permeability (md)	Porosity (fraction)	Irreducible water saturation	Residual oil saturation	Crossover water saturation	Crossover relative permeability
3V	3512	0.369	0.19	0.32	0.43	0.11
3H	4717	0.362	0.26	0.40	0.43	0.09
3AV	4928	0.323	0.23	0.47	0.38	0.09
7H	3728	0.317	0.26	0.40	0.42	0.09
Average	4221	0.340	0.24	0.40	0.42	0.10

Sample	End-point oil relative permeability	End-point water relative permeability
3V	1.00	0.25
3H	0.88	0.18
3AV	1.00	0.18
7H	1.00	0.22
Average	0.97	0.21

are therefore preferable to vertical wells to achieve higher oil rates.

FLUID SATURATION DETERMINATION

Because of the fractional-flow behavior of the oil and water at Arecuna field, fluid-saturation parameters needed to be determined accurately, especially water resistivity (Rw), which is known from measured water samples to vary greatly across the field. The lateral and vertical distribution of water resistivity was interpreted across Arecuna field by (1) investigating the stratigraphic and spatial position of the 24 available water samples, (2) calibrating Rw of the measured samples with water resistivities apparent (Rwa) on the well logs where water samples were available, and (3) extrapolating Rwa values calculated from the remaining well logs throughout the field area.

A good calibration between measured Rw and log-derived Rwa was achieved using an "a" of 0.81 and "m" and "n" values of 2.0. For example, the average Rwa in the stratigraphic interval FS 60–68 (R1–5, S1–5 sandstones) was 0.73 ohm-m, which compares to the average Rw of 0.69 from the measured samples. Similarly, the average Rwa in the stratigraphic interval FS 70–90 (U1–U3 sandstones) was 0.96 ohm-m, whereas Rw from the measured samples averaged 0.92 ohm-m.

Rw is a temperature-dependent measurement that normally decreases with depth because temperature increases with depth. Reversal of the normal Rw trend with depth at Arecuna provides evidence of stratigraphic control on Rw variation. For example, the average Rw measured from water samples in the stratigraphically shallow FS 60–68 (R1–5, S1–5 sandstones) was 0.69 ohm-m, but the average measured Rw for the stratigraphically deeper

FS 70–90 (U1 to U3 sandstones) was 0.92 ohm-m. This same trend of increasing Rw with depth is evident in Rwa observed from well logs (Figure 19). The presence of lower-salinity water in the stratigraphically deeper reservoir sandstones suggests recharge from groundwater of the modern Orinoco River. The deeper reservoir units compose thick, blocky, amalgamated channel fills of braided rivers and braid-plain associations that, because of sandstone continuity, provide opportunity for extensive, interconnected aquifer systems promoting basinward groundwater flow.

Faulting further complicates the distribution of Rws. In the stratigraphic interval FS 60–62, for example, Rwa varies from 0.35–1.2 ohm-m and, although generally consistent within individual fault blocks, changes strong-

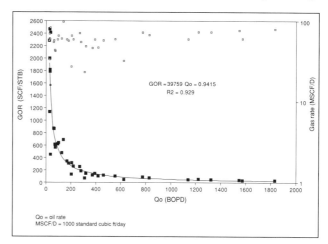

FIGURE 18. For wells in and adjacent to the study area, initial potential rates show that oil rate decreases as GOR increases. The higher GOR values are from vertical wells, whereas the GOR values at the solution GOR are from horizontal wells.

ly from fault block to fault block (Figure 20). Rw distribution in the stratigraphic interval FS 62–63 displays similar trends. Rw in the northwesternmost fault block varies from 0.35 to 0.4 ohm-m, whereas Rw in the fault block juxtaposed to the southeast varies from 0.7 to 0.8 ohm-m (Figure 21). The variability of water resistivities from fault block to fault block suggests that the faults can act either as conduits for basinward groundwater flow (high Rws) or as aquitards (low Rws), effectively sealing the reservoir from groundwater flow.

The effect of highly variable water resistivity on well-log evaluation of these heavy oils and ultimately successful horizontal drilling is profound. The unsuccessful horizontal well 1 illustrates the issue. Well 1 targeted the stratigraphic interval FS 70–80 (U1 sandstone) in an area where an Rw of 0.4 ohm-m was initially assumed. Using an Rw of 0.4 ohm-m and water saturation (Sw) of 25%, the petrophysical evaluation predicted that a response of 50 ohm-m on the resistivity log would be oil productive. When the well was drilled, resistivity on the logging-while-drilling (LWD) tool averaged 50 ohm-m along the horizontal section, but on completion, the well produced 100% water. Postmortem analysis indicated that a more appropriate Rw was 1.3 ohm-m, and using the 50 ohm-m

resistivity log response, results in an Sw calculation of 44%. The fractional-flow curve predicts that Sw of greater than 30% will produce 100% water (Figure 16). Prior to this well result, 50 ohm-m on the resistivity tool was assumed to be a reliable cutoff fieldwide for defining oil-productive zones.

COMPARTMENTALIZATION AT ARECUNA FIELD

Integration of the geologic architecture and fluid-flow trends in the Arecuna field demonstrated the highly compartmentalized character of the reservoir units. Initial fluid levels, oil-water contact (OWC), lowest known oil (LKO), and highest known water (HKW) were analyzed for all genetic stratigraphic units; the strong variability of these fluid levels across the field corresponds to the complex faulting that has compartmentalized the reservoir units into discrete fault-bounded blocks. Results of the fluid-level analysis are illustrated for Merecure Unit A and Oficina Units A and B in Figures 22, 23, and 24.

Lateral and vertical facies heterogeneities introduced by sedimentary processes are another important cause

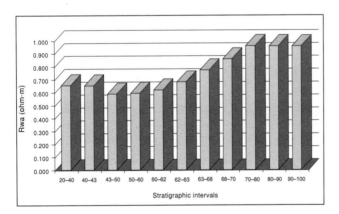

FIGURE 19. Average Rwa for all stratigraphic intervals in the Arecuna field.

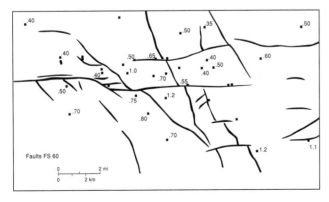

FIGURE 20. Rwa distribution in Oficina Unit FS 60–62.

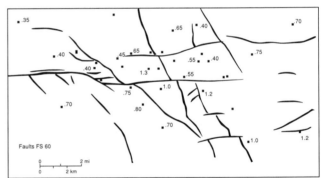

FIGURE 21. Rwa distribution in Oficina Unit FS 62–63.

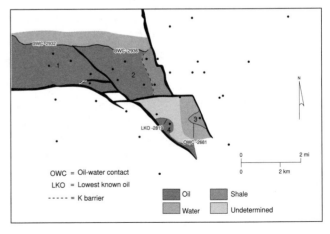

FIGURE 22. Compartmentalization in Merecure Unit A (FS 90–100).

of compartmentalization. Reservoir quality (sandstone thickness) diminishes at a distance from the axial drainage trends of the Arecuna fluvial systems at most genetic stratigraphic levels, and when the economically imposed net sandstone cutoff of 6 m (20 ft) is applied, lateral facies variability becomes a significant cause of segmentation (Figure 23).

The thin but widespread genetic sequence boundaries introduce barriers to vertical fluid flow, creating vertical stratification in the Arecuna field. Variability of fluid levels in successive genetic units, particularly water above oil, provides obvious evidence of the vertical stratification. The presence of water overlying oil in the stratigraphic unit between FS 70 and 72 illustrates that even very thin shale markers can vertically stratify the reservoir (Figure 25).

COMPARTMENTALIZATION IN MERECURE UNIT A

Four oil-bearing fluid-flow compartments are defined in Merecure Unit A (FS 90–100) by the variability of initial fluid levels (Figure 22). Faults define the updip trapping mechanism for all compartments. Compartment 1 in the northwest part of the field lies downdip from a major west-east fault and crosscutting a north-northwest–south-southeast-oriented fault. The lower limit of the compartment is defined by an OWC at −880 m (−2932 ft) subsea. This OWC is slightly higher than the OWC at −641 m (−2136 ft) subsea in the adjacent compartment 2. Compartment 2 is located in the west-central part of the field and also lies downdip of a major west-east fault. This compartment shares a faulted boundary with compartment 1 to the west and is defined on the east by pinch-out of the reservoir facies. Compartments 3 and 4 are located in the center of the field area. Compartment 3 is defined by an OWC of −804 m (−2681 ft) subsea,

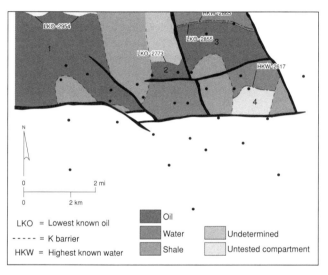

FIGURE 23. Compartmentalization in Oficina Unit A (FS 68–70).

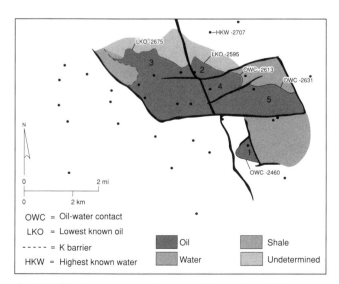

FIGURE 24. Compartmentalization in Oficina Unit B (FS 63–68).

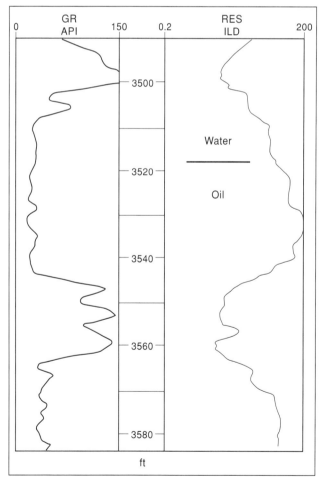

FIGURE 25. Petrophysical analysis shows water above oil in well M. The fluids are separated by a thin shale not resolvable on the gamma-ray tool.

whereas compartment 4 has an LKO of −846 m (−2819 ft) subsea (Figure 22).

COMPARTMENTALIZATION IN OFICINA UNIT A

Oficina Unit A (FS 68–70) contains four oil-bearing compartments, the lower limits of which are defined by three LKOs and an HKW (Figure 23). LKO values are generally not definitive of barriers unless they lie deeper than an adjacent OWC or HKW, as is the case in definition of compartment 4, which has an HKW of −785 m (−2617 ft) subsea that lies structurally higher than the LKO of −886 m (−2954 ft) subsea of compartment 1, the LKO of −832 m (−2773 ft) subsea in compartment 2, and the LKO of −856 m (−2855 ft) subsea of compartment 3. In addition, compartment 3 has an HKW of −860 m (−2865 ft) subsea that lies structurally higher than the LKO of −886 m (−2954 ft) subsea in compartment 1. Evidence that compartment 2 is separate from either compartment 1 or 3 is inferred largely by the faulting pattern (Figure 23). Compartment 4 is untested at this stratigraphic level, but the geologic analysis predicts good-quality sandstone updip from the water-saturated sandstones in well Q, and the structure map defines a large rise in structural level.

COMPARTMENTALIZATION IN OFICINA UNIT B

Oficina Unit B (FS 63–68) contains five oil-bearing compartments (Figure 24). Compartment 1 is defined at its updip limit by a major west-east-oriented fault and crosscutting north-northwest–south-southeast-oriented faults. The lower limit of the compartment is defined by an LKO of −803 m (−2675 ft) subsea that is structurally lower than both the OWC of −784 m (−2613 ft) subsea in the adjacent compartment 4 and the OWC of −789 m (−2631 ft) subsea in neighboring compartment 5. Similar to compartments 4 and 5, compartment 3 is defined by a distinct OWC (−792 m [−2460 ft] subsea). Compartment 2 is the only inferred compartment; its LKO of −779 m (−2595 ft) subsea is nondefinitive.

VOLUMETRIC ANALYSIS

The hydrocarbon volumetrics at Arecuna field were determined by (1) mapping storage capacity (porosity thickness [phih]) in each well for each genetic unit, and then (2) interpolating phih values between wells using the gross sandstone maps to control the interpolation trend. Once the distribution of storage capacity was mapped for each unit, fluid levels were superimposed on each compartment map, and then average oil saturation was applied

to calculate original oil in place (OOIP) for each compartment in each genetic unit. The OOIP is estimated to be 2.8 billion barrels residing in 77 discrete compartments. A field-development plan was constructed to maximize exploitation of this resource. The strategy for the horizontal-wellbore design and placement was to, wherever possible, (1) optimize drainage area; (2) drill parallel to structure and perpendicular to depositional axis; (3) start in thickest, cleanest sandstones; and (4) start near well-defined geologic boundaries. The rationale for drilling parallel to structure was to avoid drilling too close to water. Drilling perpendicular to the depositional axes was intended to penetrate any inherent lateral heterogeneity and maximize access to the reservoir. An example of the idealized well-drilling plan is illustrated in Figure 26.

POSTMORTEM RESULTS/ PROBLEM-WELL DIAGNOSIS

The 46 horizontal-well drilling program that followed the reservoir characterization study was highly successful; the accuracy of the reservoir model was confirmed by the drilling results. In some cases, geologic predictions were reliable as much as 1050 m (3500 ft) from the nearest well data. Moreover, the vast majority of the wells produced at target flow rates of 1000 BOPD or more. Despite the success, eight of the 46 wells drilled experienced water-production problems, and a detailed follow-up was requested by the operator to help identify the source of the water production. The trajectory of each well was plotted on cross sections using the LWD recorded directional data and was compared to the predicted structural, depositional, and fluid framework on the integrated compartment maps. An example of a typical well

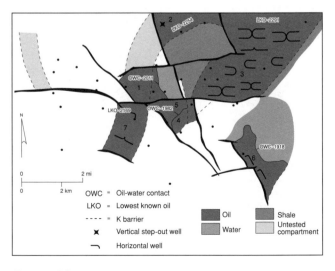

FIGURE 26. Example of an idealized well plan for genetic unit FS 43–45.

trajectory is illustrated in Figure 27 for well 8. Three main problems were identified that resulted in water production: (1) mechanical problems during drilling and completion, (2) coning of the aquifer, and (3) uncertainty in defining pay-resistivity cutoffs as a result of variable Rw. Results of the wells are summarized in Table 4.

Mechanical problems

Failure of several of the wells was clearly attributable to mechanical problems. Well 7, for example, measured high resistivity on the LWD tool from the casing shoe to total depth, indicating that the well had been successfully steered in the oil sand. There was a slight increase in gamma-ray response and an accompanying decrease in resistivity at total depth, but Sw analysis indicated that the well was still in the oil zone. The well produced 800 BOPD for 15 days before any water production. Subsequently, however, the production rapidly changed to 100% water. The plot of the well's trajectory indicated that the well had drilled to within 1.5 m (5 ft) of the underlying bounding shale that separates the target-oil reservoir from an underlying water-saturated sandstone. Pressure drawdown during production is interpreted to have caused the thin shale to break down, coning water from the underlying sandstone unit.

Well 4 also measured high resistivity on the LWD tool from the casing shoe to total depth, indicating that it too had been successfully steered in the oil sand, but the well produced 100% water on completion. No water contacts were predicted below the sandstone by the integrated study. The petrophysical analysis did, however, indicate two high-permeability water sandstones above the casing shoe, and an ultrasonic cement-evaluation tool confirmed poor isolation from these water zones.

High resistivity on the LWD tool indicated that well 5 successfully steered through the targeted oil-saturated sandstone. The 200,000 barrels of water-free oil produc-

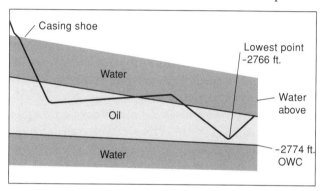

FIGURE 27. Trajectory of well 8 compared to the predicted reservoir model.

TABLE 4. Results of the wells that experienced water-production problems.

Well	Mechanical problems	Coning of aquifer	Uncertainty in pay-cutoff determination
1			Produced 100% water. LWD measured 50 ohm-m in horizontal section. Rw = 0.4 gives Sw = 25% and predicts oil production. Rw = 1.3 gives Sw = 44% and predicts water production.
2		Produced 360,000 bbl of oil, then coned water. Well drilled to 18 m (61 ft) above the OWC.	
3		Produced 100% water. Well drilled to 5 m (16 ft) above OWC.	
4	Produced 100% water. Poor isolation from overlying water sandstone.		
5	Produced 200,000 bbl of oil, then 100% water. No obvious source for water, but probably poor isolation from overlying water sandstone.		
6	Produced 100% water. Drilled through overlying water sandstone.		
7	Produced 800 BOPD for 15 days, then 100% water. Drilled within 1.5 m (5 ft) of underlying shale, which broke down because of pressure drawdown.		
8	Produced 100% water. Drilled through overlying water sandstone then drilled near to underlying OWC.		

tion in the ensuing five months also confirmed the technical success of the well. After five months, the well rapidly transformed to 100% water production. In this case, there was no obvious source for the water production. We speculate that poor isolation from overlying, low-permeability, water-saturated sandstone was the cause of the water production.

Two wells, 6 and 8, passed through low-resistivity zones along their horizontal section, indicating that they had deviated from their proposed trajectories during the drilling process. Well 6 passed through a low-resistivity/high-gamma-ray zone near total depth that when plotted against the proposed well path, indicated that the well had penetrated the shale horizon above the target pay zone into an overlying highly permeable water-saturated sandstone. Well 6 produced 100% water from the outset. Well 8 drilled through high-resistivity strata for most of the horizontal section but intersected two zones of low resistivity. Plotting the actual well trajectory against the proposed well path showed that the well first penetrated the overlying highly permeable, water-saturated sandstone and then drilled too steeply, approaching the OWC near the toe of the well. Well 8 also produced 100% water from the outset.

CONING OF THE AQUIFER

Water-production problems in two wells, 2 and 3, were attributed to coning of the aquifer. Both had high resistivity on the LWD log for the entire length of the wellbore, and no apparent shales or water zones were detected. Well 2 produced 360,000 barrels of water-free oil during the first six months of production, but once water production began, the well rapidly changed to 100% water. This well was drilled in the downdip direction, counter to the recommended strike-parallel trajectory, but was still 18 m (61 ft) above the OWC at the heel of the well. Well 3 was drilled in the same fault block as well 2 and in the same downdip direction. The trajectory plot indicated that the heel of the well was only 5 m (16 ft) above the OWC, and the well produced water immediately on completion. Relative permeability of the reservoir to water is much greater than oil, yet it was surprising that as much as 18 m (61 ft) above OWC was insufficient to allow sustained, water-free oil production.

Both wells were completed with uncemented slotted liners, which restricted possible remedial workovers in the horizontal section such as plugbacks, and both were subsequently recompleted uphole in the vertical section. Future wells drilled in proximity to water contacts could be completed with either cemented perforations or with some form of external casing packer to allow isolation of water-producing zones.

UNCERTAINTY IN PAY-CUTOFF DETERMINATION

A resistivity of 50 ohm-m was assumed to be a reliable cutoff for defining oil-productive sandstone fieldwide. The failure of horizontal well 1, however, highlighted the need for more rigorous petrophysical analysis in determining pay cutoffs because Rw varies so greatly from fault block to fault block, from stratigraphic unit to stratigraphic unit, and within individual stratigraphic units. When well 1 was drilled, resistivity on the LWD tool averaged 50 ohm-m along the horizontal section, and it was assumed that the well would be oil productive. The well produced 100% water on completion. Because of the fractional-flow behavior of the oil and water, Sw must be 25% or less for a well to be oil productive. A response of 50 ohm-m on the resistivity log would yield an Sw of 25% if Rw was 0.4 ohm-m. In contrast, an Rw of 1.3 ohm-m would yield an Sw calculation of 44% for the same 50 ohm-m resistivity log response. Investigation of offset wells in the follow-up study indicated that an Rw of 1.3 ohm-m was a more appropriate value for pay-cutoff evaluation.

CONCLUSIONS

Horizontal-well drilling technology offers a cost-effective approach to optimize commercial recovery of the multibillion barrel heavy-oil resource of the Faja region that is not possible with conventional vertical-well completions. Reservoir complexity caused by faulting, lateral and vertical facies heterogeneity, and fluid-flow dynamics, however, introduces considerable risk to successful horizontal completions. Multidisciplinary, integrated reservoir characterization that resolves the complexity of the reservoir architecture and fluid-flow behavior is essential for accurate forecasting of horizontal-well trajectories.

The horizontal drilling experience at Arecuna highlights the potential hazards for horizontal-well drilling in the Faja region. Interbedding of oil and water-saturated sandstones dictates the need for precision in drilling and careful zone isolation during cementing and completion. The fractional-flow behavior of the oil and water so strongly favors the transmission of water that the greatest possible separation from the OWC is desirable in well design. Aquifer recharge from the modern Orinoco River has introduced lower-resistivity waters in the region, but in a complex pathway controlled by faulting and facies continuity. Distribution of Rw must be resolved so that reliable petrophysical models (resistivity pay cutoffs) can be applied in the interpretation of LWD responses as drilling proceeds.

REFERENCES CITED

Audemard, F. E., I. Azpiritxaga, P. Bauman, A. Isea, and M. Latreille, 1985, Marco geologico del Terciario de la Faja Petrolifera del Orinoco de Venezuela: 6th Congreso Geofisica, Venezuela, v. I, p. 70–109.

Burkill, G. C. C., and L. A. Rondon, 1991, Steam soak pilot project in the Zuata area of the Orinoco belt, Venezuela: Fifth U.N. Institute for Training and Research International Conference for the 21st century, v. 2, p. 105–112.

Erlich, R. N., and S. F. Barrett, 1992, Petroleum geology of the eastern Venezuela Foreland Basin, *in* R W. Macqueen and D. A. Lechie, eds., Foreland basins and fold belts: AAPG Memoir 55, p. 341–362.

Galloway, W. E., and D. K. Hobday, 1983, Terrigenous clastic depositional systems: Applications to petroleum, coal, and uranium exploration: New York, Springer-Verlag, 423 p.

Hamilton, D. S., and N. Z. Tadros, 1994, Utility of coal seams as genetic stratigraphic sequence boundaries in nonmarine basins: An example from the Gunnedah Basin, Australia: AAPG Bulletin, v. 78, no. 2, p. 267–286.

Leckie, D. A., 1994, Canterbury Plains, New Zealand: Implications for sequence stratigraphic models: AAPG Bulletin, v. 78, no. 8, p. 1240–1256.

Miall, A. D., 1992, Alluvial deposits, *in* R. G. Walker and N. P. James, eds., Facies models: Response to sea level change: Geological Association of Canada, p. 119–142.

Hurley, N. F., A. A. Aviantara, and D. R. Kerr, 2003, Structural and stratigraphic compartments in a horizontal well drilled in the eolian Tensleep Sandstone, Byron field, Wyoming, *in* T. R. Carr, E. P. Mason, and C. T. Feazel, eds., Horizontal wells: Focus on the reservoir: AAPG Methods in Exploration No. 14, p. 143–159.

9

Structural and Stratigraphic Compartments in a Horizontal Well Drilled in the Eolian Tensleep Sandstone, Byron Field, Wyoming

N. F. Hurley

Colorado School of Mines
Golden, Colorado, U.S.A.

D. R. Kerr

The University of Tulsa
Tulsa, Oklahoma, U.S.A.

A. A. Aviantara

The University of Tulsa
Tulsa, Oklahoma, U.S.A.

ABSTRACT

The Tensleep Sandstone is a major oil and gas producer in the Bighorn Basin, Wyoming. Cores, borehole images, and outcrop descriptions have been used to characterize the formation. Lithofacies include: (1) porous and permeable eolian cross-stratified units; (2) lower permeability, finely laminated interdune units; and (3) relatively impermeable marine carbonate and clastic units. Individual eolian dunes range in thickness from 2 to 10 m (6 to 30 ft), with first-, second-, and third-order bounding surfaces identified in outcrops, cores, and logs.

Byron field is a northwest-southeast-trending asymmetric anticline. In 1992, a medium-radius lateral hole was drilled from the southwest flank toward the crest of the fold. The horizontal well was cased from the surface to the top of the Tensleep Sandstone. The remainder of the well, which was drilled for approximately 150 m (500 ft) on an uphill slant, stayed in the uppermost Tensleep Sandstone in a 6-m- (20-ft-) thick stratigraphic interval. This part of the borehole was left uncased. The general shape of the borehole is that of a fishhook, where the structural high is at total depth.

Borehole-image log interpretation in the horizontal well showed two sets of roughly orthogonal fractures. Average spacing, corrected for borehole geometry, is 2.3 m (7.5 ft) for Set 1 fractures, which lie approximately perpendicular to the borehole path. Set 2 fractures, which lie roughly parallel to the borehole path, have a spacing corrected for borehole geometry of 0.9 m (3.0 ft).

Using prior knowledge of the paleowind direction, the horizontal well was drilled approximately perpendicular to the trend of the eolian dunes. Interpretations of bed boundaries suggest that the borehole crossed at least five eolian-facies architectural elements in 96 m (315 ft) of lateral distance. This suggests an average spacing of 19 m (63 ft) for dune-related compartments. Outcrop and core studies show that flow barriers or baffles commonly exist at bounding surfaces between such compartments.

The last 46 m (150 ft) of the borehole was full of oil. The oil-water contact in the borehole, which was stable during several days of logging, may represent the height of the oil-water contact in the fractures. Note that this level is hundreds of meters (or feet) above the original oil-water contact in the intergranular porosity. This well suggests a novel but untested way to complete a horizontal well: (1) run tubing to total depth, and (2) use the borehole as a downhole oil-water separator. This completion technique could reduce water cuts and prolong the life of Byron field and other fractured reservoirs with strong water drives or active waterfloods.

INTRODUCTION

Byron field is one of many anticlinal traps that produce hydrocarbons from the Tensleep Sandstone in the Bighorn Basin, northwestern Wyoming (Figure 1). According to Peterson (1990), Byron is the sixth-largest oil field in the basin, with 130 MMBO (million barrels of oil) recoverable.

This project began when a horizontal well, the I. Lindsay #3H, was drilled in 1992. Log interpretations suggested that fractures and boundaries between eolian dune bodies had an influence on compartmentalization in this reservoir. Aviantara (2000) did subsequent outcrop and subsurface studies, including descriptions of 59 m (194 ft) of core from four wells and construction of a 65-well grid of cross sections around the horizontal well. The purpose of this paper is to describe compartmentalization in this eolian reservoir as revealed by the results of the 1992 horizontal well. Special emphasis is placed on the occurrence of flow barriers and conduits and the development of a novel completion technique—the use of the borehole as a downhole oil-water separator—that could be applied to similar reservoirs.

GEOLOGIC SETTING

STRATIGRAPHY AND SEDIMENTOLOGY

The Tensleep Sandstone, which consists of various eolian and marine units, was deposited on a broad coastal plain during the Middle Pennsylvanian to Lower Permian (Kerr and Dott, 1988). Erosion occurred at the sub–Goose Egg unconformity on top of the Tensleep Sandstone (Wheeler, 1986; Kerr and Dott, 1988; Simmons and Scholle, 1990). As a result, the unit is variable in thickness. The Tensleep, which is approximately 34 m (110 ft) thick in Byron field, is unconformably overlain by red beds, shales, evaporites, and dolomites of the Goose Egg member of the Phosphoria Formation. The Tensleep conformably overlies red beds and cherty carbonates of the Amsden Formation. The Tensleep, Phosphoria, and Madison intervals are the main hy-

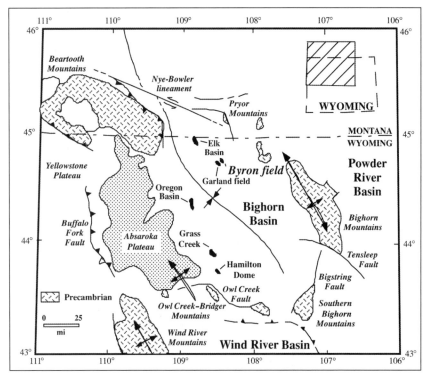

FIGURE 1. Tectonic setting of the Bighorn Basin. Note the location of Byron and Garland fields and other significant reservoirs that produce hydrocarbons from the Tensleep Sandstone. Hachured area on the inset map of Wyoming (upper right) shows the location of this index map. Modified from Johnson and Lindsley-Griffen (1986).

drocarbon-producing formations in the Bighorn Basin (Figure 2).

Most current workers believe the Tensleep was deposited in both eolian and marine environments (Fox et al., 1975; Morgan et al., 1978; Desmond et al., 1984; Kerr et al., 1986; Wheeler, 1986; Rittersbacher, 1985; Kerr and Dott, 1988; Kerr, 1989; Shebl, 1995a, b; Carr-Crabaugh and Dunn, 1996). Outcrop and subsurface studies show that the Tensleep consists from bottom to top of repetitive parasequences of dolomitic sandstone, marine sandstone, and eolian sandstone (Kerr, 1989). Each parasequence corresponds to sea-level fluctuations in which the dolomitic sandstones represent sea-level rise and the eolian sandstones are the result of sea-level fall (Kerr et al., 1986; Kerr and Dott, 1988; Kerr, 1989).

The eolian portion of the Tensleep Sandstone consists of fine to very fine-grained, rounded, well-sorted quartz arenite. Kerr and Dott (1988) described and interpreted the dune types encountered in the Tensleep Sandstone in outcrops along the west flank of the Bighorn Mountains. The dunes are oblique, slightly sinuous, crested features that support smaller crescentic dunes and lee-slope spurs. Paleocurrent studies suggest that the main bedforms migrated to present-day south-southwest (Opdyke and Runcorn, 1960; Mankiewicz and Steidtmann, 1979; Kerr and Dott, 1988). Superimposed bed forms migrated to present-day west (Kerr and Dott, 1988).

STRUCTURE

The Bighorn Basin is bordered by the Bighorn Mountains to the east, the Owl Creek Mountains to the south, the Absaroka and Beartooth Mountains to the west, and the Nye-Bowler fault zone to the north (Figure 1). The basin is characterized by elongate, doubly plunging, symmetric and asymmetric anticlines along its margins (Stone, 1967; Hoppin and Jennings, 1971; Paylor et al., 1989).

Byron anticline is asymmetric (Figure 3), with a steep limb on the northeast flank and a gently dipping limb on the southwest flank (Wyoming Geological Association, 1989). Outcrop studies at nearby Sheep Mountain anticline (Hennier, 1984) and the Thermopolis

anticline (Paylor et al., 1989) suggest that fault-propagation folding (Berg, 1962) is likely in Bighorn Basin structures. Underlying faults probably reflect preexisting weaknesses in crystalline basement (Hoppin and Jennings, 1971; Maughan and Perry, 1986).

HYDROLOGY

Groundwater hydrology is important in the Bighorn Basin because of the strong pressure support provided by the Madison, Tensleep, and Phosphoria aquifers. Hubbert (1953) cited examples of Bighorn Basin oil fields in his classic study of hydrodynamic trapping. Huntoon (1985) clarified the effects of Laramide orogenic structures on aquifer permeability. Potentiometric maps (e.g., Doremus, 1986) show broad pressure support from the aquifer and significant pressure sinks caused by fluid withdrawal from oil fields such as Byron.

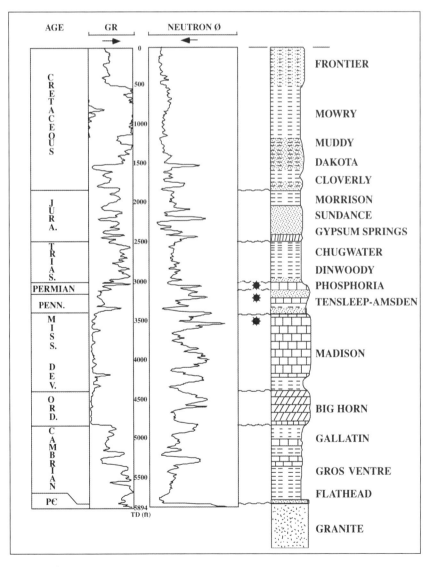

FIGURE 2. Typical log and stratigraphic units in a basement well drilled in nearby Garland field. GR = gamma ray. Modified from Demiralin et al. (1994).

FIELD HISTORY

Byron field, discovered in 1930, is a doubly plunging anticline that trends northwest-southeast (Figure 3). The Tensleep producing area is approximately 5.6 km (3.5 mi) long and 2.4 km (1.5 mi) wide. Structural closure on the anticline is about 180 m (600 ft). Although the Tensleep is the main reservoir, the Phosphoria Formation also produces oil. In 1989, the Tensleep produced intermediate-gravity crude (36° API) from 58 active wells with an average pay thickness of 30 m (97 ft). Average porosity is 14% (ranging from 10% to 20%), and average permeability is 78 md. The presence of an active water drive is indicated by relatively high reservoir pressures and water cuts that commonly exceed 90%. Cumulative production from the Tensleep is approximately 119 MMBO (Wyoming Geological Association, 1989).

Figure 3 shows the location of the cored wells, wells with borehole images, and a cross section constructed for the field. The I. Lindsay #3H well is currently the only horizontal well in Byron field.

RESERVOIR CHARACTERIZATION

CORE DESCRIPTIONS AND FACIES

Aviantara (2000) described core from four wells in Byron field and studied the Tensleep Sandstone in several exposures on the west flank of the Bighorn Mountains. The E. Jones #1 well offers the most complete core coverage of the Tensleep in Byron field (Figures 3 and 4). Based on these observations, the Tensleep has five main facies.

Tensleep eolian sandstones are composed of rounded, moderately to well sorted, fine to very fine quartz arenites. Where stained with oil in core, these sandstones have a distinctive brown color. In outcrop, they are light gray, medium gray, or yellow. Dolomite nodules are common throughout the section, and anhydrite nodules are associated with the interdune facies. Mankiewicz and Steidtman (1979) provided a detailed assessment of Tensleep diagenesis.

Tabular planar cross-stratified sandstone facies are characterized by large-scale cross-strata sets that typically achieve a maximum thickness of 10 m (30 ft). Locally confined in these large sets are pods of trough cross-stratified eolian sandstones (later referred to as ***trough cross-***

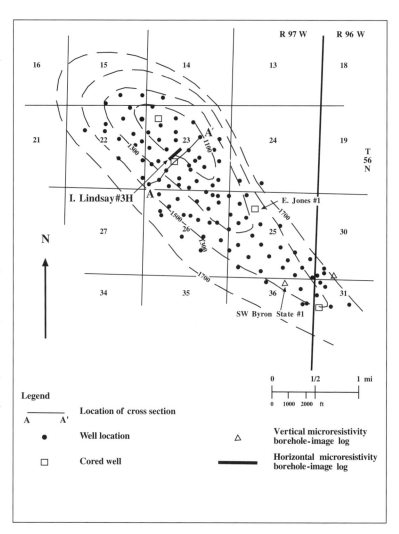

FIGURE 3. Structure map on top of the Tensleep Sandstone for Byron field. Note locations of cored well (E. Jones #1), vertical well with FMS log (SW Byron State #1), and horizontal well with FMS log (I. Lindsay #3H). Cross section A-A′ is shown in Figure 8. Contour interval is 200 ft (61 m) subsea. Modified from Wyoming Geological Association (1989).

stratified intrasets). Wind-ripple cross strata dominate the lower parts of a given crossstata set. Dips range from nearly horizontal to as much as 12°. Grainfall strata dominate the upper parts of cross-strata sets, and they typically dip up to 27°. The sandstone framework of wind-ripple strata is characteristically more tightly packed when compared to grainfall strata. This results from the grain-packing condition at the time of sedimentation (Hunter, 1977).

Interdune sandstone facies are characterized by crinkle or wavy lamination, and less commonly by horizontal wind-ripple strata. Kerr (1989) also noted the local development of subaqueous physical structures. The interdune sandstone facies occurs as broad lenses between tabular-planar sandstone facies sets or, less commonly, as sheets resting on marine sandstones.

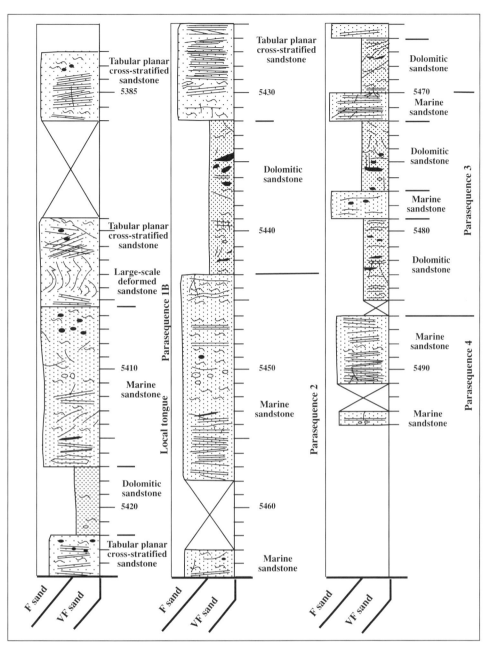

FIGURE 4. Core description from E. Jones #1 well (Sec. 25-T56N-R97W), Byron field. See Figure 3 for location. Parasequences are shown. Interval 5405.5–5422 ft is a local tongue of marine deposits in parasequence 1B. F = fine; VF = very fine. Depth scale is in feet.

The tabular-planar cross-stratified sandstone facies in this part of the Bighorn Basin was deposited probably by large south-southwest-advancing oblique dunes with superimposed crescentic dunes (Kerr and Dott, 1988; Aviantara, 2000). The interdune sandstone facies probably represents sand accumulation between dunes under intermittently wet conditions (Kerr, 1989).

Large-scale deformed sandstone facies are associated with the tabular-planar sandstone facies and are comprised of contorted eolian cross strata. Based on the common occurrence of this facies below Tensleep marine deposits, Kerr (1989) believed the deformation was caused by marine incursion into the eolian dune field.

The gamma-ray log signature for eolian sandstones has a blocky profile. Grainfall-dominated sandstones have the highest porosity values. In contrast, interdune sandstone facies have the lowest eolian porosity values. Porosity values through the lower parts of the tabular-planar cross-stratified sets are intermediate as a result of the gradation from wind-ripple-dominated strata at the base to grainfall-dominated strata at the top of a set.

Tensleep marine sandstones have textural characteristics similar to those of the eolian sandstones. Two marine facies are recognized in this study, although Kerr (1989) reported additional marine sandstone facies in the area.

Marine sandstone facies include wavy lamination, horizontal lamination, and low-angle cross stratification. Locally, there are biogenic traces and thin laminations of green claystone. The inferred depositional environment is shoreface to foreshore.

Dolomitic sandstone facies are characterized by gray, green, or red color and contain pervasive dolomite cement with localized pyrite cement and anhydrite nodules. Wavy lamination predominates with subordinate structureless intervals and biogenic traces. The inferred depositional environment is siliciclastic sabkha, although peritidal deposits also occur.

Log signatures for the marine sandstones are variable but differ from those of the eolian sandstones. Gamma-ray values are slightly to markedly higher in the marine sandstones. Porosity values are lower, and the dolomitic sandstone facies has high resistivity values.

PARASEQUENCES AND BOUNDING SURFACES

Tensleep parasequences are made up of eolian and marine sandstone couplets. The characterisitic repetition throughout the Tensleep Sandstone is traceable in outcrop (Kerr, 1989) and throughout the subsurface of the Bighorn Basin. The marine sandstones (marine sandstone facies and dolomitic sandstone facies) represent the flooding of a preexisting eolian dune field with a decrease in the ratio of sediment supply to accommodation space. The eolian sandstones (tabular-planar, large-scale deformed, and interdune sandstone facies) represent the outbuilding of the eolian dune field at a time when the ratio of sediment supply to accommodation space increased.

In the subsurface, parasequence identification is accomplished through well-log calibration to cores and well-log correlation. The contrasting well-log character of the marine sandstones from the eolian sandstones, especially the tabular-planar sandstone facies, serves to delineate parasequence boundaries. Parasequence boundaries are placed at the base of the low-porosity values from neutron or density log traces (Figure 5). In Byron field, six parasequences (1A, 1B, 2, 3, 4, and 5) are recognized, with designations roughly corresponding to operator zones. Figure 5 illustrates the parasequences in Byron field. Parasequence 1A is absent in some wells because of erosion below the sub–Goose Egg unconformity.

Bounding surfaces related to eolian sedimentation processes are recognized between parasequence boundaries. These include first-, second- and third-order eolian bounding surfaces (Brookfield, 1977, 1992; Kocurek, 1996). In Aviantara's (2000) analysis of Tensleep facies architectural elements, these surfaces were designated as 1.0-, 2.0-, and 3.0-bounding surfaces, respectively. He also recognized 0.0- and 0.1-bounding surfaces.

0.0-bounding surfaces occur at the contact between underlying marine facies and overlying eolian facies in a parasequence. These surfaces are flat, are regionally extensive in outcrop, and can be correlated in the subsurface. The 0.0-bounding surfaces represent the initiation of eolian sediment accumulation. The 1.0-bounding surfaces, locally accompanied by interdune facies, rise upward from a given 0.0-bounding surface.

1.0-bounding surfaces (first-order of Brookfield, 1977, 1992) are regionally extensive and locally exhibit low relief. The 1.0-bounding surfaces truncate lower-rank (2.0, 3.0, and 0.1) surfaces and eolian cross strata. Tensleep outcrops along the west flank of the Bighorn Mountains have 1.0-bounding surfaces that separate very large scale tabular-planar eolian cross-strata sets (Kerr and Dott, 1988; Aviantara, 2000). Typically, Tensleep eolian cross strata are in tangential contact above and discordant contact below a 1.0-bounding surface. Locally, 1.0-bounding surfaces are overlain by interdune deposits. Such surfaces are attributed to the migration and accumulation of the largest eolian bedforms (dunes or draas).

2.0- and 3.0-bounding surfaces (second- and third-order of Brookfield, 1977, 1992) separate bundles of eolian cross strata in a set. Along the western flank of the Bighorn Mountains, Tensleep Sandstone 2.0-bounding surfaces conform to eolian strata dip directions high in tabular-planar cross-strata sets. They sweep downward and define trough cross-strata axes that plunge west-southwestward at the base of tabular-planar cross-strata sets (Kerr and Dott, 1988). The 3.0-bounding surfaces occur mostly high in a tabular-planar set. They separate eolian cross strata with minor variations in dip direction across the surface. In the Tensleep, 2.0-bounding surfaces probably formed when lower lee-slope spurs migrated obliquely across the larger host dunes (Kerr and Dott, 1988). The 3.0-bounding surfaces are reactivation surfaces that developed during minor variations in wind velocity across the lee slope.

0.1-bounding surfaces have not been described in other eolian facies architectural hierarchies. They encapsulate medium- to large-scale trough cross-strata cosets with foreset dip directions and axes that plunge westward (Kerr and Dott, 1988). These elements occur as eolian pods of varying dimensions in the 1.0-bounding surface tabular-planar sets (Aviantara, 2000). Thus, they are called intrasets. At their base, 0.1-bounding surfaces truncate 2.0- and 3.0-bounding surfaces and eolian cross strata. The 0.1-bounding surfaces are downlapped from above by 2.0-bounding surfaces and eolian cross strata. Wind-ripple strata dominate the cross strata, although thicker sets also include avalanche strata. Kerr and Dott (1988) suggested that these intrasets represent a temporary domination of the wind field by a westward-directed wind component, perhaps related to monsoon circulation.

The described eolian bounding surfaces can be recog-

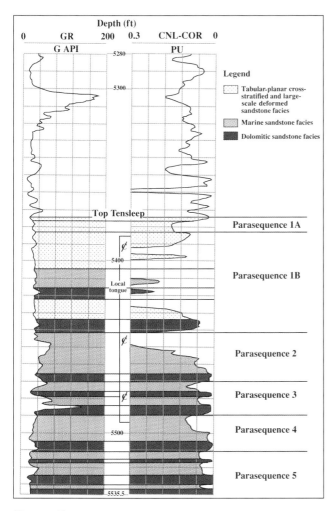

FIGURE 5. Well-log profile and lithology distribution at E. Jones #1 well (Sec. 25-T56N-R97W), Byron field. Parasequences are shown. GR = gamma ray; CNL-COR = neutron porosity; ¢ = core.

nized in the subsurface (core and borehole images; Figure 6) by truncation and hierarchical rank, association with eolian strata types, and cross-strata dip orientation. Carr-Crabaugh et al. (1996) provided expected tadpole-plot vertical trends for the bounding surfaces and cross-strata dip orientations. Later in this paper, we discuss applications of borehole image analysis to Tensleep bounding surfaces in a horizontal well in Byron field.

PETROPHYSICAL PROPERTIES

Figure 7 is a profile of porosity and permeability for the E. Jones #1 well. Facies zonations and parasequence numbers are superimposed on this plot. Table 1 shows the average values of porosity, permeability, water saturation, and oil saturation for each parasequence from all cores in the field. Mean porosities range from 1.5% to 14.8%, with the highest mean value in eolian sandstones of parasequence 1B. Permeability ranges from 0.3 to 197 md, with the highest mean value in eolian sandstones of para-

sequence 2. According to this table, parasequences 1B, 2, and 3 are the best reservoirs in terms of porosity and permeability.

Among the five facies, the tabular-planar cross-stratified sandstones and large-scale deformed sandstones have the highest porosities and permeabilities (Table 2). Dolomitic sandstones have the lowest porosity and permeability values. This relationship can be confirmed by looking at core porosity and permeability profiles (Figure 7) and well-log responses (Figure 5). In the core-porosity and permeability profile, the eolian facies shows that porosity generally ranges from 10% to 20%, with a range in permeability from 10 to several hundred md. The marine facies shows porosity that generally ranges from 5% to 10%, with a range in permeability from 1 to 100 md.

In the tabular-planar cross-stratified sandstone facies, grainfall stratification has higher porosity and permeability than wind-ripple stratification. Above 1.0-bounding surfaces, intervals are dominated by wind-ripple lamination. Below 1.0-bounding surfaces, they are dominated by grainfall stratification. Thus, wind-ripple-dominated strata associated with 1.0-bounding surfaces represent potential baffles to fluid flow (see also Carr-Crabaugh and Dunn, 1996).

Other studies of eolian deposits have shown directional anisotropy in permeability as a result of spatial variation or heterogeneity in facies and stratification (Emmett et al., 1971; Lindquist, 1988; Goggin et al., 1992; Shebl, 1995a, b; Carr-Crabaugh and Dunn, 1996; Iverson et al., 1996). Previous studies have shown that the highest permeabilities are oriented parallel to the dune crests. Directional permeability ratios of 3:1 to 10:1 are commonly reported. Aviantara (2000) confirmed these results using samples from Byron field.

CROSS SECTIONS

Facies distribution has a great impact on reservoir heterogeneity, and each facies behaves as a unit of distinctive fluid-flow behavior (Lindquist, 1988; Chandler et al., 1989; Shebl, 1995a, b). Reconstruction of facies architecture in specific mapping units (i.e., eolian facies in each parasequence) has implications for geologic modeling. Aviantara (2000) constructed a grid of cross sections around the I. Lindsay #3H horizontal well to help constrain dune geometries.

Outcrop observations had a major impact on this part of the study. Rather than correlate using a layer-cake model (Figure 8a), outcrop studies showed that a parasequence-oriented approach is needed (Figure 8b). For example, outcrop work showed that 1.0-bounding surfaces rise to the southwest at an angle of 0.9° to 2.7° from 0.0-bounding surfaces. Using a simple calculation, this sug-

FIGURE 6. FMS images from SW Byron State #1 well (Sec. 36-T56N-R97W), Byron field. Lines trace the intersection of foresets and fractures in the borehole. Tadpoles indicate dip magnitude of the foresets, and tick marks point in the dip direction. The labels 1.0, 2.0, and 3.0 represent first-, second-, and third-order bounding surfaces, respectively.

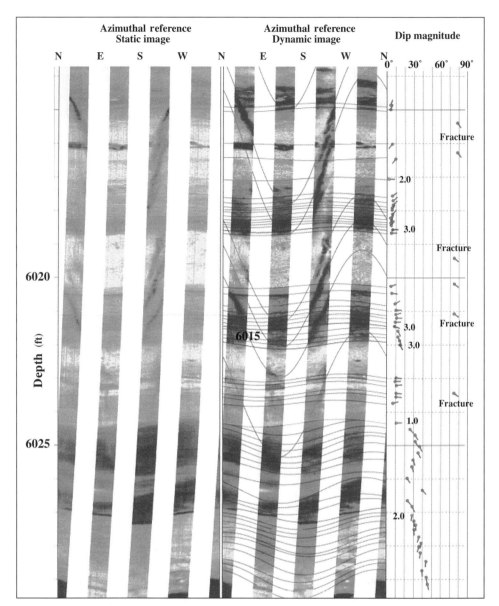

gests that a 1° angle of rise of the 1.0-bounding surface in a distance of 300 m (1000 ft) will translate to a 5.3-m (17.5-ft) rise. Figure 8 shows a comparison between stratigraphic correlations to the southeast of the I. Lindsay #3H horizontal well. The 1.0-bounding surfaces rise from the 0.0-bounding surfaces and are erosionally truncated by the sub–Goose Egg unconformity or by the next parasequence boundary (marine flooding surface) in the cross section.

I. LINDSAY #3H HORIZONTAL WELL

DRILLING FACTS

Marathon Oil Company drilled and logged the I. Lindsay #3H horizontal well in December 1992. The well had conventional openhole logs (gamma ray, resistiv-ity, density, neutron) and two runs of Schlumberger's FMS (Formation MicroScanner) and AMS (Auxiliary Measurement Sonde). Deviation surveys from the FMS log runs closely matched the measurement-while-drilling (MWD) survey. The purpose of the AMS was to determine mud resistivity, mud temperature, and tool compression. Water saturations determined from resistivity logs were at or near virgin saturations (K. Byerly, personal communcation). This is despite the fact that the well had two nearby offset wells, and the field had been on production for more than 60 years, including a recent waterflood.

The medium-radius lateral hole was drilled along a northeast trend (Figure 3). This trend was chosen because it was roughly perpendicular to the Tensleep dunes, and it was approximately parallel to the known major fracture orientation in the field. The goal of the well was to intersect as few northeast-trending fractures as possible in an

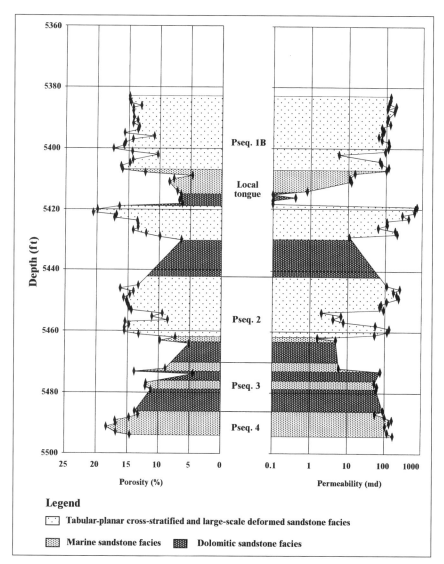

FIGURE 7. Core porosity and permeability profiles versus facies and parasequence number from E. Jones #1 well (Sec. 25-T56N-R97W), Byron field. Interval 5405.5–5422 ft is a local tongue of marine deposits in parasequence 1B. Pseq. = parasequence.

attempt to minimize water cut. The well reached 90° deviation where it encountered the Tensleep Sandstone on the southwest flank of the Byron anticline. Casing was set from the surface to this point. The remainder of the well, which was drilled for about 150 m (500 ft) on an uphill slant, stayed in the uppermost Tensleep Sandstone in a 6-m- (20-ft-) thick stratigraphic section. This part of the well was left uncased. The borehole path for the I. Lindsay #3H is shaped somewhat like a fishhook (Figure 9). Most of the lateral section is open hole, and the structurally highest point is at total depth.

LOG INTERPRETATION

Fractures

FMS log interpretation in 96 m (315 ft) of logged interval (measured depth 1681–1777 m [5515 to 5830 ft]) revealed two roughly orthogonal open-fracture sets. Set 1, the most abundant fracture set (n = 41), was oriented approximately perpendicular to the borehole path. The

mean strike and dip of this set were N49° W/88° SW (Figure 10). The intensity of Set 1 fractures was about one fracture per 2.3 m (7.5 ft) of measured depth. Set 2 fractures (n = 7), which were nearly parallel to the borehole path, had a mean strike/dip of N43° E/69° NW (Figure 10). The intensity of Set 2 fractures was about one fracture per 13 m (44 ft) of measured depth. A set of healed fractures, Set 3, had a mean strike/dip of N85° E/41° SE.

Using simple geometric calculations (Figures 11 and 12), it is possible to determine a corrected average spacing between fractures of Sets 1 and 2. Even though Set 1 fractures were encountered much more commonly in the borehole (n = 41), Set 2 (n = 7) fractures had a 0.9-m (3.0-ft) spacing that is 2.5 times lower than the 2.3-m (7.5-ft) spacing for Set 1 fractures.

Eolian Facies Architecture

FMS log interpretation, using the approach outlined by Hurley et al. (1994), was used to pick bed boundaries. Figure 13 is a stick plot of the apparent dips of each ob-

TABLE 1. AVERAGE DISTRIBUTION OF POROSITY AND PERMEABILITY, WATER SATURATION (SW), AND OIL SATURATION (SO) FOR EACH PARASEQUENCE AT BYRON FIELD. THE DATA SET CONSISTS OF 670 CORE-PLUG ANALYSES FROM 32 WELLS.

Parasequence	Facies	Average permability (md)	Average porosity (%)	Average Sw (%)	Average So (%)
1A	Eolian	103.0	14.6	38.3	25.6
	Marine	2.0	7.1	19.4	57.2
1B	Eolian	162.0	14.8	32.9	28.6
	Marine	2.8	6.6	28.5	26.9
2	Eolian	197.0	13.5	33.8	27.2
	Marine	2.5	6.9	24.1	32.3
3	Eolian	158.0	14.1	34.7	28.6
	Marine	3.7	7.2	17.7	38.9
4	Eolian	54.0	11.5	24.0	39.2
	Marine	0.3	4.6	15.1	45.9
5	Eolian	64.0	11.6	27.8	39.4
	Marine	0.4	1.5	8.3	9.0

TABLE 2. AVERAGE DISTRIBUTION OF POROSITY AND PERMEABILITY VALUES VERSUS FACIES AND STRATIFICATION TYPES FROM THE HOSKINS #A1 WELL, BYRON FIELD.

Facies	Stratification	No.	Average porosity (%)	Average permeability (md)
1) Tabular-planar cross-stratified sandstones	Grain fall	6	16.5	216.0
	Wind ripple	7	15.7	167.0
2) Large-scale deformed sandstones		2	17.7	420.0
3) Marine sandstones		3	12.1	54.3
4) Dolomitic sandstones		4	5.7	4.6

No. = number of core plug analyses.

served bed boundary projected into the plane of the horizontal well. In this well, at least five bounding surfaces (1.0, 2.0, and 0.1) were crossed in a lateral distance of about 96 m (315 ft).

Figure 14 shows a crossplot of cumulative dip magnitude of bedding planes versus an arbitrary bedding-plane sample number (a function of depth). The bedding planes are numbered consecutively from top to bottom of the logged interval. Inflection points visible between dip domains may indicate faults, parasequence boundaries, bounding surfaces (1.0, 2.0, and 0.1), or unconformities (Hurley, 1994). Figure 15a shows the vector plot of the dip domains interpreted from FMS bedding-plane dip directions. A vector plot is a projection in the horizontal plane

of oriented unit vectors that correspond to the observed bedding planes. Each vector points in the dip direction of the corresponding bed. When vectors are plotted end to end from deepest to shallowest reading, inflection points that occur between straight-line segments can be used to define tops and bottoms of dip domains or similar groups of dips. The interpreted arrangement of 0.1-bounded elements cutting across a 2.0-bounded element and erosion by a 1.0-bounded element has been observed in the outcrop study of Aviantara (2000). The schematic diagram of the Tensleep Sandstone in the horizontal well (Figure 15b) shows our interpretation of the eolian dune geometries intersected by the I. Lindsay #3H well.

Figure 16 shows a stereoplot of poles and a rose diagram of the FMS bed-boundary distribution for the I. Lindsay #3H well. Vector means have been corrected for structural dip. Note that most beds are flat lying, or they dip to the southwest. This is consistent with published results of Tensleep paleowind directions (Kerr and Dott, 1988). This is also consistent with cross-bed orientations observed in the SW Byron State #1, a vertical well in Byron field that was logged with the FMS tool (Aviantara, 2000).

Hydrocarbons

The last 46 m (150 ft) of the borehole was full of oil (Figure 9). The FMS log in this interval was of very poor quality because of oil in the well. The mud-resistivity log for both AMS passes showed high values. Finally, the dual laterolog showed anomalously high resistivity kicks in this portion of the borehole. This oil-water contact in the borehole was relatively stable during multiple log runs over a period of several days.

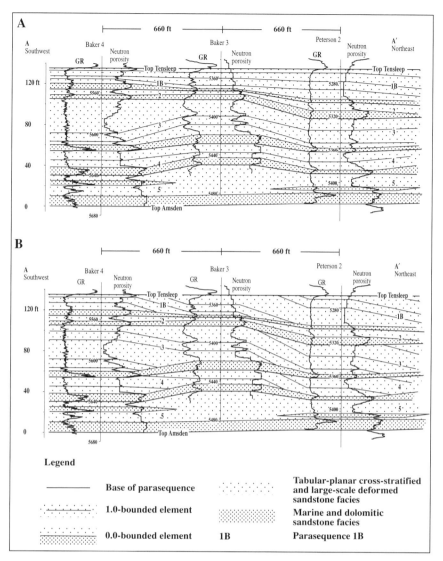

FIGURE 8. Well-log correlations show alternate views of 1.0-bounding surfaces. (a) Conventional, layer-cake correlation. (b) Correlation, thought to be more valid, based on outcrop studies by Aviantara (2000). The 1.0-bounding surface rises at an angle of about 1° in the southwest direction. Location of cross section A-A′ is shown in Figure 3.

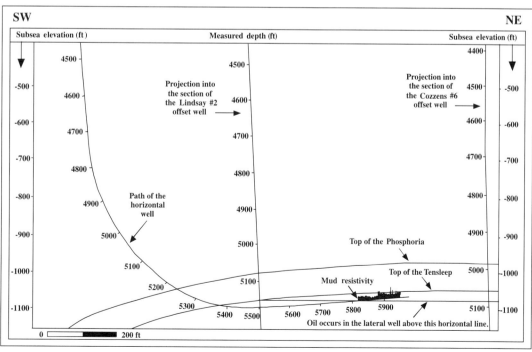

FIGURE 9. Structural cross section of the I. Lindsay #3H horizontal well, Byron field. Two nearby offset wells are projected into the plane of the cross section. These wells were used to help pick formation tops. The increase on the mud-resistivity log corresponds to oil in the borehole. There is no vertical exaggeration.

DISCUSSION

Log interpretations from the I. Lindsay #3H horizontal well show that the Tensleep reservoir is compartmentalized by at least two sets of open fractures. Fracture spacings, corrected for borehole geometry, show that northeast-trending fractures (Set 2) are much more closely spaced than northwest-trending fractures (Set 1). Note that this well was purposely drilled to the northeast to

avoid Set 2 fractures and the resulting high water cut. This is because (1) Set 2 fractures are parallel to maximum horizontal in situ stress, (2) directional permeabilities from pressure interference tests tend to be very high along a northeast-strike azimuth, and (3) Set 2 fractures are suspected of being major carriers of water to the boreholes in this field (Haws and Hurley, 1992). The well, when put on production, had a very high water cut. This suggests that it may be impossible to avoid northeast-trending Set 2 fractures with a horizontal well in a field like this. Furthermore, the scatter in orientations shown in Figure 10 confirms that these fractures are not simple, parallel planes.

FMS evidence further shows that the borehole crossed at least five eolian facies architectural elements in roughly 96 m (315 ft) of lateral distance. Therefore, the average length of a bounded element is 19 m (63 ft). Based on our knowledge of a southwest-trending dune-migration direction and the fact that this well was drilled to the northeast, the borehole path is nearly perpendicular to the dunes. Therefore, the spacing we compute is close to a perpendicular spacing of dune facies elements. Permeability baffles commonly occur between dune facies elements, which suggests significant compartmentalization in this reservoir. Results imply that a given vertical well would inefficiently drain a reservoir such as this. Horizontal wells may be the best way to connect compartments.

A stable oil-water contact was observed at a measured depth of 1778 m (5832 ft) in the horizontal well. It is not

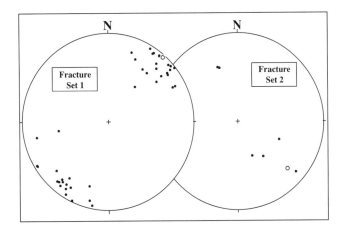

FIGURE 10. Stereoplots of poles to fracture planes for Fracture Set 1 and Set 2. Vector means are shown as open circles. The plots are lower hemisphere, Schmidt equal-area projections. The mean strike/dip of Set 1 fractures is N48.8°W/87.9°SW. The mean strike/dip of Set 2 fractures is N42.9°E/69.2°NW.

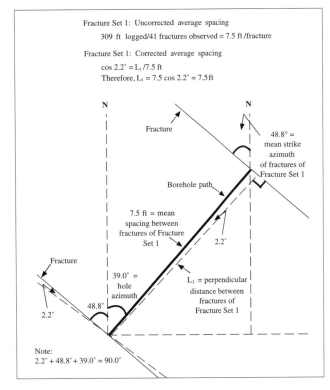

FIGURE 11. Diagram showing Set 1 fracture spacing, corrected for borehole geometry.

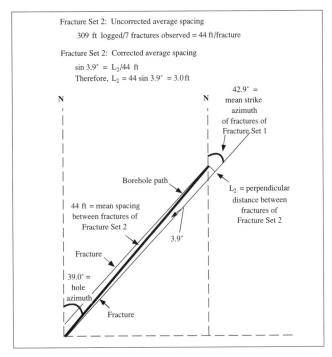

FIGURE 12. Diagram showing Set 2 fracture spacing, corrected for borehole geometry.

known whether that contact would have been stable over a period of months to years. We infer that the oil-water contact observed in the borehole represented the height of the oil-water contact in the fractures. If so, this would put the oil-water contact in the fractures at an elevation several hundred meters (or feet) above the oil-water contact in

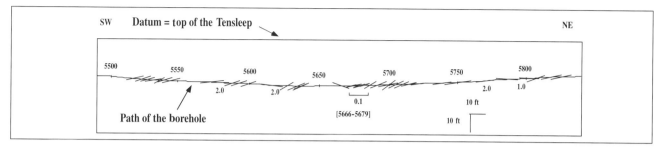

FIGURE 13. Apparent dips of FMS bed boundaries projected into the plane of the I. Lindsay #3H horizontal well, Byron field. Note the alternation between flat and southwest-dipping beds. The borehole crossed five eolian architectural elements and their associated bounding surfaces (labeled 0.1, 1.0, 2.0). The datum for the figure is the top of the Tensleep. There is no vertical exaggeration.

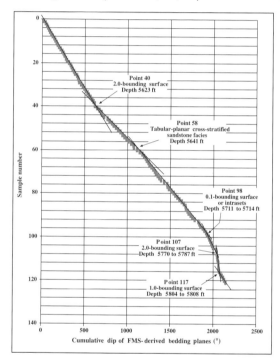

FIGURE 14. Cumulative dip plot of FMS-derived bedding planes for all lithologies, I. Lindsay #3H horizontal well, Byron field. Inflection points between dip domains are interpreted as bounding surfaces.

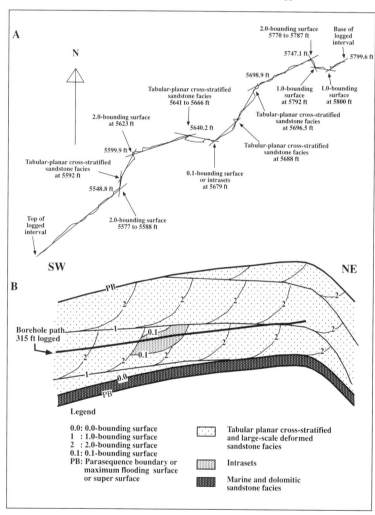

FIGURE 15. I. Lindsay #3H horizontal well, Byron field. (a) Vector plot shows dip domains interpreted from FMS bedding-plane dip directions for all lithologies. The plot is read from the deepest point (upper right) to the shallowest point (lower left). Most beds in this well have present-day dips to the southwest. Inflection points between dip domains are interpreted as bounding surfaces. (b) Schematic diagram illustrates the inferred relationship between the borehole path and the eolian bounding surfaces in the Tensleep Sandstone, I. Lindsay #3H horizontal well.

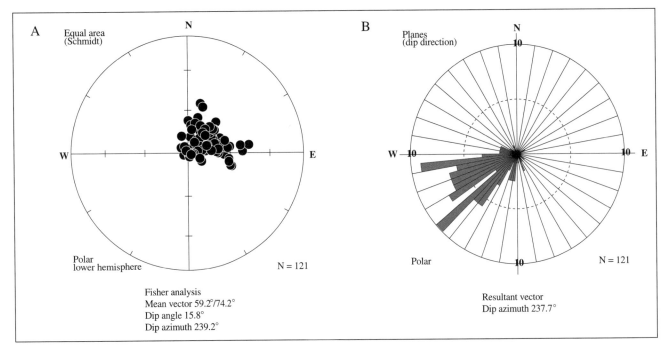

FIGURE 16. Bed boundaries observed in the I. Lindsay #3H horizontal well, Byron field. (a) Stereoplot of poles to FMS bed boundaries. This is a lower-hemisphere, Schmidt equal-area projection. The vector mean is stated, corrected for structural dip. (b) Rose diagram of FMS bed boundaries. This is a polar plot of the dip directions. The vector mean is stated, corrected for structural dip.

the rock matrix itself. This may have been caused by water coning into the very permeable fractures.

NOVEL COMPLETION IDEA

The possibility that a differential oil-water contact exists in the intergranular porosity versus the fractures suggests an innovative, although untested, method of completing a well such as the I. Lindsay #3H. As water imbibes into the matrix and expels oil along a horizontal well, or as oil flows into the well from fractures that intersect the borehole, the oil will gravity-segregate toward the top (total depth) of a fishhook-shaped well. Tubing could be run to a point near total depth, a packer could be placed on the outside of tubing at the bottom of casing, and a rod pump could be set through tubing to the deepest practical point (Figure 17). The rate at which the borehole is charged with oil can be determined using step-rate pump tests. If the determined rate is economically viable, the well could be produced at that rate with minimal water production. Note that water cuts are very high in this and other Bighorn Basin fields because of the presence of fractures, waterfloods, and strong water drives. If this completion method works, the upside potential for similar horizontal wells could be very high in Bighorn Basin fields and other fractured anticlines. Water cuts could be reduced and reservoir lives could be extended significantly.

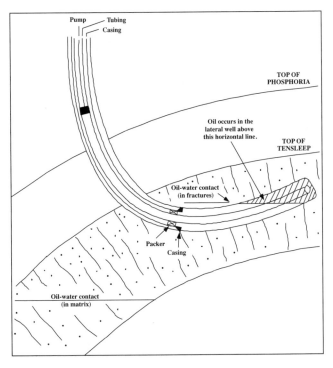

FIGURE 17. Diagram shows a proposed completion technique that could cause a fishhook-shaped horizontal well to act as a downhole oil-water separator in a fractured reservoir. Note the elevation difference between the oil-water contact in the fractures versus that in the rock matrix. Tubing is run to total depth, and a rod pump is installed to produce oil at low water cuts.

CONCLUSIONS

1) We have used reservoir characterization from core, outcrop studies, and log correlations to define the facies architecture at Byron field, Wyoming. In this context, a horizontal well was drilled roughly perpendicular to the trend of dunes and parallel to the major fracture orientations. We have used borehole image interpretations in the horizontal well to help define structural and stratigraphic compartments in the field.

2) The I. Lindsay #3H horizontal well stayed in the uppermost Tensleep Sandstone along the entire logged portion of the lateral borehole. The well stayed in a Tensleep interval approximately 6 m (20 ft) thick.

3) The last 46 m (150 ft) of the borehole, from about 1778 m (5832 ft) to total depth, was full of oil. The FMS log in this interval was of very poor quality because of oil in the well. The mud-resistivity log for both FMS passes also showed high values. Finally, the dual laterolog showed anomalously high resistivity kicks in this portion of the borehole.

4) Two roughly orthogonal sets of open fractures were observed. Set 1, the most abundant fracture set, was oriented approximately perpendicular to the borehole path. The mean strike and dip of this set were N49° W/88° SW. The intensity of Set 1 fractures was about one fracture per 2.3 m (7.5 ft) of measured depth. Set 2 fractures, which were nearly parallel to the borehole path, had a mean strike/dip of N43° E/69° NW. The intensity of Set 2 fractures was about one fracture per 13 m (44 ft) of measured depth. After correcting for borehole geometry, the mean spacing of Set 1 fractures is still 2.3 m (7.5 ft), whereas for Set 2 fractures, mean spacing is 0.9 m (3.0 ft).

5) The borehole path for the I. Lindsay #3H is shaped somewhat like a fishhook. Casing was set at the point where the well turned horizontal. Therefore, most of the lateral section is open hole, and the structurally highest point is at total depth. A relatively stable oil-water contact existed in the borehole at about 1778 m (5832 ft). The stability of this contact could be determined with future runs of the AMS log after a more prolonged shut-in. It is inferred that this level is an oil-water contact in the fractures. If so, the oil-water contact in the fractures is several hundred meters (or feet) above the oil-water contact in the rock matrix. This may have been caused by water coning into very permeable fractures.

6) A novel method of pumping this well has been proposed but not tested. Because the oil is lighter than water, any free oil in the well will gravity-segregate toward the total depth. This oil may be coming from the intergranular pores along the borehole path or from fractures that intersect the borehole. Tubing could be set at a point near total depth, a packer could be placed on the outside of tubing at the bottom of casing, and a rod pump could be run through tubing to the deepest possible point. The rate at which the borehole is being charged with oil can be determined with step-rate pump tests. It is hoped that an economical rate can be determined using this approach in which water production is minimized. If so, the well represents an unconventional application of horizontal drilling in a fractured reservoir. The potential is high for horizontal wells used in this manner to reduce water cuts and prolong reservoir life in Byron and other fractured, water-producing anticlinal reservoirs throughout the world.

REFERENCES CITED

Aviantara, A. A., 2000, Facies architecture of the Tensleep Sandstone, Bighorn Basin, Bighorn County, Wyoming: Ph.D. dissertation, The University of Tulsa, Tulsa, Oklahoma, 249 p.

Berg, R. R., 1962, Mountain flank thrusting in Rocky Mountain foreland, Wyoming and Colorado: AAPG Bulletin, v. 46, p. 2019–2032.

Brookfield, M. E., 1977, The origin of bounding surfaces in ancient eolian sandstones: Sedimentology, v. 24, p. 303–332.

Brookfield, M. E., 1992, Eolian system, in R. G. Walker and N. P. James, eds., Facies model response to sea level change: Geological Association of Canada, p. 143–156.

Carr-Crabaugh, M., and T. L. Dunn, 1996, Reservoir heterogeneity as a function of accumulation and preservation dynamics, Tensleep sandstone, Bighorn and Wind River Basins, Wyoming, in M. W. Longman and M. D. Sonnenfeld, eds., Paleozoic systems of the Rocky Mountain region: Rocky Mountain Section, Society for Sedimentary Geology (SEPM), Denver, Colorado, p. 305–320.

Carr-Crabaugh, M., N. F. Hurley, and J. L. Carlson, 1996, Interpreting eolian reservoir architecture using borehole images, in J. A. Pacht, R. E. Sheriff, and B. F. Perkins, eds., Stratigraphic analysis: Gulf Coast Section, Society for Sedimentary Geology (SEPM) Foundation, 17th Annual Research Conference, p. 39–50.

Chandler, M. A., G. Kocurek, D. J. Goggin, and L. W. Lake, 1989, Effects of stratigraphic heterogeneity on permeability in eolian sandstone sequence, Page Sandstone, northern Arizona: AAPG Bulletin, v. 73, p. 658–668.

Demiralin, A. S., N. F. Hurley, and T. W. Oesleby, 1994, Influence of karst fabrics on reservoir heterogeneity, Madison Formation, Garland field, Wyoming, in J. C. Dolson, M. Hendricks, K. Shanley, and B. Wescott, eds., Unconformity related hydrocarbon accumulation and exploitation in clastic and carbonate settings: Rocky Mountain Association of Geologists, 1994 Field Conference Guidebook, p. 219–230.

Desmond, R. J., J. R. Steidtmann, and D. F. Cardinal, 1984,

Stratigraphy and depositional environments of the middle member of the Minnelusa Formation, central Powder River Basin, Wyoming, *in* The Permian and Pennsylvanian geology of Wyoming: Wyoming Geological Association, 35th Annual Field Conference Guidebook, p. 213–239.

Doremus, D. M., 1986, Groundwater circulation and water quality associated with the Madison aquifer, northeastern Bighorn Basin, Wyoming: Master's thesis, University of Wyoming, Laramie, 81 p.

Emmet, W. R., K. W. Beaver, and J. A. McCaleb, 1971, Little Buffalo Basin, Tensleep heterogeneity—Its influence on drilling and secondary recovery: Journal of Petroleum Technology, v. 23, p. 161–168.

Fox, J. E., P. W. Lambert, R. F. Mast, N. W. Nuss, and R. D. Rein, 1975, Porosity variation in the Tensleep and its equivalent, the Weber Sandstone, western Wyoming: A log and petrographic analysis: Rocky Mountain Association of Geologists Guidebook, p. 185–216.

Goggin, D. J., M. A. Chandler, G. Kocurek, and L. W. Lake, 1992, Permeability transects of eolian sands and their use in generating random permeability fields: Society of Petroleum Engineers Formation Evaluation, v. 7, p. 7–16.

Haws, G. W., and N. F. Hurley, 1992, Application of pressure-interference data in reservoir characterization studies, Bighorn Basin, Wyoming: Society of Petroleum Engineers 67th Annual Technical Conference and Exhibition, Washington, D. C., SPE Preprint 24668, p. 53–62.

Hennier, J. H., 1984, Structural analysis of the Sheep Mountain anticline, Bighorn Basin, Wyoming: Master's thesis, Texas A&M University, College Station, 119 p.

Hoppin, R. A., and T. V. Jennings, 1971, Cenozoic tectonic elements, Bighorn Mountain region, Wyoming-Montana: Wyoming Geological Association Guidebook, 23rd Annual Field Conference, p. 39–47.

Hubbert, M. K., 1953, Entrapment of petroleum under hydrodynamic conditions: AAPG Bulletin, v. 37, p. 1954–2026.

Hunter, R. E., 1977, Basic types of stratification in small eolian dunes: Sedimentology, v. 24, p. 361–387.

Huntoon, P. W., 1985, Fault severed aquifers along the perimeters of Wyoming artesian basins: Groundwater, v. 23, no. 2, p. 176–181.

Hurley, N. F., 1994, Recognition of faults, unconformities, and sequence boundaries using cumulative dip plots: AAPG Bulletin, v. 78, p. 1173–1185.

Hurley, N. F., D. R. Thorn, J. L. Carlson, and S. L. W. Eichelberger, 1994, Using borehole images for target-zone evaluation in horizontal wells: AAPG Bulletin, v. 78, p. 238–246.

Iverson, W. P., T. L. Dunn, and I. Ajdari, 1996, Relative permeability anisotropy measurements in Tensleep Sandstone: 1996 Society of Petroleum Engineers/Department of Energy 10th Symposium on Improved Oil Recovery, v. 2, SPE Preprint 35435, p. 317–324.

Johnson, J. S., and N. Lindsley-Griffen, 1986, New interpretation of Clark's Fork field, northern Bighorn Basin, Montana, *in* J. H. Noll and K. M. Doyle, eds., Rocky Mountain oil and gas fields: Wyoming Geological Association Symposium, Casper, Wyoming, p. 159–165.

Kerr, D. R., 1989, Sedimentology and stratigraphy of Pennsylvanian and Lower Permian strata (Upper Amsden Forma-

tion and Tensleep Sandstone) in north-central Wyoming: Ph.D. dissertation, University of Wisconsin–Madison, 381 p.

Kerr, D. R., and R. H. Dott Jr., 1988, Eolian dune types preserved in the Tensleep sandstone (Pennsylvanian-Permian), north-central Wyoming: Sedimentary Geology, v. 56, p. 383–402.

Kerr, D. R., D. M. Wheeler, D. J. Rittersbacher, and J. C. Horne, 1986, Stratigraphy and sedimentology of the Tensleep sandstone (Pennsylvanian and Permian), Bighorn Mountains, Wyoming: Wyoming Geological Association Earth Science Bulletin, v. 19 (1–2), p. 61–77.

Kocurek, G. A., 1996, Desert aeolian systems, *in* H. A. Reading, ed., Sedimentary environments: Processes, facies and stratigraphy, 3rd ed.: Oxford, England, Blackwell Science Ltd., p. 125–153.

Lindquist, S. J., 1988, Practical characterization of eolian reservoirs for development: Nugget Sandstone, Utah-Wyoming thrust belt: Sedimentary Geology, v. 56, p. 315–339.

Mankiewicz, D., and J. R Steidtmann, 1979, Depositional environments and diagenesis of the Tensleep Sandstone, eastern Bighorn Basin, Wyoming: Society for Sedimentary Geology (SEPM) Special Publication 26, p. 319–336.

Maughan, E. K., and W. J. Perry Jr., 1986, Lineaments and their tectonic implications in the Rocky Mountains and adjacent plains region, *in* J. A. Peterson, ed., Paleotectonics and sedimentation in the Rocky Mountain region, United States: AAPG Memoir 41, p. 41–53.

Morgan. J. T., F. S. Cordiner, and A. R. Livingston, 1978, Tensleep reservoir, Oregon Basin field, Wyoming: AAPG Bulletin, v. 62, p. 609–632.

Opdyke, N. D., and S K. Runcorn, 1960, Wind direction in the western United States in the late Paleozoic: Geological Society of America Bulletin, v. 71, p. 959–972.

Paylor, E. D., H. L. Muncy, H. R. Lang, J. E. Conel, and S. L. Adams, 1989, Testing some models of foreland deformation at the Thermopolis anticline, southern Bighorn Basin, Wyoming: Mountain Geologist, v. 26, p. 1–22.

Peterson, J. A., 1990, Petroleum potential outlined for northern Rockies, Great Plains: Oil & Gas Journal, v. 88, July 30, p. 103–110.

Rittersbacher, D. J., 1985, Facies relationship of the Tensleep Sandstone and Minnelusa Formation, Western Powder River basin, Johnson County, Wyoming: Master's thesis, Colorado School of Mines, Golden, 188 p.

Shebl, M. A., 1995a, The impact of reservoir heterogeneity on fluid flow in the Tensleep Sandstone of the Bighorn Basin: 1995 Field Conference Guidebook, Wyoming Geological Association, p. 343–359.

Shebl, M. A., 1995b, Geochemical and diagenetic investigations of the Tensleep Sandstone reservoirs, Bighorn Basin, Wyoming: Ph.D. dissertation, University of Wyoming, Laramie, 285 p.

Simmons, S. P., and P. A. Scholle, 1990, Late Paleozoic uplift and sedimentation, northeast Bighorn Basin, Wyoming: Wyoming Geological Association, 41st Field Conference Guidebook, p. 39–55.

Stone, D. S., 1967, Theory of Paleozoic oil and gas accumulation in the Bighorn Basin, Wyoming: AAPG Bulletin, v. 51, p. 2056–2114.

Wheeler, D. M., 1986, Stratigraphy and sedimentology of the Tensleep Sandstone, Southeast Bighorn Basin, Wyoming: Master's thesis, Colorado School of Mines, Golden, 169 p.

Wyoming Geological Association, 1989, Byron field, *in* Wyoming Oil and Gas Fields Symposium, Casper, Wyoming, p. 76–78.

Chalk
Reservoirs

10

Feazel, C. T., and H. H. Nielsen, 2003, Reservoir characterization, well planning, and geosteering in the redevelopment of Ekofisk Field, North Sea, *in* T. R. Carr, E. P. Mason, and C. T. Feazel, eds., Horizontal wells: Focus on the reservoir: AAPG Methods in Exploration No. 14, p. 163–172.

Reservoir Characterization, Well Planning, and Geosteering in the Redevelopment of Ekofisk Field, North Sea

Charles T. Feazel

Phillips Petroleum Company
Bartlesville, Oklahoma, U.S.A.

Hardy H. Nielsen

Phillips Petroleum Company
Bartlesville, Oklahoma, U.S.A.

ABSTRACT

After 28 years of oil and gas production from chalk of Danian and Maastrichtian age, the giant Ekofisk field is only halfway through its economic life. New surface facilities installed during 1998 offer opportunities to plan and drill as many as 50 new wells in a field with a mature waterflood under way. Locations of the new wells are based on a history-matched reservoir flow-simulation model derived from a 3-D geologic model.

Attributes (e.g., permeability, initial fluid saturation, facies type, rock composition) are distributed in the 23-million-cell geologic model from a database of well logs, cores, and 3-D seismic data inverted for porosity and thickness. An upscaled 39,000-cell flow-simulation model is used to predict the movement of injected water and to optimize the paths of wells that vary from near-vertical to long-reach horizontal, and from simple slant-holes to complex multilaterals.

Planning the new Ekofisk wells is an iterative exercise: A preliminary well path defined in the geocellular and flow-simulation models is displayed against a seismic backdrop to ensure that it avoids major faults and penetrates reservoir units with best properties. The proposed well then is compared with nearby wells to avoid collision problems and to ensure correlation of major flow units, and is plotted through the seismic volume using an automatic fault-picking routine to guard against operational or production problems resulting from minor faults. Throughout this planning process, interactions among the field geoscientists, operations geoscientists, reservoir engineers, and directional drillers ensure optimal well paths and value-based data collection programs.

Particular successes attributed to this teamwork include wells producing more than 12,000 BOPD, wells with total depths exceeding 7772 m (25,500 ft) (including horizontal reservoir sections more than 2377 m [7800 ft] long), and a total contribution to Ekofisk production from partial completions in the first 22 wells of more than 135,000 BOPD and 175 MMCFGD with minimal water production.

Additional challenges to this redevelopment include (1) a gas-charged, overpressured overburden, which obscures surface seismic returns; and (2) production-induced reservoir compaction, which causes both wellbore collapse and seafloor subsidence.

INTRODUCTION

Ekofisk, a giant oil and gas field (Van den Bark and Thomas, 1981; Thomas et al., 1987; Feazel et al., 1990) in the center of the North Sea (Figure 1), has been producing since 1971 and has been under waterflood since 1987. A major redevelopment of the field is under way after the 1997–1998 installation of new surface facilities. Planning and drilling of as many as 50 new wells—many of them horizontal—present particular challenges in avoiding collisions with approximately 225 existing wellbores while simultaneously avoiding areas of the reservoir swept by injected seawater. Beginning with an intensive reservoir characterization effort (1994–1996), followed by an interdisciplinary approach to well planning and execution (1996 and continuing), a team of geoscientists, reservoir engineers, petrophysicists, flow simulation experts, and drilling engineers has created a detailed view of the Ekofisk reservoir and has designed innovative wells to optimize the remaining reserve potential during the field's next 30 years of anticipated production.

Ekofisk field produces from chalk of Danian and Maastrichtian age (Ekofisk and Tor Formations, respectively). Deposited as coccolith-foraminiferal ooze and

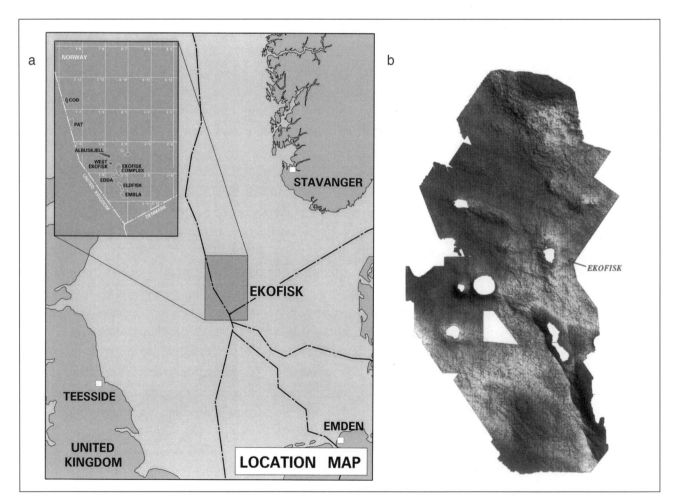

FIGURE 1. (a) Location map showing the Ekofisk area in the center of the North Sea. (b) Map of the top of the chalk group of the area compiled from merged 3-D seismic volumes, central North Sea.

commonly resedimented as debris flows and turbidites (Hancock and Scholle, 1975; Watts et al., 1980; Feazel et al., 1985; Schatzinger et al., 1985; Kennedy, 1985; Feazel and Farrell, 1988), the soft sediment had an initial porosity approaching 80%. Dewatering by early mechanical compaction (enhanced by water escape through the burrows of an active infauna) resulted in accumulation of calcareous skeletal grains with about 50% interparticle porosity. The fact that 45–48% porosity remains in some chalk intervals in the field means that little diagenesis or further compaction occurred in these rocks despite the intervening 65 million years and burial to depths exceeding 3 km (10,000 ft). This remarkable porosity preservation at depth has invited much speculation about the timing of hydrocarbon entry and the role of overpressure (Scholle, 1977; D'Heur, 1984; Feazel et al., 1985; Brasher and Vagle, 1996). The resulting rock drills easily, although it contains tight intervals and horizons rich in nodular chert, which occasionally cause steering problems if encountered at sufficiently shallow angles that the drill bit glances off instead of cutting through them.

The Ekofisk reservoir is highly fractured. Recent advances in 3-D seismic processing and interpretation have allowed mapping of hundreds of previously unsuspected faults in the field (Figure 2). Fractures recognized in cores and in borehole-image logs can be related to the orientation of the field's main faults, but the gap in scale between borehole and seismic resolution makes detailed correlation problematic. Although wellbore fault penetrations can be minimized in the well-planning process, it is impossible to design a well at Ekofisk that will avoid faults. This limitation has negative consequences in terms of intersecting possible conduits for breakthrough of injected water, but it has positive implications as well for crossing permeability-enhancing fractures that parallel the faults.

RESERVOIR CHARACTERIZATION

The Ekofisk reservoir—described from well data (drilling histories, logs, and cores) and 3-D seismic data inverted for porosity and thickness away from well control—can be visualized in a 23-million-cell 3-D geocellular model (Figure 3). Each cell contains as many as 30 attributes, such as porosity, shale volume, silica content, or initial water saturation. Slices through this model are used to display, for example, porosities and initial fluid saturations to be expected along the path of a future well (Figure 4). The model is also used in conjunction with an extracted seismic line during the geosteering of the well after it enters the reservoir.

The geocellular model has been upscaled to a 39,000-cell flow-simulation model, which has been matched to

the field's production history and then run in forward mode to predict future production profiles. A goal of the modeling process is better prediction of permeability anisotropy, because this remains the least understood aspect of reservoir management in the North Sea chalk fields. With water injection at Ekofisk exceeding 800,000 BWPD, it is increasingly important to understand and quantify where that volume is going. The ability to map faults in the reservoir—a consequence of reprocessing 1988-vintage 3-D seismic data—provides the basis for a revised understanding of the field's historical anisotropy. New modeling techniques, particularly corner-point geometry, are being investigated to make better quantitative use of the results of the seismic interpretation. In addition, a new 3-D seismic data set was acquired in summer 1999. The 1988 and 1999 3-D seismic surveys cover 150 km², extending far beyond the productive area. Very good repeatability outside the productive area has been achieved by reprocessing both surveys and performing post-DMO (dip moveout) binning to the same bin grid (12.5 × 12.5 m). The two surveys are at present used for time-lapse analysis in the productive area as well as for planning new production wells (Table 1).

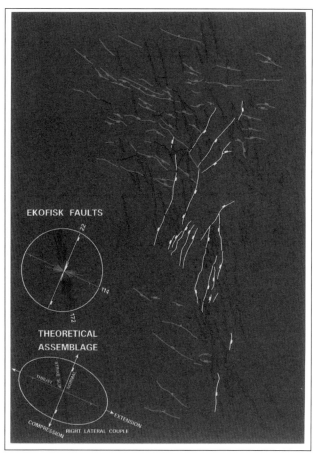

FIGURE 2. Fault systems at Ekofisk field, mapped from wells and 3-D seismic data.

FIGURE 3. Isometric view of the 23-million-cell Ekofisk geocellular model, displaying porosity.

Ekofisk 3-D reservoir characterization

FIGURE 4. Extracted line from the Ekofisk geocellular model, used to plan well path 2/4-X-7 through high-porosity layers in the chalk.

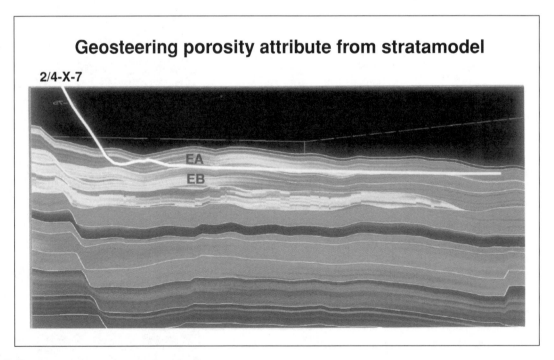

Geosteering porosity attribute from stratamodel

2/4-X-7

EA

EB

TABLE 1. EKOFISK 3-D SEISMIC SHOOTING PARAMETERS.

3-D seismic vintages	Receiver types	Number of receiver cables	Receiver-cable length (m)	Source
1988	Streamer	2 streamers	3000	Dual air-gun array
1999	Streamer/OBC for platform undershooting	4 streamers/ 2 OBC cable layouts	3000/2 × 6000	Dual air-gun array

The flow-simulation model is the basis for picking initial locations for new production and injection wells. The geocellular model is used for detailing and optimizing the well-path design to create optimum geologic positions of the wells in relation to local structural and stratigraphic variabilities. Several well-path iterations are run to create optimal well-path designs based on the two models.

WELL PLANNING

Planning new Ekofisk wells is an iterative, interdisciplinary exercise. A preliminary well path defined in the geocellular and flow-simulation models is first displayed against a seismic backdrop to ensure that it avoids major faults and penetrates reservoir units with best properties (primarily porosity and fluid saturation). Then the path of the proposed well is compared with existing nearby wells to avoid collision problems and ensure correlation of major flow units. It is subsequently plotted through the seismic volume using an automatic fault-picking routine (proprietary) to minimize operational or production problems resulting from minor faults. Throughout this process, which may require several months, interaction among the various team members ensures optimal well paths and value-based data acquisition programs. Coventurer companies—some with their own reservoir models—are asked to comment on each well path early in the process, and technical discussions continue until a final bottom-hole location and path are approved by consensus of the partnership.

Production-induced reservoir compaction and its surface expression as seafloor subsidence present an addi-

tional challenge to well planners. Some facilities standing on the seabed above Ekofisk have subsided more than 7 m (22 ft) from their preproduction elevation, necessitating a major simultaneous jack-up of several platforms in 1987 and design of considerably taller structures for the present installation, known as Ekofisk II. Subsidence has significant detrimental effects on well casing, with tensional, shear, and compressional deformation apparent in casing strings from various positions around the field and from different stratigraphic intervals. Loss of production as a result of restricted or parted casing has led to the expensive redrilling of many Ekofisk wells. Countermeasures, including the use of sliding casing joints and underreaming of the hole to allow the mobile overburden to deform around the casing string, have been part of the effort to mitigate casing collapse and thus extend the lifetimes of the new Ekofisk wells.

GEOSTEERING

Horizontal wells have become commonplace in the Ekofisk field; on average, they require 40 days to drill and complete. The upper part of the hole is usually drilled without incident through an overpressured (>2500 psi above hydrostatic) and gas-charged overburden. This part of the section is drilled "blind," using only offset well control, because both gas and fluid pressure in the overburden obscure mappable seismic returns from the crest of the field, resulting in a hole in the center of the seismic volume (Figure 5). Current efforts in borehole geophysics and 3-D reprocessing offer the promise of filling this gap and imaging the gas-affected area (Figure 6).

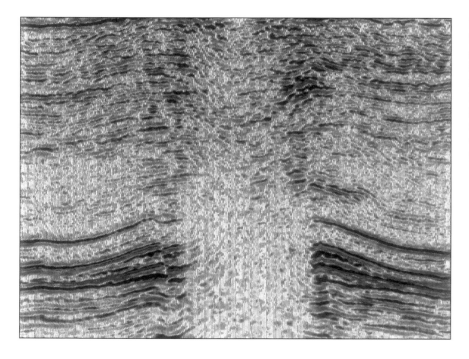

FIGURE 5. Gas chimney above Ekofisk field, a zone of poor seismic returns above the crest of the reservoir, caused by overpressure and gas in the overburden. The original interpretation of the Ekofisk structure included a collapse zone at the crest of the dome. Subsequent drilling allowed mapping of reservoir layers across the top of the structure.

A critical step in the drilling process is picking the 9⅝-in. casing point, ideally 3 m (10 ft) true vertical depth (TVD) above the top of the chalk reservoir. An experienced well-site operations geologist comparing logging-while-drilling (LWD) readings with drill rate and cuttings descriptions can usually make this decision within 1 m (3 ft) TVD of the intended position, even at considerable hole-deviation angle. Having the casing shoe in the correct position is essential to the conduct of all subsequent operations in the reservoir section, as shales of the Våle Formation just above the reservoir commonly cave or deform plastically. Having minimal Våle section exposed allows for building hole angle and multiple tripping, whereas too much Våle exposed creates problems with torque and drag. Setting the casing too deep—in the uppermost chalk—exposes the well to the risk of lost circulation because of the presence of abundant fractures and a downward pore-pressure drop on entering the reservoir. The judgment of the well-site geologist is thus a critical element for both technical and economic success in these wells, which commonly cost $10 million or more to drill and complete.

Once a deviated well exits the gas chimney and enters the area of good seismic imaging, the well path through the 3-D seismic volume is updated with position information as the bit advances, as a complement to well-site biostratigraphy for real-time geosteering. A biostratigraphic model of anticipated microfossil and nannofossil assemblages is prepared for each well in advance of drilling,

and cuttings samples are processed as soon as they reach the shaker, resulting in a biostratigraphic interpretation usually less than two hours behind the drill bit. At-bit inclination tools offer real-time positioning information but require sacrificing the at-bit position of some other logging tool in the bottom-hole assembly (Figure 7). It is often more important to know where the drill bit is headed next than to measure resistivity or porosity at its present location. Well-site operations geologists, in communication with onshore geophysical interpreters, coordinate the flow of these diverse data types and recommend steering changes to the directional driller (Figure 8). At Ekofisk, it has been possible to keep wells within 10 m (30 ft) TVD of their intended paths (the approximate resolution of the seismic data) while drilling horizontal reservoir sections more than 2377 m (7800 ft) long (Figure 9).

RESULTS

Although we are less than halfway into the redevelopment of Ekofisk field and many wells are completed only partially and are still on choke, the first 22 wells drilled in this program (Figure 10) have added more than

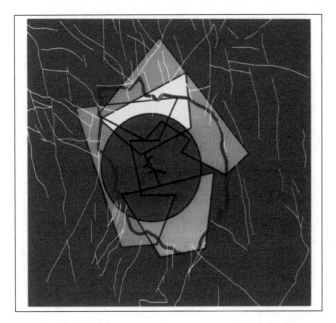

FIGURE 6. Outline (red) of the gas-affected area above the crest of Ekofisk field, shown with trapezoidal outlines of 3-D walkaway seismic profiles with the potential to undershoot the overburden gas, and the gray outline of a circular vertical seismic profile acquired in an attempt to image the top of the structure.

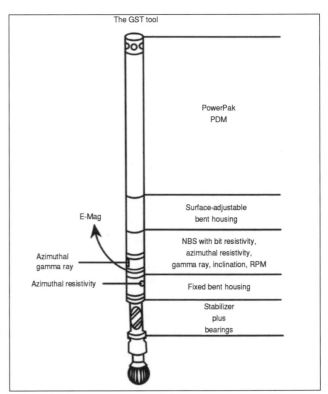

FIGURE 7. Bottom-hole assembly of GST tool used at Ekofisk field. Geosteering requires compromises among the various logging tools that might be desirable close to the bit. In this illustration (courtesy of Schlumberger), resistivity is measured at the bit. In Ekofisk wells, the at-bit resistivity measurement commonly is sacrificed in favor of at-bit inclination tools.

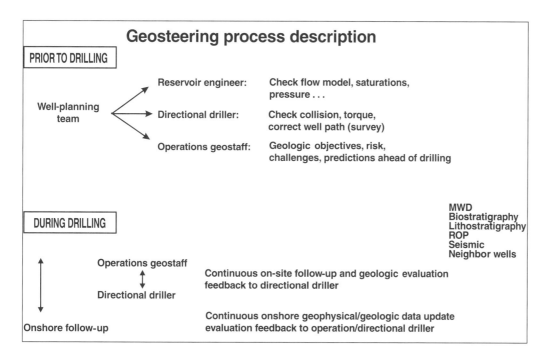

Geosteering process description

PRIOR TO DRILLING

Well-planning team →
- Reservoir engineer: Check flow model, saturations, pressure . . .
- Directional driller: Check collision, torque, correct well path (survey)
- Operations geostaff: Geologic objectives, risk, challenges, predictions ahead of drilling

MWD
Biostratigraphy
Lithostratigraphy
ROP
Seismic
Neighbor wells

DURING DRILLING

Operations geostaff ⇕ Directional driller
Continuous on-site follow-up and geologic evaluation feedback to directional driller

Onshore follow-up
Continuous onshore geophysical/geologic data update evaluation feedback to operation/directional driller

FIGURE 8. Geosteering process description, prior to and during drilling. Well planning and geosteering are iterative, interdisciplinary team activities that balance geologic risk and operational risk with objectives for each well (safety, production, injection, data collection, etc.). A highly efficient but flexible workflow evolved among the team members during the first two years of Ekofisk redevelopment.

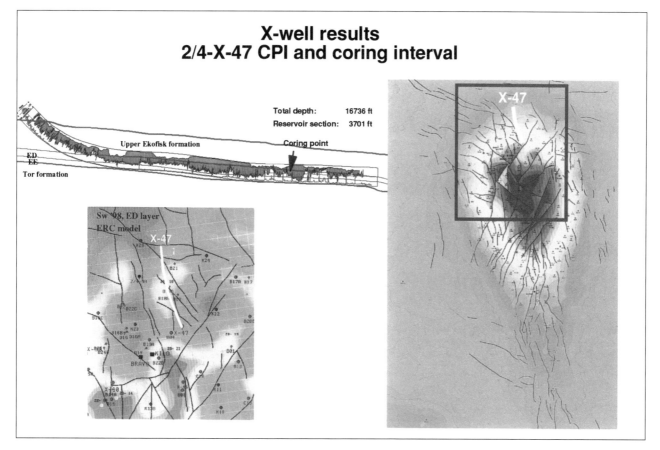

X-well results
2/4-X-47 CPI and coring interval

Total depth: 16736 ft
Reservoir section: 3701 ft

Upper Ekofisk formation
Coring point
ED
EE
Tor formation

Sw 98, ED layer
ERC model
X-47

FIGURE 9. Detailed reservoir characterization and 3-D seismic data allow steering of Ekofisk wells with long horizontal sections, which are drilled with sufficient accuracy to permit horizontal coring across faults for assessment of the width of the associated fracture zones. In the 2/4-X-47 well, a fault was cored to capture such information and to confirm the location of the fault in the geocellular model.

135,000 BOPD and 175 MMCFGD to the production stream, with minimal water production. Most wells have met their forecast initial flow rates (usually 2000–

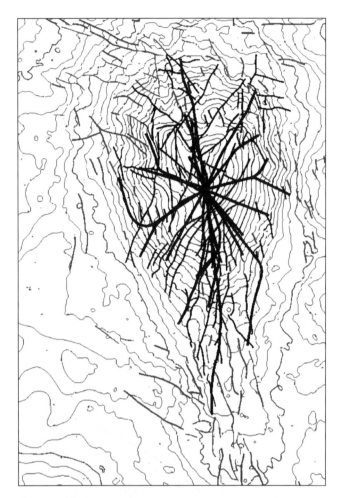

FIGURE 10. Map of the top of the chalk reservoir at Ekofisk field, showing faults (gray) and paths of the new wells (black).

8000 BOPD, depending on location), but a few have exceeded 12,000 BOPD (Figure 11). Drilling has been possible to measured depths exceeding 7800 m (25,600 ft), and horizontal sections longer than 1.5 km (5000 ft) are common. Multilateral wells also have been drilled, with objectives in each of the two chalk reservoir layers that are fitted with a sliding sleeve to isolate each lateral, if necessary, during workover operations (Figure 12). Perforations are selective, to avoid tight spots and possible water conduits. Acid-fracture stimulations performed after perforating are followed by periodic acid washes to maintain production rates.

The very detailed models have made it possible to plan and drill wells longer than 7625 m (25,000 ft) measured depth, keeping within pay zones 21–27 m (70–90 ft) thick. Layer tops are normally penetrated within 8 m (25 ft) of their predicted depths, which is a high level of predictive accuracy in a field that is subject to dynamic subsidence. Attributes such as porosity and fluid saturation in rocks penetrated by the new wells have been remarkably close to modeled values (Figure 13). On average, oil production has been slightly higher than predicted, whereas water production in this initial phase has been lower than expected. Pressures were slightly higher than predicted, and gas production was lower.

As a result of the efficiency of the waterflood, improvements in well placement and design, and a license extension granted by the Norwegian authorities when the Ekofisk II plan was approved, recoverable reserves from Ekofisk have increased significantly. As a result of the reservoir-characterization effort, estimates of original oil in place have increased from 6.3 to 7.1 billion barrels (Landa et al., 1998).

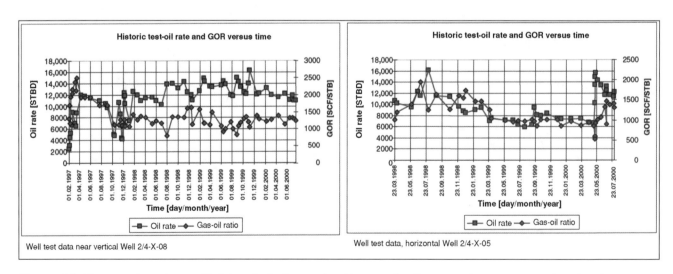

FIGURE 11. Flow rates from the new Ekofisk wells. Both vertical and horizontal wells commonly meet or exceed predicted rates.

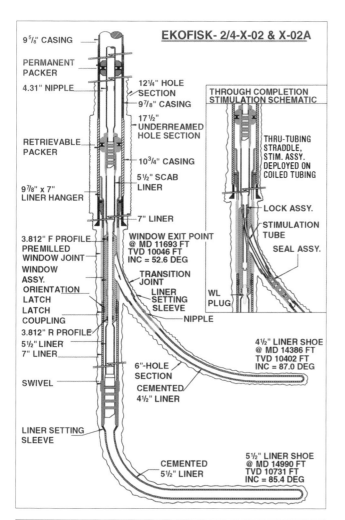

FIGURE 12. Wellbore schematic diagram for a dual-lateral well at Ekofisk field.

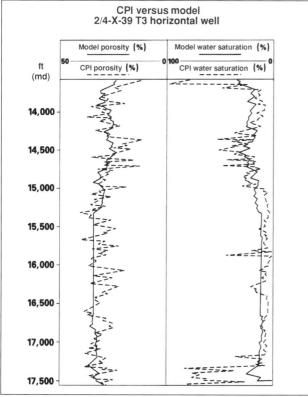

FIGURE 13. Results of the first 22 wells drilled in the redevelopment of Ekofisk field have confirmed model predictions in close detail, as shown by porosity and water saturation along horizontal well 2/4-X-39.

ACKNOWLEDGMENTS

The work processes and results described in this paper represent a true team approach to a complex set of problems. Members of the Ekofisk Reservoir Characterization and Reserves Optimization teams have had a major impact, both technologically and financially, on redevelopment of the field, and we thank them for their efforts. The authors are solely responsible for the content of this paper. We acknowledge permission to publish from Phillips Petroleum Company and its Ekofisk coventurers, including TotalFinaElf Exploration Norge, Norsk Agip, Norsk Hydro, Statoil, and Saga Petroleum.

REFERENCES CITED

Brasher, J. E., and K. R. Vagle, 1996, Influence of lithofacies and diagenesis on Norwegian North Sea chalk reservoirs: AAPG Bulletin, v. 80, p. 746–769.

D'Heur, M., 1984, Porosity and hydrocarbon distribution in the North Sea chalk reservoirs: Marine and Petroleum Geology, v. 1, p. 211–238.

Feazel, C. T., and H. E. Farrell, 1988, Chalk from the Ekofisk area, North Sea: Nannofossils + micropores = giant fields, in A. J. Lomando and P. M. Harris, eds., Giant Oil and Gas Fields, A Core Workshop: Society for Sedimentary Geology (SEPM) Core Workshop 12, p. 155–178.

Feazel, C. T., J. Keany, and R. M. Peterson, 1985, Cretaceous and Tertiary chalk of the Ekofisk field area, central North Sea, in P. O. Roehl and P. W. Choquette, eds., Carbonate petroleum reservoirs: New York, Springer-Verlag, p. 495–507.

Feazel, C. T., I. A. Knight, and L. J. Pekot, 1990, Ekofisk field, Norway, Central Graben, North Sea, in E. A. Beaumont and N. H. Foster, eds., AAPG Treatise of petroleum geology, Structural traps IV: p. 1–25.

Hancock, J. M., and P. A. Scholle, 1975, Chalk of the North Sea, in A. W. Woodland, ed., Petroleum and the continental shelf of northwest Europe: New York, John Wiley & Sons, p. 413–425.

Kennedy, W. J., 1985, Sedimentology of the Late Cretaceous and Early Paleocene Chalk Group, North Sea Central Graben: North Sea Chalk Symposium, Book I, p. 1–35.

Landa, G. H., W. H. Holm, and E. V. Hough, 1998, Ekofisk area management and redevelopment: OTC 8654, presented at the 1998 Offshore Technology Conference, Houston, Texas, May 4–7, 1998.

Schatzinger, R. A., C. T. Feazel, and W. E. Henry, 1985, Evidence of resedimentation in chalk from the Central Graben, North Sea, in P. D. Crevello and P. M. Harris, eds., Deep water carbonates: Society for Sedimentary Geology (SEPM) Core Workshop 6, p. 342–385.

Scholle, P. A., 1977, Chalk diagenesis and its relation to petroleum exploration: Oil from chalk, a modern miracle?: AAPG Bulletin, v. 61, p. 982–1009.

Thomas, L. K., T. N. Dixon, C. E. Evans, and M. E. Vienot, 1987, Ekofisk pilot waterflood: Journal of Petroleum Technology, v. 39, p. 221–332.

Van den Bark, E., and O. D. Thomas, 1981, Ekofisk: First of the giant oil fields in western Europe: AAPG Bulletin, v. 65, p. 2341–2363.

Watts, N. L., J. F. LaPre, F. S. Van Schijndl-Goester, and A. Ford, 1980, Upper Cretaceous and Lower Tertiary chalks of the Albuskjell area, North Sea: Deposition in a base-of-slope environment: Geology, v. 8, p. 217–221.

Jørgensen, O., and N. W. Petersen, 2003, Interpreting natural fracture directions in horizontal wells from wellbore image logs, *in* T. R. Carr, E. P. Mason, and C. T. Feazel, eds., Horizontal wells: Focus on the reservoir: AAPG Methods in Exploration No. 14, p. 173–182.

Interpreting Natural Fracture Directions in Horizontal Wells from Wellbore Image Logs

O. Jørgensen

Mærsk Olie og Gas AS
Copenhagen, Denmark

N. W. Petersen

Mærsk Olie og Gas AS
Copenhagen, Denmark

ABSTRACT

A method is presented to discriminate on wellbore image logs natural open fractures from far more abundant drilling-induced fractures in horizontal wells in the Dan chalk field, offshore Denmark. The method recognizes that drilling-induced fractures are predominantly shear failures in a preferred orientation relative to the wellbore, whereas natural fractures are confined to a limited strike range determined by the regional stress field.

The strike of natural fractures derived from the image logs agrees well with core analyses and with the strike of hydraulic fractures generated during injection and stimulation. Valuable information on likely injection fracture propagation direction can thus be derived.

INTRODUCTION

The Dan field is situated in the southern part of the Danish North Sea (Figure 1) and contains oil and gas in Upper Cretaceous (Maastrichtian) and Lower Paleocene (Danian) chalk. The structure is dome shaped and divided into two blocks (A and B) by a northeast-striking, northwest-dipping normal fault, with a displacement of as much as 91 m (300 ft) (Figure 2).

The Maastrichtian chalks are the main hydrocarbon reservoir, and most productive reservoir units are located in the uppermost part. Porosity averages 29% (ranging from 24% to 41%), with low permeabilities in the 1–5-md range. Core studies have shown that the frequency of natural open fractures is lower than in most other Central Graben chalk fields.

A waterflood development using horizontal production wells with multiple sand-propped fractures has been implemented. Waterflooding at high rates has produced large vertical fractures of as much as several hundred me-ters (several thousand feet) in length, thereby improving injectivity and reducing the number of injection wells required. Knowledge of the fracture direction is key to optimizing the well pattern for maximum areal sweep efficiency, and tools capable of predicting the extent and direction of injection-driven fractures are therefore valuable. Existing fracture growth models (Carter, 1957; Hagort, 1981; Koning, 1988; Ovens, 1997) can predict both the extent of a given fracture and the size and shape of the zone swept by the injection water as a function of injection time, but they contribute nothing to understanding the propagation direction.

The large number of horizontal wells drilled in the Dan field has provided information about the sweep pattern when narrow, flushed zones seen on open-hole logs can be related geometrically to the injection points. This knowledge has been used to calibrate fracture growth models (Ovens, 1997), but the interpreted fracture-propagation direction stands alone, without a model or independent physical verification.

Current earth stress strongly controls the fracture-propagation direction. Several methods of estimating the maximum horizontal stress direction exist, based on observations of stress-driven deformation and fracturing in vertical and moderately deviated wells (Barton et al., 1997; Brudy and Zoback, 1993; Peska and Zoback, 1995).

This present work is based on the analysis of drilling-induced rock deformation and fractures revealed on borehole image logs. It will be proposed that the sum of open fractures seen on the image logs consists of drilling-induced shear failures along the horizontal wellbore axis, and natural fractures characterized by a distinct strike parallel to that of hydraulic fractures.

FIGURE 1. Danish-sector map of the North Sea.

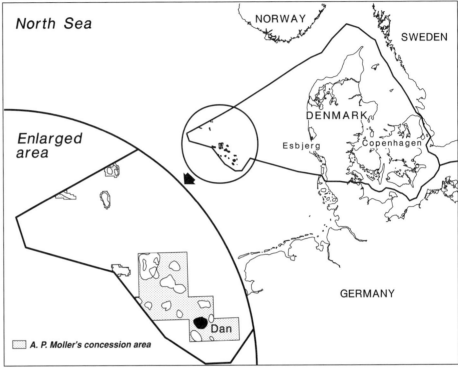

FIGURE 2. Dan field structure map with locations of wells with image logs.

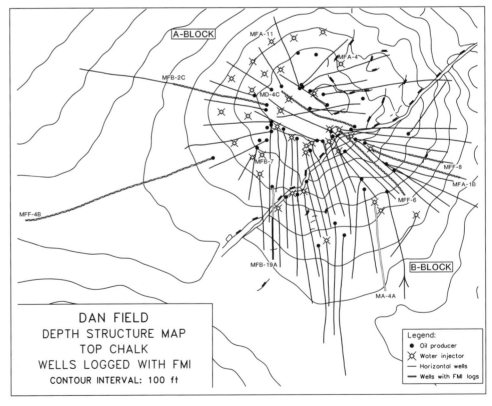

We propose a method of recognizing drilling-induced fractures that allows them to be separated from natural fractures. We present a series of examples from the Dan field that demonstrates the classification and how the fracture direction deduced from image logs compares to other field evidence.

GROUPING FRACTURES ACCORDING TO ORIGIN

THE APPEARANCE OF DRILLING-INDUCED FRACTURES ON FORMATION MICROSCANNER IMAGE (FMI) LOGS OF HORIZONTAL WELLS

The stress changes induced by drilling extend only a few wellbore diameters into the surrounding rock, and drilling-induced fractures therefore are very localized. In vertical holes, fractures that run parallel to the borehole and that strike in the minimum in-situ stress direction commonly are interpreted as having been induced during drilling (Ma et al., 1993). In horizontal wells, vertical fractures running along the top of the hole are considered

drilling-induced (Lehne and Aadnoy, 1993). These conventions do not, of course, rule out the possibility that fractures intersecting a wellbore at any other angle are also drilling-induced, as evidenced on the image log from well MD-4C (Figure 3). Clearly, the strike of open fractures along horizontal-well sections correlates with the azimuth of the well trajectory. As the azimuth of the well changes, the strike of the fractures plotted along the trajectory changes accordingly. This could be a coincidence, but most likely it indicates that a large fraction of the interpreted fractures were induced during drilling of the well.

To understand the rock-failure mode in the horizontal wells studied here, we must consider the stress distribution around the wellbores, as well as the failure envelope of the surrounding rock. This issue is discussed in the next subsection.

The mode of fracturing during drilling, however, can be further quantified based on the log data alone. Across the Dan field, the angle, α, between a vector normal to the plane of the fracture and a unit vector defining the tangent of the well trajectories (Figure 4) shows a symmetrical distribution centered on 45° (Figure 5). The dis-

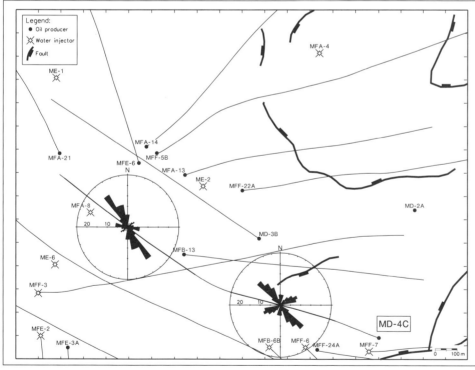

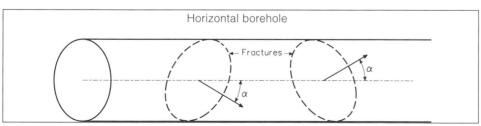

FIGURE 3. Strike of logged fractures along the path of well MD-4C. Note the apparent correlation with well azimuth.

FIGURE 4. Definition of angle α.

tribution shows that most fractures identified on image logs are formed preferentially along planes at 45° to the horizontal-well axis. When considering the stress field around the wellbore, these fractures are denoted as shear fractures. Although the exact mechanism giving rise to the shear fracturing is not known, the fundamental degrees of freedom of the fracturing process have been identified from the observed distribution. This is an important result. A failure mechanism that may explain the phenomenon is proposed in the following subsection.

PROPOSED MECHANISM FOR GENERATING FRACTURES AT 45° TO THE WELLBORE

The stress distribution around a horizontal wellbore is controlled largely by the orientation of the borehole relative to the in-situ principal stresses (Zhou et al., 1994; Kirsch, 1898). It is assumed that one of the principal stress directions is vertical. In this situation, the principal stresses line up with the horizontal wellbore as depicted in Figure 6. At any point on the periphery of the borehole, the generator, tangent, and radius are directions of principal stresses, regardless of which type of deformation applies to the surrounding rock. With this orientation of the stress field, the observed fractures at 45° to the wellbore (principal-stress direction) are shear fractures.

The deformations at the very end of the hole have to be considered in order to envisage the formation of shear failures during drilling. At point A (Figure 7), the rock is subject to large compressive stresses from the geometric stress concentration caused by the hole end. Hence, the rock in the vicinity of the hole end will deform plastically. Plastic deformation causes the lowering of the stress deviator, $q = \sigma_1 - \sigma_3$, which is the difference between the largest and smallest principal stresses; it causes smoothing of the stress field around the circumference toward a more axis-symmetrical distribution. On the failure envelope, point A is characterized by high compressive-mean stress and less high deviatoric stress (Figure 8). When drilling farther ahead, material at point A moves away from the hole end toward B. It thereby moves through a transition zone where the hole diameter, supported by the closed end, changes from the drilled diameter to the resulting hole diameter. At point B, the hoop stress (σ_1) on the sides of the hole are larger than the vertical stress prior to drilling. By contrast, the horizontal-stress component parallel to the wellbore axis is reduced (it oscillates along the path A-B) because the wall of the hole is bent. The state of stress at point B is characterized by a decline in mean stress relative to that of point A, but by an enhanced deviatoric stress (Figure 8). If, during the passage from A to B, the state of stress meets the horizontal tangent of the failure envelope, which is denoted by the dotted line on Figure 8, then a shear failure will form at an angle of 45° to the wellbore.

THE APPEARANCE OF NATURAL FRACTURES ON IMAGE LOGS

Obviously, drilling-induced fractures need to be eliminated from further analysis of fracture-enhanced permeability estimates, fracture-density estimates, and fracture-propagation orientation.

From core analysis, natural fractures are known to present one dominant strike. The 45° symmetry of drilling-induced fractures allows sorting of the open fractures into drilling-induced and natural fractures. If a dis-

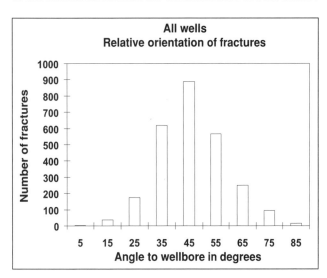

FIGURE 5. I-distribution of the entire Dan field.

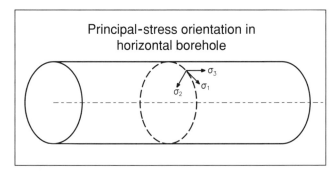

FIGURE 6. Principal-stress coordinate system on the wall of the horizontal borehole.

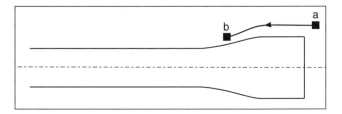

FIGURE 7. Schematic of deformation at the end of the hole.

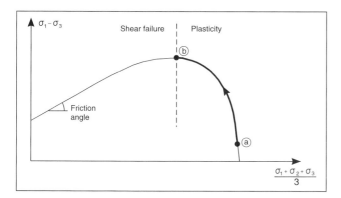

FIGURE 8. Schematic failure/yield envelope.

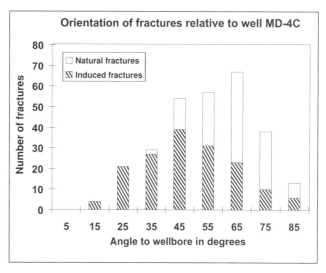

FIGURE 9. α-distribution of well MD-4C, with and without filtering.

tribution shows a distinct skewing and it is possible to restore the 45° symmetry by filtering out fractures with a certain strike, then the data set has been divided into one group representing drilling-induced shear fractures (45° symmetry) and another group of fractures characterized by a confined strike indicative of natural fracture direction.

EXAMPLE CASES FROM THE DAN FIELD

Image-log interpretation identifies a variety of features, including bedding planes, closed fractures, and open fractures. In this work, focus is on the geometry of open fractures only. Because the orientation of each fracture plane needs to be well determined, only fractures recognized by at least three of the four pads on the image logging tool were included in the analysis. Also, because the strike of a fracture is poorly determined if its dip is close to horizontal, open fractures with a dip of less than 45° were excluded from analysis. These low-dip fractures, which are not easily explained, represent less than 15% of the total amount of open fractures seen on the log.

WELL MD-4C

This well is situated on the crest of the Dan field A block, where relatively unambiguous information on hydraulic-fracture propagation exists from a variety of logged swept zones in the region. In this well, a symmetrical fracture distribution of about 45° can be restored by filtering out all fractures with a strike in the range of 127–162° (Figure 9). The strike of each of the natural fractures filtered out is indicated as a line fragment intersecting the well path on the map (Figure 10). The map also shows the fracture paths interpreted from flushed zones logged in neighboring wells. Clearly, the large water-injection fractures grown from the injectors ME-1, ME-2, ME-6,

MFA-4, and MFA-8 have propagated along azimuths that are near-parallel to the strike of the natural fractures interpreted in MD-4C.

WELL MA-4A

The α-distribution of the data in well MA-4A (Figure 11) lacks the 45° symmetry found in well MD-4C. Symmetry can be restored by filtering out fractures with a strike in the range of 130–145°. Supporting evidence for the propagation direction of induced hydraulic fractures in this area consists of mud-log data from the nearby well MA-5C, where frac sand was observed in ditch cuttings when crossing well MD-9 (Figure 12). A fracture-propagation direction of 125° can be estimated on the basis of the placement of the perforations in the reservoir.

WELL MFB-2C

In this well, the image log α-distribution (Figure 13) is not as asymmetrical as the previous two distributions studied, but a more symmetrical distribution still can be obtained by removing all fractures with a strike of 135–150° (Figure 13). Evidence of hydraulic fracture-propagation direction in the region consists of the four swept zones seen, which can be tied back to respective injection points (Figure 14). The strikes of the natural fractures are reasonably consistent with the range of strikes of the hydraulic fractures grown from the high-rate injection points, although the strikes of the long injection fractures interpreted as originating from MFB-5, MFA-9A, and ME-4 are all oriented more north-south than the fractures interpreted in well MFB-2C. This discrepancy may

FIGURE 10.
Areal map showing strike of natural fractures in well MD-4C and fracture paths of wells ME-1, ME-2, ME-6, MFA-4, and MFA-8.

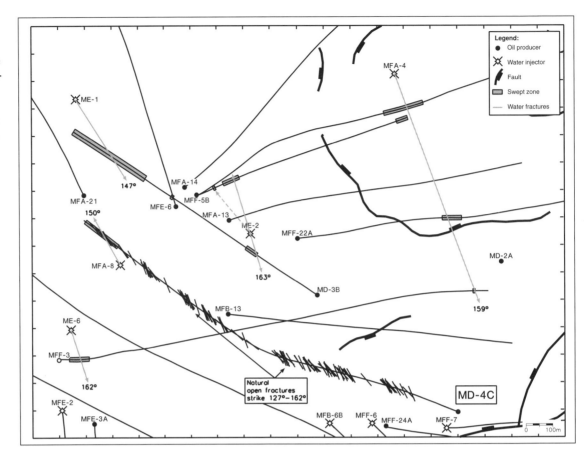

reflect local variations in propagation direction as well as the accuracy of the method.

WELLS MFA-1B AND MFF-8

The image-log data sets for wells MFA-1B and MFF-8 both show a distinct 45° peak for the α-distribution, but the symmetry of the respective distributions still can be improved by filtering out fractures striking 130–150° from the MFA-1B data (Figure 15), and likewise by filtering out fractures striking 150–170° from the MFF-8 data (Figure 16).

The hydraulic fracture propagated from the injector MFA-1B was seen in MFF-8 and nearby well MA-1C as swept zones. MFA-1B is completed with five isolated injection zones, and a production log has shown that most of the water is injected into zones 5 and 4. Combining this information with the image analysis gives the fracture paths shown in Figure 17.

All three remaining image logs shown in Figure 2 (wells MFF-4B, MFB-19A, and MFF-6) show α-distributions with a symmetry of 45° without filtering. The absence of a skewed distribution may be caused by the natural fractures having been oriented at a 45° angle to the wellbore, or it may mean simply that the wells were drilled in regions with few natural fractures.

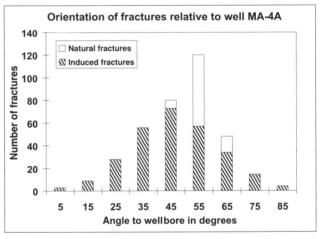

FIGURE 11. α-distribution of well MA-4C, with and without filtering.

DISCUSSION

The probability-density distributions of the filtered and unfiltered data are shown in Figure 18. It is clear that the fractures filtered out have an angle to the wellbore of more than 45°. This reflects the fact that the wells are drilled along azimuths oriented approximately east-southeast–west-northwest and south-southeast–north-northwest, whereas the fractures filtered out are predominantly northwest-southeast-bound vertical fractures. The proba-

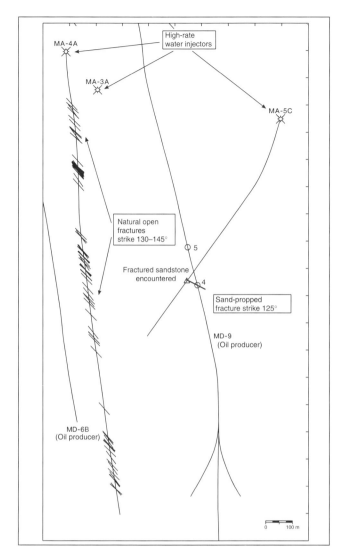

FIGURE 12. Map of wells MA-4A, MA-5C, and MD-9 showing strike of natural fractures in well MA-4A and the anticipated path of the sand-propped fracture logged during drilling of well MA-5C.

bility-distribution function of the filtered data can be considered a characteristic of the reservoir rock. The symmetrically distributed drilling-induced fractures account for more than 65% of the open fractures found. This finding supports the general perception, based on core observations and well tests, of a very low natural-fracture density in the Dan field (Jørgensen, 1992). Moreover, it emphasizes that image logs can be used only as a measure of reservoir fracture density, once observed fractures are properly classified.

COMPARISON OF IMAGE-LOG METHODOLOGY TO CORE DATA

The strike of natural open fractures measured in cores from three deviated wells (MFA-4, MFB-7, and MFA-11) coincides with the groups of open fractures filtered out from the image logs (Figure 19). Further, the strike of core fractures in MFA-4 coincides with the propagation

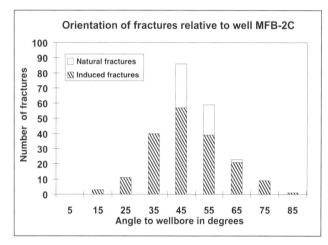

FIGURE 13. I-distribution of well MFB-2C, with and without filtering.

FIGURE 14. Areal map showing strike of natural fractures in well MFB-2C and fracture paths of wells MFB-5, MFA-9A, MFB-2B, and ME-4.

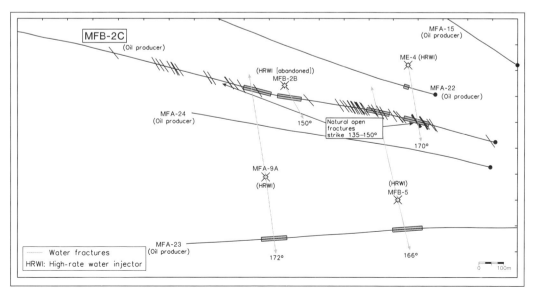

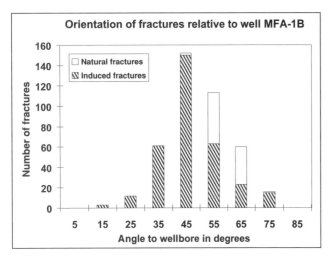

FIGURE 15. I-distribution of well MFA-1B, with and without filtering.

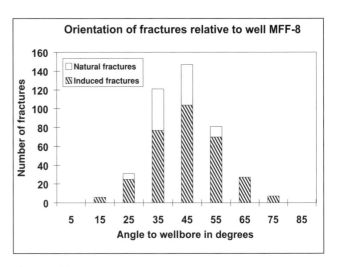

FIGURE 16. I-distribution well MFF-8, with and without filtering.

FIGURE 17. Areal map showing strike of natural fractures in wells MFF-8 and MFA-1B and fracture paths as interpreted from swept zones in well MA-1C and MFF-8.

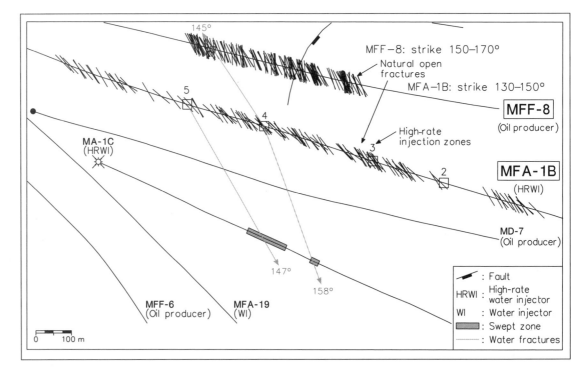

FIGURE 18. Probability-density functions based on the eight wells subject to image logging (MD-4C, MA-4A, MFB-2C, MFA-18, MFF-8, MFF-6, MFF-4B, and MFB-19A), before and after filtering.

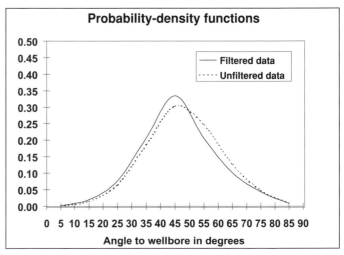

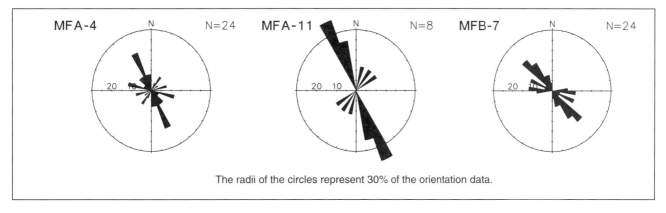

The radii of the circles represent 30% of the orientation data.

FIGURE 19. Strike of natural open fractures observed in cores from wells MFA-4, MFA-11, and MFB-7.

direction of the hydraulically induced fracture responsible for flushed zones observed in four wells—MFF-5, MFF-5B, MFF-22A, and MFF-3 (Figure 10).

The number of observed open fractures in the cores is limited, which agrees with the general assumption of a sparse distribution of natural open fractures in the Dan field.

CONCLUSIONS

Four key conclusions arise from this study.

1) The open fractures seen on the image logs in the Dan field predominantly are drilling-induced shear fractures that formed preferentially along slip planes oriented 45° to the wellbore axis.

2) Natural fractures detected on the image logs are of a confined strike, which agrees reasonably well with the strike of hydraulically propagated fractures and with the strike of open fractures seen on cores.

3) A method has been developed that divides logged open fractures into one group of drilling-induced fractures and another group of natural fractures.

4) The generic nature of the method makes it likely that the technique can be applied to other oil fields.

The reported method for discriminating natural fractures from induced fractures is a tool for predicting the propagation direction of hydraulic fractures. The method predicts hydraulic-fracture orientation along the entire extent of a horizontal well, as well as in areas where hydraulic-fracture orientation has not been observed by other methods. The predictive capability will be exploited in the placement of high-rate injection points and further infill drilling in the Dan field.

ACKNOWLEDGMENTS

The authors would like to thank Mærsk Olie og Gas AS and its partners in the Danish Underground Consortium (DUC), A.P. Møller, Shell Olie og Gasudvinding Danmark G.V. (Holland), and Texaco Denmark Inc., for permission to publish this paper.

REFERENCES CITED

Barton, C. A., D. Moos, P. Peska, and M. D. Zoback, 1997, Utilizing wellbore image data to determine the complete stress tensor: Application to permeability anisotropy and wellbore stability: The Log Analyst, v. 38, no. 6, p. 21–33.

Brudy, M., and M. D. Zoback, 1993, Compressive and tensile failure of borehole arbitrarily inclined to principal stress axes: Application to KTB boreholes, Germany: International Journal of Rock Mechanics and Mineral Science, v. 30, p. 1035–1038.

Carter, R. D., 1957, Derivation of the general equation for estimating the extent of the fracture area, *in* G. C. Howard and C. R. Fast, eds., Appendix of optimum fluid characteristics for fracture extension: American Petroleum Institute, Drilling and Production Practice, p. 261–268.

Hagort, J., 1981, Waterflood-induced hydraulic fracturing: Ph.D. dissertation, Technical University of Delft, the Netherlands.

Jørgensen, L. N., 1992, Dan Field—Denmark, Central Graben, Danish North Sea, *in* N. H. Foster and E. Beaumont, comps., Structural traps VI: AAPG Treatise of petroleum geology—Atlas of oil and gas fields, p. 199–215.

Kirsch, G., 1898, Die theorie der elasticitaet und die beduerfnisse der festigkeitslehre: Zeitschrift VDI, v. 29, p. 797–807.

Koning E., 1988, Waterflooding under fracturing conditions: Ph.D. dissertation, Technical University of Delft, the Netherlands, 253 p.

Lehne, K. A., and B. S. Aadnoy, 1993, Quantitative analysis of stress regimes and fractures from logs and drilling records of a North Sea chalk field: The Log Analyst, v. 33, no. 4, p. 351–361.

Ma, T. A., V. Lincerum, R. Reinmiller, and J. Mattner, 1993, Natural and induced fracture classification using image analysis: Society of Professional Well Log Anaysts 34th Annual Logging Symposium, Calgary, Manitoba, Canada, June 13–16, 1993.

Ovens, J. E. V., 1997, Making sense of water injection fractures in the Dan field: Society of Petroleum Engineers 1997 Annual Technical Conference and Exhibition, San Antonio, Texas, October 5–8, 1997, SPE paper 38928, p. 887–901.

Peska, P., and D. Zoback, 1995, Compressive and tensile failure of inclined wellbores and determination of in situ stress and rock strength: Journal of Geophysical Research, v. 100(B7), p. 12791–12811.

Zhou, S., R. Hillis, and M. Sandiforf, 1994, A study of the design of inclined wellbores with regards to both mechanical stability and fracture intersection, and its application to the Australian North West Shelf: Journal of Applied Geophysics, v. 32, p. 293–304.

12

Petersen, N. W., J. Toft, and P. M. Hansen, 2003, Appraisal of an extensive thin chalk reservoir with long horizontal wells: Cyclostratigraphy as a complementary steering tool, *in* T. R. Carr, E. P. Mason, and C. T. Feazel, eds., Horizontal wells: Focus on the reservoir: AAPG Methods in Exploration No. 14, p. 183–189.

Appraisal of an Extensive Thin Chalk Reservoir with Long Horizontal Wells: Cyclostratigraphy as Complementary Steering Tool

N. W. Petersen

Mærsk Olie og Gas AS
Copenhagen, Denmark

J. Toft

Mærsk Olie og Gas AS
Copenhagen, Denmark

P. M. Hansen

Mærsk Olie og Gas AS
Copenhagen, Denmark

ABSTRACT

An extensive flank reservoir extension has been appraised and production has been tested by drilling horizontal wells of record length for the North Sea, drilled from the existing Dan field production installation.

The main target reservoir unit with producible hydrocarbons is a thin (<3 m [10 ft]), highly porous zone that forms part of a sequence of cyclic beds in the low-permeable Maastrichtian chalk. The cyclic sequences can be correlated stratigraphically across the field.

It was planned to steer the wells with the help of biostratigraphy and logging while drilling (LWD). Steering proved more problematic than anticipated because the biostratigraphic zones in the target unit changed along the hole.

To prevent jeopardizing well reach and thus the appraised area, severe doglegs and extensive open-hole sidetracking were avoided. This resulted in a gently undulating well trajectory that crossed the target zone several times.

At times, however, it proved difficult to determine whether the well was above or below the highly porous target zone. It was found that the LWD log could be used to determine the position of the well relative to the highly porous zone by means of cyclostratigraphic correlation.

Long-term production testing has been performed in the wells using existing production facilities, with encouraging results.

INTRODUCTION

The Dan field was discovered in 1971 and was the first Danish oil field to be developed. The field is located in the southern part of the North Sea Central Graben (Figure 1).

The main trap is a structural dome (Figure 2) with the bulk of oil and gas trapped in chalk reservoirs of Maastrichtian (Upper Cretaceous) and Danian (lower Tertiary) age. A major fault divides the field into a northwestern A block and a southeastern B block. The reservoir chalk is characterized by high porosities but very low permeabilities. Natural fracturing appears to be limited (Jørgensen, 1992). The initial delineation phase was concluded in 1983 with the vertical exploration well M-10X (Figure 2), which appraised the southwestern flank of the A block.

The field was put on stream in 1972 and was developed until 1987 by conventional wells. Since 1987, development has been based on horizontal wells with multiple, induced, sand-propped fractures (Figure 2). Oil production increased accordingly, proving that horizontal drilling is a viable method for development of low-permeable chalk reservoirs.

To position the horizontal-development wells optimally in the Maastrichtian reservoir, a high-resolution biostratigraphic zonation was developed. This zonation has been refined during the years, and biosteering has been the main tool for positioning of horizontal wells at the crest and in the proximal flank areas.

The horizontal-well development campaign in 1990–1992 on the western flank of the A block (Figure 2) proved that oil saturations in the upper Maastrichtian in this part of the field were higher than anticipated.

Detailed geologic studies performed in 1994 on cores from all parts of the Dan field indicated the presence of a repetitive or cyclic bedding development in the upper Maastrichtian (Scholle et al., 1998) (Figure 3). Each cycle consists of a high- and low-porosity interval. Characteristic cycles can be correlated stratigraphically on cores, as well as on wireline logs, across the field (Toft et al., 1996) and thereby provide an additional steering tool. The thickness of the high-porosity portions of the cycles associated with higher oil saturations increases toward the west, leading to increased hydrocarbon volumes on the western flank as compared to previous models.

The development campaign initiated in 1995 included two infill, high-rate water injectors on the proximal

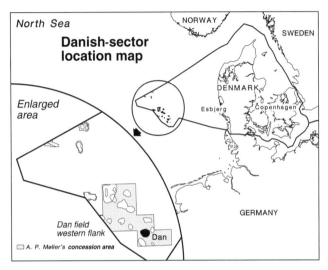
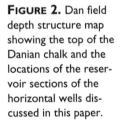

FIGURE 1. Location map showing the Dan field in the Danish North Sea.

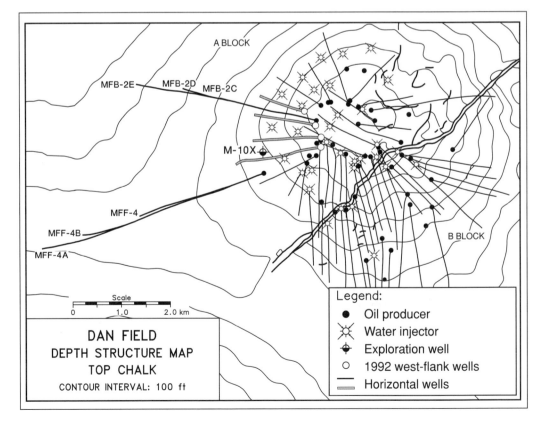

FIGURE 2. Dan field depth structure map showing the top of the Danian chalk and the locations of the reservoir sections of the horizontal wells discussed in this paper.

western flank (MFB-2C and MFF-4, Figure 2). With the improved horizontal-drilling techniques now available, these injectors were planned to be extended toward the distal western flank to appraise further the earlier identified oil accumulation. Both wells were drilled from the existing Dan field production installation to record horizontal lengths for the North Sea.

This paper describes the appraisal of the distal western flank with these horizontal wells. The main target was a highly porous unit, M1b1, identified in the M-10X well (Figure 4). This target unit has a vertical thickness of less than 3 m (10 ft). Steering of the wells was planned as a combination of logging while drilling (LWD) and biostratigraphy. However, cyclostratigraphic correlation proved invaluable when the biostratigraphy became inconclusive.

By avoiding severe doglegs and extensive open-hole sidetracking, gently undulating well trajectories were obtained, including several drop-and-build sections through

the upper Maastrichtian suitable for long-distance correlation of cyclostratigraphy.

DAN FIELD GEOLOGIC SETTING

The main formation of the Dan field structural dome was initiated in the Triassic by minor movements of Permian salt, which was mobilized again at the end of the Jurassic, resulting in the formation of a salt pillow. Transgression, especially through the Late Cretaceous, resulted in a reduced influx of clastic material and deposition of pronounced chalk strata into the early Tertiary. The Dan salt pillow experienced several growth episodes from the Late Jurassic to the late Tertiary, with major structural growth occurring in the Eocene.

The Maastrichtian chalks are the main hydrocarbon reservoir and have the most productive reservoir units in

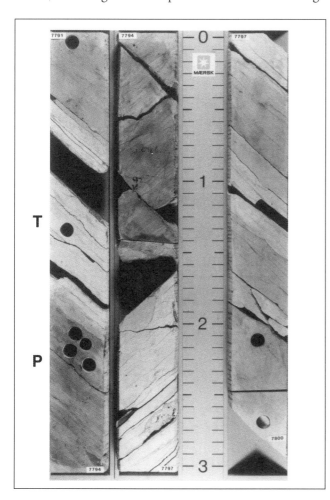

FIGURE 3. Core photo illustrating the pronounced cyclicity in the upper Maastrichtian chalk. Porous (P) laminated intervals show dark oil staining. Tight (T) bioturbated intervals are light colored. Example shown is from the Dan field well MFB-7. Depth scale is in feet. (After Toft et. al., 1996.)

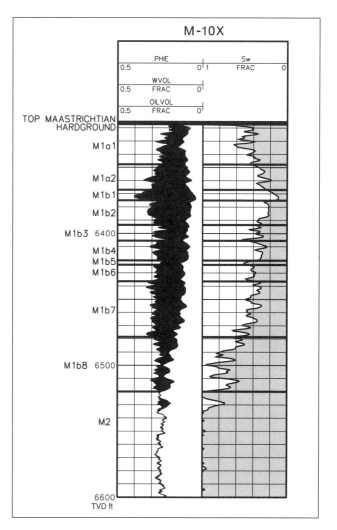

FIGURE 4. Petrophysical log of the vertical exploration well M-10X. True-vertical-depth scale is in feet. PHIE = effective porosity (the porosity calculation includes a correction for shale content); WVOL = water volume; OIL VOL = oil volume; Sw = water saturation; FRAC = fraction.

the uppermost part of the Dan field (M1a and M1 units, Figure 4). The chalks in these units consist of foraminiferal wackestone grading to coccolithic mudstone. Clay and silica content is generally low (less than 3 wt.%). The porosity in the uppermost Maastrichtian is, on average, 29% (ranging from 24% to 41%). Corresponding permeabilities, however, are low—approximately 1–5 md (Jørgensen, 1992).

Sedimentological core studies (Scholle et al., 1998) demonstrated that the entire reservoir section is characterized by a distinct cyclic development (Figure 3). Individual cycles are typically 1–1.5 m (3–5 ft) thick and consist of a porous and a tight interval. The porosity variation across a cycle is 2–8%.

Porous intervals of a cycle were formed by relatively continuous deposition of penecontemporaneous, remobilized chalk that was transported by a combination of continuous current and gravitationally induced processes. The tight intervals were formed primarily by slow pelagic deposition. The factor controlling the cyclicity is sedimentation rate.

Average thickness of the porous intervals increases from east to west. On the western flank, tight intervals constitute 10–15% of a cycle; on the eastern flank, they form 50% of a cycle (Scholle et al., 1998).

This cyclic bedding is considered generally to have formed in response to orbital forcing (De Boer and Smith, 1994) and thus represents chronostratigraphic surfaces. Individual beds can be correlated across the field and thus offer the possibility of a significantly improved chronostratigraphic correlation in otherwise monotonous chalk reservoirs (Toft et al., 1996).

This cyclostratigraphic model predicted that the pattern of increasingly thickening cycles, with thicker porous intervals and thin tight intervals, would continue toward the west, indicating a volumetric upside on the distal western flank of the A-block.

BIOSTRATIGRAPHY AND CYCLOSTRATIGRAPHY AS STEERING TOOLS

Horizontal wells in the Dan field traditionally have been positioned in the reservoir by extensive use of high-resolution biostratigraphic analysis of cores and cuttings at the well site.

The biostratigraphic zonal scheme is based on local assemblages and principally reflects changes in environment rather than evolution. This could cause problems with diachronous events, but until recently had not caused problems in the Dan field. Some events could be correlated to other fields in the area.

The biostratigraphic zones are defined from the top of the Maastrichtian chalk as DM1 (youngest), DM2 (with subzones A, B, and C), and DM3. These zones cover approximately the uppermost reservoir unit M1a at the crest of the Dan field. Older zones correspond to deeper reservoir units.

However, in conjunction with drilling of the two new wells on the distal western flank, biostratigraphy was found to be of limited value because the biostratigraphic zones changed along the hole relative to the target unit. This is believed to have been caused by one or more of the following:

1) sample contamination escalation in proportion to the length of the horizontal section drilled
2) depositional facies changes in the zones
3) diagnostics and identification of individual biozones, complicated by increases in the distance to the reference well sections in the crest of the field

When biostratigraphy failed as an effective steering tool, steering on cyclostratigraphy was implemented. Toft et al. (1996) demonstrated that correlation based on logs from other horizontal wells may be possible if the logs cover a certain stratigraphic interval. This possibility is dependent on the thickness and appearance of characteristic cycles; i.e., the wellbore must be discordant to stratigraphy.

MFB-2C/D/E WELL

Drilling of MFB-2C (Figure 5) was initiated in autumn 1996. The main objective was to cover three downhole locations for high-rate water injection. The secondary objective was to appraise oil accumulation in the M1b1 unit on the western flank. MFB-2C was unintentionally dropped through the oil-water contact (OWC), which was encountered significantly deeper than expected. To meet the main objective and secure the hole, a 7-in. liner was run.

Comparison of the logs from the discordant build-and-drop sections in MFB-2C with the vertical log from M-10X indicated distinct correlatable cycles and promising oil saturations, especially at the base of the target zone—the M1b1 unit, shown in Figure 4. The horizontal sidetrack MFB-2D was kicked off and drilled with a 6-in. bit, which excluded the use of LWD because of the limited number of tools available at that time. The steering was therefore based on hydrocarbon shows and biostratigraphy, because it was known that M1b1 was related to Dan field biozones DM2C-DM3 in the MFB-2C drop section. M1b1 was followed for more than 305 m (1000 ft); the well trajectory was then dropped because biostratigraphy indicated that the well was approaching the overlying Danian chalk. Oil saturations decreased immediately, and the drop was continued until the OWC was encountered.

The subsequently obtained log indicated that the sudden decrease in oil saturation was caused by exiting the M1b1 unit. Consequently, the well was sidetracked (MFB-2E) to further appraise M1b1. Again, a comparison of the log from the MFB-2D discordant drop section and the M-10X log indicated correlatable cycles (Figure 6) and promising oil saturations in M1b1. The longer stratigraphic interval covered by the drop section in MFB-2D improved the correlation to M-10X.

The MFB-2E sidetrack indicated that the biostratig-

raphy diagnostic for the M1b1 unit changes at the most distal part of the flank. No drop section was drilled in MFB-2E, which excluded cyclostratigraphic correlation.

Drilling of MFB-2C/D/E demonstrated that M1b1 is the main reservoir unit on the western flank. Further, appraisal of such a thin unit is very difficult without LWD, especially when the biostratigraphic age changes. The diagnostic cyclic-stacking pattern recognized in the discordant drop sections in MFB-2C and MFB-2D was succesfully correlated to M-10X, a lateral distance of more than 2438 m (8000 ft), with the undulating horizontal sections between them indicating a coherent stratigraphy. The well then was tied back to existing infrastructure for long-term production tests.

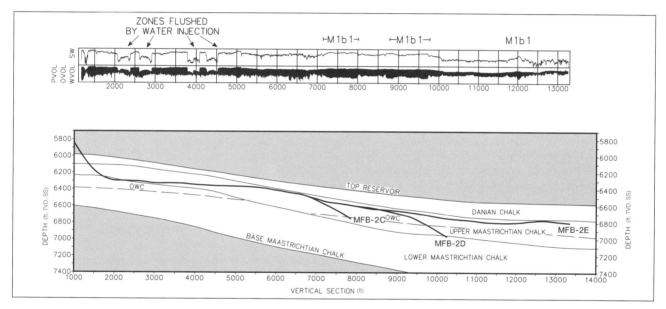

FIGURE 5. Geologic cross section of well MFB-2E and petrophysical log of well MFB-2C/D/E. PVOL = pore volume; OVOL = oil volume; WVOL = water volume; SW = water saturation.

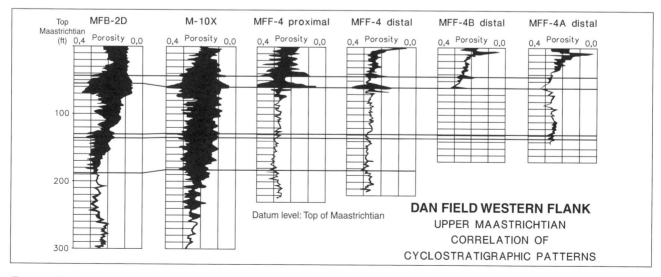

FIGURE 6. Cyclostratigraphic correlation among M-10X, MFB-2D, and MFF-4/A/B. True-vertical-depth scale is in feet.

MFF-4/A/B WELL

Drilling of MFF-4 was initiated in January 1997. The main objective was to cover a downhole location for high-rate water injection. The secondary objective was to further appraise oil accumulation on the southern part of the western flank. The three sidetracks all were drilled with LWD, and steering was initiated by a combination of bio- and cyclostratigraphy.

After reaching the planned injection point, drilling continued in the uppermost Maastrichtian, targeting the M1b1 unit (Figure 7). The cyclostratigraphic sequence in the initial build section was correlatable to M-10X (Figure 6), with M1b1 initially related to Dan field biozones DM3-DM4 and later to DM2C.

At 4075-m (13,369-ft) MDRT (measured-depth rotary table), the oil zone could not be identified. Biostratigraphic analysis indicated a position in the uppermost Maastrichtian, and drilling continued downward until a level 46 m (150 ft) true vertical depth (TVD) below the predicted position of the M1b1 unit. The trajectory was turned upward and drilling continued, penetrating M1b1 and terminating at the top of the Danian chalk. Oil was limited to M1b1; the chalk reservoir above and below was 100% water saturated. An evaluation of the cyclostratigraphic pattern obtained along MFF-4 established the correlation to M-10X, and the structural position of M1b1 was geometrically interpolated between the MFF-4 drop-and-build sections. The MFF-4A sidetrack subsequently could be positioned accurately in this target unit.

MFF-4A was planned to further appraise M1b1 beyond total depth in MFF-4 (Figure 7); doglegs were avoided from kickoff to avoid jeopardizing the potential length of the hole. By drilling MFF-4A beyond total depth in MFF-4, the structural position of M1b1 became

uncertain. With water-saturated reservoir above and below and the biostratigraphy becoming inconclusive, it proved difficult to determine whether the bit was above or below the target zone. The well trajectory was dropped in order to penetrate a stratigraphic section suitable for cyclostratigraphic correlation. Characteristic porosity peaks in this section indicated a position below the target unit; therefore, the trajectory was turned upward, penetrating a water-saturated M1b1. Drilling terminated in the Danian chalk at a horizontal-record length for the North Sea of 7553 m (24,781 ft) MDRT.

Again, the cyclostratigraphic sequence obtained in the distal discordant build section of MFF-4A demonstrated an excellent match to M-10X (Figure 6). The cyclostratigraphy thus could be correlated laterally for more than 4572 m (15,000 ft) (Figure 2).

MFF-4A was sidetracked to appraise the position of the M1b1 oil-water contact (OWC), which consequently should be positioned somewhere along MFF-4A. MFF-4B encountered M1b1, showing an oil saturation of 40%. In the two latter sidetracks, real-time LWD readings were in periods of poor quality because of background noise. Significant problems with sample quality also masked the biostratigraphy. M1b1 was found to be related to different biozones along the hole, and cyclosteering was consequently the primary steering tool.

DISCUSSION

The appraisal of the western flank of the Dan field has demonstrated that the lithologic cycle pattern can be correlated across the entire western flank (Figures 2 and 6). The proximal- and distal-build sections of MFB-2D, MFF-4A, and M-10X represent the apexes of a polygon

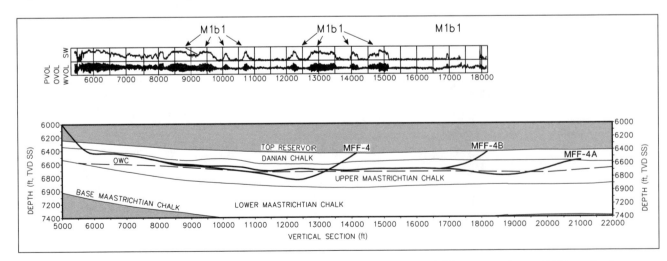

FIGURE 7. Geologic cross section of well MFF-4B and petrophysical log of MFF-4/A/B. PVOL = pore volume; OVOL = oil volume; WVOL = water volume; SW = water saturation.

enclosing an extensive area of the western flank, in which it is possible to predict, based on cyclostratigraphy, the probable structural position of the M1b1 unit.

As demonstrated in the distal part of the MFF-4A sidetrack, extrapolation of the position of the target unit beyond well control is very uncertain. This is caused by the position of M1b1 being directly related to the position of the top of the reservoir. The position of this top is interpreted from seismic data, on which the uncertainty significantly exceeds the thickness of the target unit. When structural control is provided by wells, this uncertainty is reduced considerably, and interpolation of the position of the target unit is possible.

Using correlation of log patterns, undulating horizontal wells provide information on thickness and reservoir properties of the target unit across a large area. Acquisition of similar information with conventional appraisal wells would require more wells and would be costly. A further advantage is the opportunity for long-term production tests by tying the wells to the existing production facilities. However, the use of sophisticated horizontal-drilling techniques, including LWD, is required.

Paradoxically, when aiming for target-unit horizontal intervals as long as possible, correlation of cycles requires discordant well sections. Such sections are normally avoided in horizontal production wells; consequently, they rarely provide information on thickness of the pay zone. When appraising with horizontal wells, however, such sections are required for subsequent reservoir modeling and positioning of horizontal production wells in a development situation.

CONCLUSIONS

Drilling of MFB-2C/D/E and MFF-4/A/B has demonstrated that a thin reservoir (< 3 m [10 ft]) unit can be appraised with horizontal wells in a very large area.

The cyclostratigraphy provides information for detailed correlation along the wells, thus proving a consistent and coherent stratigraphic pattern.

Gently undulating well trajectories must be part of the appraisal phase to obtain discordant well sections suit-

able for cyclostratigraphic correlation and subsequent cyclosteering. At the same time, doglegs must be minimized to avoid jeopardizing well length.

Extrapolation of the structural position of the target unit is directly related to the uncertainty about the position of the top of the reservoir. Extrapolation beyond well control was found to be difficult, whereas inside the distal build-and-drop sections, the position of the target unit could be identified accurately across significant distances.

The appraisal technique presented here requires the use of sophisticated horizontal-drilling techniques, including LWD.

Long-term test-production experience is obtained by appraisal drilling with horizontal wells from existing infrastructure.

ACKNOWLEDGMENTS

The authors wish to express their appreciation to Mærsk Olie og Gas AS and the partners in Danish Underground Consortium (A. P. Møller, Shell Olie og Gasudvinding Danmark G.V. [Holland], and Texaco Denmark Inc.) for permission to publish this paper.

REFERENCES CITED

De Boer, P. L. and D. G. Smith, eds., 1994, Orbital forcing and cyclic sequences: International Association of Sedimentologists, Special Publication Number 19, p. 1–14.

Jørgensen, L. N., 1992, Dan Field—Denmark, Central Graben, Danish North Sea, in N. H. Foster and E. A. Beaumont, eds., Structural Traps VI: AAPG Treatise of Petroleum Geology—Atlas of Oil and Gas Fields, p. 199–215.

Scholle, P. A., T. Albrechtsen, and H. Tirsgaard, 1998, Formation and diagenesis of bedding cycles in uppermost Cretaceous chalks of the Dan Field, Danish North Sea: Sedimentology, v. 45, p. 223–243.

Toft, J., T. Albrechtsen, and H. Tirsgaard, 1996, Use of cyclostratigraphy in Danish chalks for field development and appraisal, in Joint Chalk Research Program, Topic V: Fifth North Sea Chalk Symposium, Reims, France, October 7–9, 1996, p. 1–8.

Carbonate Reservoirs

13

Pearce, L. A., C. M. Hewitt, and L. M. Corder, 2003, Horizontal drilling in the Northern Reef Trend of the Michigan Basin, *in* T. R. Carr, E. P. Mason, and C. T. Feazel, eds., Horizontal wells: Focus on the reservoir: AAPG Methods in Exploration No. 14, p. 193–203.

Horizontal Drilling in the Northern Reef Trend of the Michigan Basin

Lester A. Pearce

Shell Western E&P Inc., Houston, Texas, U.S.A.

Lisa M. Corder

Shell Noreast Co., Houston, Texas, U.S.A.

Christine M. Hewitt

Shell Noreast Co., Houston, Texas, U.S.A.

ABSTRACT

For a company operating in a mature province, it is extremely important continuously to seek new ways to improve recovery from known hydrocarbon accumulations. As the mechanical process of drilling horizontal wells matured in the mid-1990s, Shell Western E&P Inc. (SWEPI) was quick to apply this new technology to just such a province, the Northern Silurian Niagaran Pinnacle Reef Trend (NRT) in the Michigan Basin. To date, SWEPI has drilled more than 60 horizontal wells in this trend, which accounts for more than 55% of all horizontal drilling in the NRT and continues to be the industry leader. Much has been learned in the past decade, and it appears that the application of lateral drainhole technology can be an effective tool in the NRT in the right geologic and reservoir situations.

SWEPI first attempted to apply horizontal technology to the NRT in 1987. Those early attempts failed mechanically because of the inability to maintain a horizontal path through a significant section of reef. In only one well, the Bancroft-Bisard 2-1, was a horizontal path achieved, and that lateral was less than 30 m (100 ft) in length. No further horizontal activity occurred until 1994, when a review of the decline curve from that well indicated that not only had there been a production increase of 30 BOPD after the lateral, but the rate of production decline had been reduced from 37% to 18%. It was estimated that an additional 100 MBO would be recovered from that well as a result of the lateral. This information, coupled with improvements in drilling technology and the knowledge that great improvements in production were being made worldwide by the application of lateral technology, caused SWEPI to initiate a new lateral program in the NRT.

Five major opportunity types were evaluated with horizontal wells. SWEPI's first attempt, in 1987, was a large gas cap over a thin oil leg. This type has continued to provide the most consistently positive results, and it accounts for 47% of the total laterals drilled. The second-largest opportunity type, with 36% of the total, is the heterogeneous reservoirs. These opportunities closely resemble exploratory drilling in which the success rate is low but the risk capacity is high. Only two low-pressure reservoirs, which depend on pure gravity drainage, have been drilled. Neither was successful. Another type is the improved kh in a low-porosity reservoir. Although SWEPI has drilled only four, these can be profitable in certain circumstances. The most disappointing type has been the thin oil column over a large water leg. We have drilled three thin-oil-column horizontals with no successes.

Approximately 40% of the 55 horizontal wells that had been drilled by May 1998 were considered economically successful. Ultimately, SWEPI has forecast that the horizontals will recover an additional 3.4 MMSTBO and 5.7 bcfg.

Issues that remain a concern include lost returns, wellbore damage, vertical position of the lateral, and the application of laterals in settings that have been unsuccessful to date. Significant oil volumes remain in these environments. We are attempting to address some of these issues as the program continues to mature.

INTRODUCTION

In 1994, Shell Noreast Company, a former division of SWEPI, undertook a horizontal-drilling program in the Northern Reef Trend (NRT) of Michigan (Figure 1). This trend is comprised of Niagaran Silurian pinnacle reefs that form a swath across the northern Lower Peninsula of the state, and it extends from Lake Michigan on the west to Lake Huron on the east. All of the reservoirs in which horizontal drilling was applied were depleted partially. Several types of horizontal-drilling candidates were identified and tested. This discussion addresses the results of the first 55 horizontal wells drilled in this program.

FIGURE 1. Map of Michigan, with the Northern Reef Trend indicated in relation to other active portions of the Michigan Basin.

BACKGROUND

The Michigan Basin is essentially bowl shaped and was filled with mixed clastic and carbonate sediments between the Precambrian and Late Devonian. The Silurian-Devonian is dominantly a carbonate sequence. During Niagaran (Silurian) time, pinnacle reefs grew in a ring around the entire basin, with heights ranging from 61 to 183 m (200 to 600 ft). The reef density varies widely around the perimeter of the basin; the NRT has the highest concentration. Discovery of the NRT dates to 1968, with the highest activity levels occurring in the early to mid-1970s. Activity has continued at a modest pace to the present. More than 700 reefs have been discovered in the NRT, and well over half of these continue to produce. Cumulative production in the trend through mid-1998 was 380 MMSTBO and 2.1 tcfg.

The depositional and diagenetic history of the pinnacle reefs was established and described during the first decade of development (Mesolella et al., 1974; Huh et al., 1977). Among the factors that affect the porosity content and distribution and are thus critical to the evaluation of horizontal drilling opportunities are the following:

1) The primary pore network varies, depending on position in the reef.
2) Relatively planar dolomite zones caused by mixed water dolomitization occurred during a sea-level decline immediately after reef formation.
3) Fracturing, solution enhancement of fractures, and partial cementation occurred during and immediately subsequent to sea-level lowstands.
4) Hypersaline dolomitization occurred during deposition of the overlying A1 Carbonate zone, which in many cases forms part of the reservoir.
5) Salt plugging occurred during burial.

Numerous reservoir and fluid characteristics have been mapped across the NRT. Several of these characteristics change somewhat methodically when moving from a basinward to a shelfward position (Caughlin et al., 1976). Those that are important to an understanding of the horizontal drilling program include:

1) The reefs located most basinward are 100% limestone. Shelfward, the percentage of dolomite increases until it reaches 100%. The reefs located most shelfward contain excess dolomite in the form of dolomite cement.
2) The reefs located most basinward contain gas, and the hydrocarbon mix becomes more oily in a shelfward direction.
3) Basinward, reefs tend to have thicker hydrocarbon columns, whereas water levels are generally higher in the reservoir as the shelf is approached.

Little natural water support has been identified in the reef trend; the major drive mechanisms are gas-cap expansion, depletion, and gravity drainage. The improved recovery techniques of water, natural gas, or CO_2 injection have been used in only a very small fraction of the reservoirs in this trend (Tinker, 1982; Tinker et al., 1983; Pieters and Pearce, 1996). These factors have been important in assessing volume opportunities and placement of laterals in the reefs, as well as in distinguishing one candidate type from another.

PREVIOUS HORIZONTAL DRILLING

In the early 1990s, numerous reports surfaced about the successful application of horizontal drilling in various parts of the world. Shell was interested in testing horizontal drilling in those arenas where it had not yet been applied successfully. In some of these areas, however, it had been tried, unsuccessfully, as recently as 1987. Michigan's NRT was one of those areas.

During an early 1994 well review, it became apparent that the lack of previous success in Michigan might have been overstated. For several years, it was generally understood that little additional recovery

would come from the 1987 horizontal leg drilled in the Bancroft-Bisard 2-1 well. The earliest production data subsequent to the horizontal-drilling operation showed an immediate doubling of the oil-production rate (Figure 2). However, production had declined within a few months to levels that approximated the predrilling rates. With the benefit of several years of additional production, the horizontal leg in this well clearly would result in production of more than 100,000 incremental barrels of oil as a result of the lessening of the decline rate, apparently because of the drilling of the lateral.

Further investigation of the 1987 program revealed that three horizontal wells had been planned. In all cases, the gas cap had expanded such that gas was coning into the wellbore, thereby reducing the liquid component of the production stream. Thus, the horizontal was to be drilled low in the oil column of the reef to decrease the gas-oil ratio (GOR) and improve ultimate oil recovery. A plat of Mayfield 1A field shows that the Bancroft-Bisard 2-1 was originally drilled into the center of a 70-ac reef (Figure 3).

A schematic cross section through this reef shows that the horizontal leg was drilled low in the oil column (Figure 4).

However, the horizontal leg had extended only 30 m (100 ft) before drilling was abandoned. Of the other two planned wells, one resulted in a sidetrack when the drilling operation was unable to turn the corner and become horizontal, and the third was never drilled because of the mechanical difficulty of the first two.

REDEVELOPMENT TEAM

During the period from 1990 to 1992, Shell did very little drilling in the NRT. As a result, oil production de-

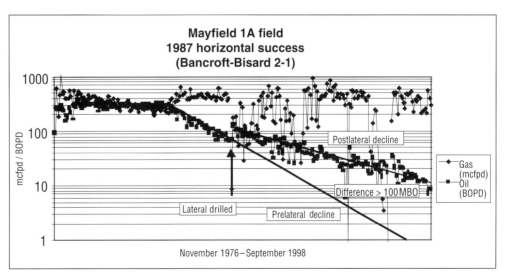

FIGURE 2. Production curve from the Shell Bancroft-Bisard 2-1 well, showing the impact of a horizontal leg drilled in 1987.

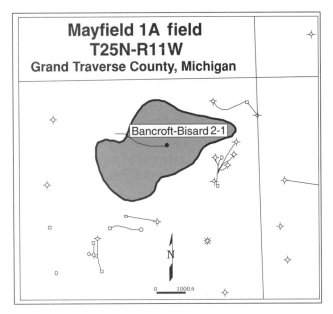

FIGURE 3. Location of the Bancroft-Bisard 2-1 well, approximately in the center of an irregularly shaped pinnacle reef.

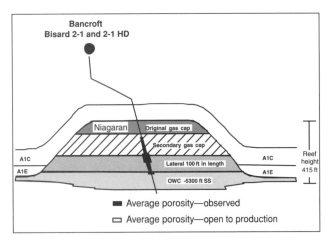

FIGURE 4. Diagrammatic cross section of Mayfield 1A reef, showing the original Bancroft-Bisard 2-1 penetration and the location of the horizontal leg at the base of the oil column.

clined severely. Partly in response to this decline, Shell decided to undertake a multidisciplinary review of the fields in which they operated wells to see if additional drilling opportunities remained. This charter was subsequently expanded to include all fields in the NRT, regardless of Shell's working interest.

When the concept of drilling horizontally in this trend was revived in 1994, it proved beneficial that a multidisciplinary team was already in place. Because the team was knowledgeable about the processes that impact porosity and hydrocarbon distribution in the NRT, it was possible quickly to develop a spreadsheet of possible opportunities to which the technology could be applied.

These opportunities were combined into groups with similar characteristics, and categories were formed. Ultimately, five of these categories were tested, the results of which are shown in Figure 5.

An initial look at Figure 5 suggests that two of the categories clearly have been profitable: "Large gas cap/thin oil leg" and "Improved kh." A third category, "Heterogeneous reservoir," stands out as being anomalous. Although only 25% of attempts in the heterogeneous reservoir category had economic outcomes, only one other category (Large gas cap/thin oil leg) contained more attempts. Therefore, each of the five categories and their outcomes warrants a closer look.

LARGE GAS CAP/ THIN OIL COLUMN

The prototypical success described above is of this type. The category is a combination of two categories from the original spreadsheet. At first, a distinction was made between those reservoirs that had an original gas cap and those that had only a secondary gas cap. This distinction proved to be important only in determining the target volume. Because it does not affect where or how a horizontal leg is drilled, the two categories were combined into one. The importance of understanding the target volume, however, should not be minimized. Among the early failures were wells that probably should not have been drilled because of insufficient target volumes. One characteristic of this type of opportunity is that the porosity is relatively well connected. Under these circumstances, a material-balance calculation can yield a reasonable estimate of the volumes in place and the remaining target.

Although the large gas cap/thin oil column opportu-

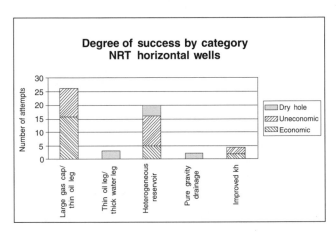

FIGURE 5. Results of the five categories of NRT horizontal-drilling opportunities distinguished by wells plugged and abandoned (dry holes), wells that produced uneconomic volumes, and wells that produced economic volumes.

nity had been successfully drilled in the past, four of the first five attempts in the 1994 and 1995 round of drilling were economic failures. In part, the early failures were a result of cost overruns that occurred during the time that techniques to drill horizontal wells in the reef trend were being established and refined. Among the earliest wells were a long-radius (~213-m [700-ft]) lateral and several very short-radius (~14-m [45-ft]) laterals. Both types proved to be prone to severe drilling problems and subsequently were abandoned. Once drilling techniques and selection criteria were refined, the economic success rate soared to the current overall economic rate of 60%. Moreover, oil was added in all cases. However, without the determination to fully and effectively test the use of horizontal drilling in this setting, the program could have been abandoned prematurely and erroneously labeled as a failed method of improving production.

SUCCESSFUL LARGE GAS CAP/ THIN OIL COLUMN

Mayfield 16 field consists of two wells, the Wurm 1-16 and the Grant Farms 2-16 (Figure 6). At the time the Wurm was entered and extended laterally low in the oil column, it had started to produce free gas; the Grant Farms well still was producing at a low GOR (Figure 7). The Wurm, one of the earlier reentries, was drilled using 14-m- (45-ft-) radius technology (Figure 8); it encountered serious drilling problems but was drilled to a distance of 259 m (850 ft) from the wellbore. Abundant water was lost to the formation, which led to an extended cleanup time. There was an immediate drop in GOR, however, and oil production ultimately climbed to the level that had been produced before gas started to cone into the original wellbore.

After this success, the field was unitized to eliminate internal barriers and to allow a horizontal leg in the Grant Farms well to be optimized. Subsequently, a 396-m (1300-ft) horizontal leg was put in the Grant Farms well, and oil production climbed to levels that had never been produced before from this field.

UNSUCCESSFUL LARGE GAS CAP/ THIN OIL COLUMN

Charlton 9 field has been problematic. On the surface, it seems to be very similar to Mayfield 16 field.

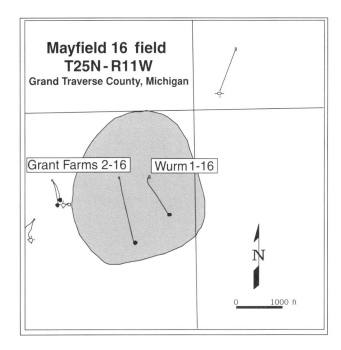

FIGURE 6. Location plat of Mayfield 16 field, showing the position of the two horizontal wells drilled in that field.

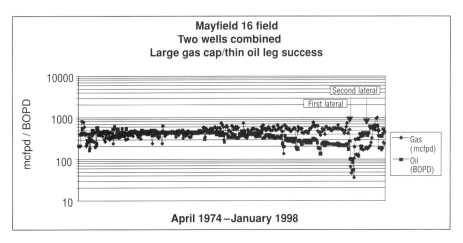

FIGURE 7. Production curve for Mayfield 16 field, with the timing of the two horizontal wells indicated.

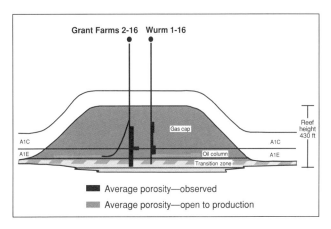

FIGURE 8. Diagrammatic cross section of the Mayfield 16 reef, showing the position of the two horizontal wells in this field.

Three wells produce from this field—the Charlton 1-9, Charlton 2-9, and Charlton 6-4 (Figure 9). When the Charlton 1-9 was extended laterally, the Charlton 2-9 was still producing at its allowable oil rate, while the Charlton 1-9 and the Charlton 6-4 were producing free gas. The Charlton 1-9 was producing at a particularly high GOR; therefore, it was reentered and extended laterally a distance of 290 m (950 ft) at a position deep in the oil column (Figure 10). Excellent porosity was encountered in the oil zone, but immediately after being put back on production, the liquids disappeared. Because the most likely

explanation seemed to be coning at the top of the curve, a tubing and open-hole packer system was used in an attempt to shut off gas. This effort was unsuccessful, and the well subsequently has produced mostly gas.

Because of the inconsistency between the apparent drilling results and subsequent production, the third well in the field, the Charlton 6-4, was reentered and extended laterally for a distance of 244 m (800 ft) (Figure 11). Porosity of as much as 20% was encountered during drilling, which is exceptionally high in a reservoir in which the nominal pay cutoff is 3% porosity. However, allowable production rates have not been obtained since the well was put back on production. The most likely explanation for the failure to produce oil in greater quantities is damage to the lateral, but repeated attempts to clean out and stimulate the lateral have failed. The oil production from this field (Figure 12) has declined particularly severely since the laterals were drilled, because Charlton 2-9 has begun to produce free gas.

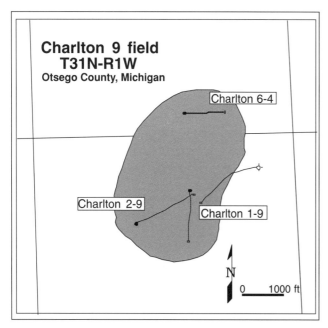

FIGURE 9. Location plat of Charlton 9 field showing the three producers in this field and the orientation of their horizontal legs.

THIN OIL LEG/THICK WATER LEG

The thin oil leg/thick water leg category of opportunity was particularly attractive because it represented an apparent chance to extend the limits of the producing trend. Regional data from the NRT indicates that the most shelfward area appears to have oil charge, but also high water levels and poor-quality pay. Oil has been produced from this area, but from only a few reefs and generally in small quantities. Occasionally, production of as much as 300,000 to 400,000 barrels of oil has been reported from a reef in this part of the trend, but production of a few thousand barrels or no production at all was more commonly reported, despite oil shows while

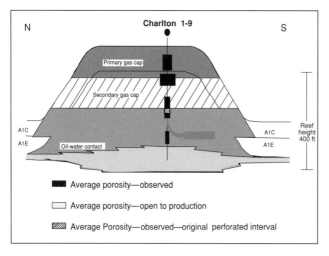

FIGURE 10. Diagrammatic cross section, in an essentially north-south orientation through the Charlton 9 reef, showing the original porosity distribution in the Charlton 1-9 well and the position of the subsequent horizontal leg.

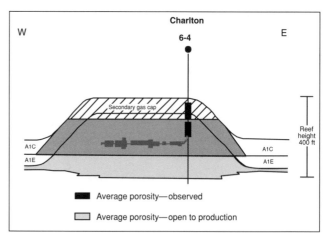

FIGURE 11. Diagrammatic cross section in a west-to-east orientation across the northern portion of Charlton 9 reef, showing orientation and porosity information from the Charlton 6-4 well.

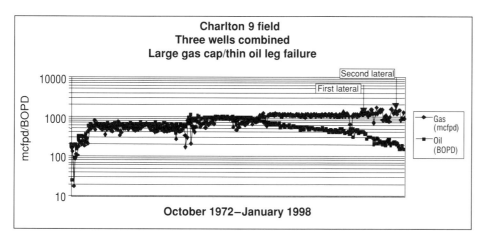

FIGURE 12. Combined production curve for Charlton 9 field, showing the minimal impact of the two horizontal-drilling operations on the oil decline curve.

drilling. Based on the few large producers, this category would seem to provide an outstanding target for horizontal drilling.

The team saw two ways in which development of this part of the trend might succeed. Because there is some degree of heterogeneity in all reefs, it is possible that pockets of higher-quality pay might be contacted by penetrating more rock in the oil column. Alternatively, flow rates might be improved by increasing the amount of formation that is open to the wellbore, thereby increasing the permeability times net-pay factor, or "kh." Although only three wells appear on the scorecard in Figure 5, other attempts have been made subsequently. Failure has resulted in all cases.

Unfortunately, no capillary pressure or relative permeability data are available from the low-porosity dolomites of which these reservoirs are comprised. However, it is not difficult to imagine the petrophysical characteristics that could lead to the results we have seen. These results consist of two types. In the first, large quantities of water were produced from laterals drilled well within the apparent oil column and which had excellent oil shows, including oil on the pits. If the rock is poor enough, the 100% water level and top of transition zone could be hundreds of feet above the free-water level. Consequently, the rock encountered might easily yield oil shows while drilling, but only water production on test. In the second type of result, production tests yielded no fluid at all. This outcome can be explained best by a relative-permeability curve in which oil and water are present in a ratio such that the permeability is reduced to a small fraction of that for oil or water alone. In this case, there may be so little flow into the lateral leg that a short test might exhibit no fluid production at all.

Although the results are easy to explain, no rock data are available to prove that this explanation is correct. Because neither possible path to success in this type of opportunity has developed in any of the attempts that have been made, the risk of working in this category is now considered to be so high that future work cannot be justified.

HETEROGENEOUS RESERVOIRS

The heterogeneous-reservoir type of opportunity comes about as a direct result of the complex diagenetic history to which these reefs have been subjected. The opportunities are difficult to identify, very difficult to quantify, and exceedingly difficult to justify. Of the 20 laterals of this type included in the data set, five were clear economic successes, 11 were uneconomic but produced additional volumes, and four were absolute failures.

In this type of opportunity, a factor that distinguishes horizontal-development drilling from traditional-development drilling is most obvious: Dry holes are relatively rare. The 55% of wells in this program that did not succeed economically still contribute to cash flow and do not detract substantially from the successes. They range from losses that approach those expected of a dry hole to the marginally economic. As a whole, the program clearly is profitable economically (Figure 13).

SUCCESSFUL HETEROGENEOUS RESERVOIR

Cleon 12 was a one-well field for approximately 20 years. The Hoekwater 1-12 had been drilled on the western flank of the field (Figure 14) and encountered, low in the oil column, porosity of as much as 10% (Figure 15). After about two years of allowable production, the rate of both oil and gas production declined rapidly, reaching a rate of about 20 BOPD. At that point, the decline slowed significantly; in 1994, the rate was about 12 BOPD. A volumetric calculation, based on the pay seen in the Hoekwater well and the mapped field size, indicated that the field was being depleted effectively by the one well. Similarly, the measured pressure decline was such that a material-balance calculation also showed effective depletion from the one well. Two curious facts were the very slow production decline in the Hoekwater and that pro-

FIGURE 13. A notional comparison between traditional-development drilling economics and horizontal-development drilling economics, based on experience from the Northern Michigan Reef Trend.

Notional comparison between traditional-development drilling-program economics and horizontal-development drilling-program economics*

Evaluation based on common success/failure criteria

	Number of wells	Cost per well	Total cost	Present value return per well	Present value profit per well	Total profit
Successful wells	5	$500,000	$2,500,000	$2,000,000	$1,500,000	$7,500,000
Dry holes	15	$500,000	$7,500,000	$0	-$500,000	-$7,500,000
Project profitability			$10,000,000			$0

Percent present value profit	0.00

Evaluation based on economic/uneconomic success/failure criteria

	Number of wells	Cost per well	Total cost	Present value return per well	Present value profit per well	Total profit
Economic wells	5	$500,000	$2,500,000	$2,000,000	$1,500,000	$7,500,000
Uneconomic wells	11	$500,000	$5,500,000	$300,000	-$200,000	-$2,200,000
Dry holes	4	$500,000	$2,000,000	$0	-$500,000	-$2,000,000
Project profitability			$10,000,000			$3,000,000

Percent present value profit	33.00

*Note: This notional example does not provide actual project figures but is representative of the level of return from the heterogenous program.

duction from this field was much less on a barrel-equivalent-per-acre basis than for fields surrounding Cleon 12.

Based on these two facts, the traditional-development well Wexford 2-7 was drilled to the eastern flank of the field in 1994. It encountered the expected gross section, but found less than 3 m (10 ft) of pay, with porosity of only 5%. Because this was poorer than the first well in the field, the volumetric oil-in-place assessment declined, suggesting that recovery from the one well was even better than had been thought. Consequently, the Wexford 2-7 well was abandoned. Shortly after drilling this dry hole, however, the horizontal-drilling program began, and poorly connected reservoirs were successfully contacted by drilling laterally. Because of this experience, the Wexford 2-7 was reentered and drilled vertically into the center of the reef and then horizontally to the reef edge at the vertical position where pay had been found previously. Hundreds of feet of oil-saturated reservoir with porosity of 5–10% were encountered, and the well was completed successfully. The field-production plot (Figure 16) shows that the second well in the field had initial production similar to that of the first well, and the Wexford is still producing at rates that exceed those of the Hoekwater.

UNSUCCESSFUL HETEROGENEOUS RESERVOIR

One of the many possible examples of the unsuccessful heterogeneous-reservoir category is Otsego Lake 35 field. This is a four-well field (Figure 17) that had three

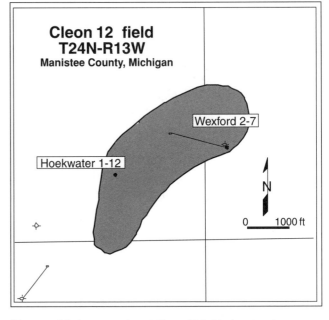

FIGURE 14. Location plat of Cleon 12 field, showing the two wells that produce from this reservoir, the Hoekwater 1-12, a vertical well, and the Wexford 2-7, a horizontal well.

flank oil producers and a single crestal well in which the only porosity development was in the gas cap (Figure 18). The crestal well, Otsego Lake 1-35, was not being used as a depletion point; thus, it appeared to be an ideal location from which to explore for undrained pockets of porosity

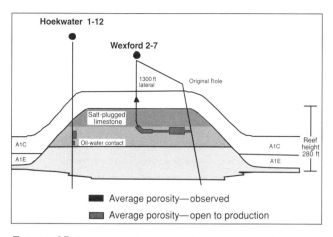

FIGURE 15. Diagrammatic cross section across the Cleon 12 reef, showing the position of the horizontal leg in the Wexford 2-7 well in the most porous part of the reef.

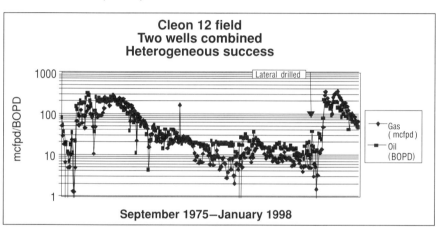

FIGURE 16. Production curve from Cleon 12 field, showing the impact that the Wexford 2-7 well made on field production.

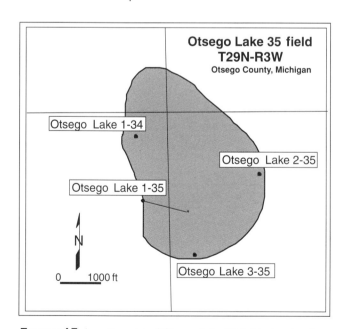

FIGURE 17. Location plat of Otsego Lake 35 field, showing the position of the four wells and the location of the horizontal leg drilled in the Otsego Lake 1-35 well.

in the field. The well was reentered and extended laterally toward the flank to a position midway between two of the earlier producers. After drilling several hundred feet of porosity in the 2–3% range, numerous 1.5–6-m (5–20-ft) stringers of reservoir with porosity of as much as 20% were encountered. Moreover, pressure was encountered that was more than 2000 psi higher than that measured at that time in producers.

Despite the understandable excitement generated by these drilling results, the new lateral never produced more than 10 BOPD and has had a negligible impact on field production. In fact, the remaining wells in the field have shown further declines in oil production, as shown on the production plot (Figure 19).

PURE GRAVITY DRAINAGE

One of the early concepts to be tested was the creation of a horizontal drainhole placed very low in the oil column of an abandoned and very pressure-depleted field (<200 psi). It was thought that gravity might be sufficient to fill the drainhole at rates that could be pumped out economically. Two attempts were made, and both experienced severe drilling problems. The drilling costs greatly exceeded the predrilling estimate. Abundant drilling mud was lost, and no oil was ever produced. Because of these poor results, no additional efforts have been made to pursue this type of opportunity.

The data available indicate that no incremental oil has been recovered from reefs that had remaining pressure less than 600 psi (Figure 20). This figure is currently used as a pressure cutoff in identifying new opportunities.

IMPROVED KH

One of the most widely described uses of horizontal drilling is simply to increase the flow rate into the wellbore by contacting more reservoir and increasing the amount of permeable rock that will contribute to flow. Not only is production accelerated, but also incremental production may occur if the well's economic production limit is deferred. As indicated in the summary chart (Figure 5), this has been applied successfully in the NRT. Although it has been successful, it is not widely applicable because few reefs in the trend exhibit the kind of homogeneity required of opportunities in this category.

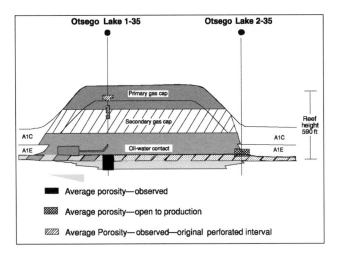

FIGURE 18. Diagrammatic cross section of the Otsego Lake 35 field, showing the position of the single horizontal leg relative to the perforated interval in the well producing on the opposite side of the field.

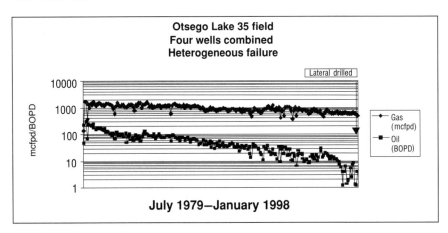

FIGURE 19. Production curve for the combined Otsego Lake 35 field. Despite encouragement while drilling, the horizontal well had virtually no impact on production when completed.

DRILLING ISSUES

In the discussion of the various opportunity types tested in the NRT, several mechanical issues have been mentioned. These issues can play a critical role in the success of a program such as the one described here.

The first major drilling issue to be addressed is the optimum radius of curvature that should be used to drill wells in this setting. Several considerations were included in the final decision to focus on two types of horizontal well, including various regulatory and mechanical issues. Ultimately, a short-radius design of 26 m (85 ft) and a medium-radius design of 91 m (300 ft) have proved applicable to most situations encountered. One of the strongest arguments for using these two designs is that the technology has been developed such that it is possible repeatedly to drill horizontal wells with few problems by using them. Of the two, the short-radius design is of lower cost when re-entering wells, and the medium-radius design is generally preferred for new wells. For wells drilled from surface, the costs of the two designs are comparable, but in the medium-radius design, the curve is cased and the pump can be lowered to a position closer to the depth at which oil is flowing into the well.

Given the state of pressure depletion of most of these reefs, the NRT setting seemed ideal for the use of underbalanced drilling. A nitrogen-based system, tried several times, led

FIGURE 20. Plot of additions to ultimate recovery (ATU) versus bottom-hole pressure measured in the NRT reservoir. Although no significant relationship was found, it is useful to observe that no reserves were added to fields in which the pressure had declined below 600 psi. mbeq = thousand barrels of oil equivalent.

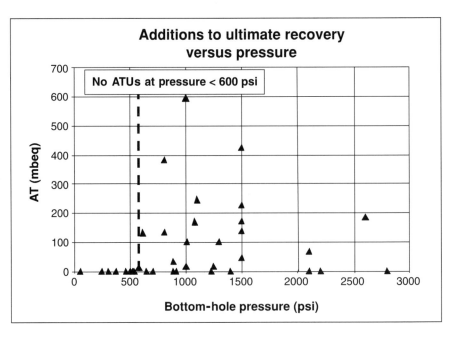

to an increase in cost of 13–20%. Unfortunately, it has not been possible to maintain an underbalanced condition through the drilling and completion processes. As a result of this and other factors, there is no conclusive evidence that production is improved sufficiently to justify the increased cost. Therefore, underbalanced drilling technology has been set aside for the present.

COMPLETION AND PRODUCING ISSUES

Clearly, considerable damage is caused by drilling a horizontal leg, including high fluid loss to the formation, poor cuttings removal, and cuttings so fine that the wellbore can be glazed. This results in significant skin effect, which may be impossible to reduce. Although perhaps not critical to success in the very best opportunities, it is believed to be an important factor for most opportunities. The effectiveness of acidizing has been increased by use of an open-hole packer system to assure that acid reaches the targeted section, but it does not always yield the expected results.

Because short-radius wells have an open hole at least 26 m (85 ft) above the target level, they are prone to early gas coning. A tubing-and-inflatable-packer system has sometimes been effective in delaying coning, but the packer failure rate has been high.

Beam pumps are used exclusively in the NRT, and significant efficiency is lost because the pump cannot be placed closer than 26 m (85 ft) from the lateral in short-radius wells. In medium-radius wells, it is possible to lower the pump to the point where the angle reaches 70°, which is about 14 m (45 ft) above the lateral. Although results are better, it still causes a serious loss of efficiency. In some short-radius wells, a retrievable whipstock was used so that a sump could be drilled, into which the pump could be lowered nearly to the level of the horizontal leg. Although this may improve efficiency, it seriously limits the ability to stimulate the lateral in the future.

SUMMARY

1) The mid-1990s NRT horizontal-drilling program has been a technical and economic success that resulted largely from the reevaluation of a 1987 lateral deemed unsuccessful at the time.

2) Persistence by both technical staff and management was required to overcome early technical problems and misconceptions.

3) An initial wide-ranging concept list was quickly winnowed down to two types of opportunities, which have been pursued aggressively. A significant factor in the speed with which the opportunity could be captured was the existence of a multidisciplinary subsurface team that was already working in the area.

4) In pinnacle reefs, horizontal wells are very different from vertical wells. This statement applies equally to identifying, justifying, drilling, completing, producing, and even conducting a look-back on a program. For such a production-enhancement project to succeed, it is not enough for geoscientists to understand how it could and should work. Management, operations, and other technical staff members must share in this understanding.

REFERENCES CITED

Caughlin, W. G., F. J. Lucia, and N. L. McIver, 1976, The detection and development of Silurian reefs in northern Michigan: Geophysics, v. 41, no. 4, p. 646–658.

Huh, J. M., L. I. Briggs, and D. Gill, 1977, Depositional environments of pinnacle reefs, Niagara and Salina groups, Northern Shelf, Michigan Basin: AAPG, Studies in Geology No. 5, p. 1–20.

Mesolella, K. J., J. D. Robinson, L. M. McCormick, and A. R. Ormiston, 1974, Cyclic deposition of Silurian carbonates and evaporites in Michigan Basin: AAPG Bulletin, v. 58, p. 34–62.

Pieters, D. A., and L. A. Pearce, 1996, An application utilizing horizontal re-entries versus waterflooding for depleting a mid-life Niagaran reef: Presented at the Society of Petroleum Engineers 71st Annual Technical Conference and Exhibition, Denver, Colorado, October 6–9, 1996, SPE 36756, p. 753–767.

Tinker, G. E., 1982, Design and operating factors that affect waterflood performance in Michigan: Presented at the Society of Petroleum Engineers 57th Annual Technical Conference and Exhibition, New Orleans, Louisiana, September 26–29, 1982, SPE 11134, 11 p.

Tinker, G. E., P. F. Barnes, E. E. Olson, and M. P. Wright, 1983, Geochemical aspects of Michigan waterfloods: Presented at the Society of Petroleum Engineers 58th Annual Technical Conference and Exhibition, San Francisco, California, October 5–8, 1983, SPE 12208, 12 p.

Bhattacharya, S., P. M. Gerlach, and T. R. Carr, 2003, Cost-effective techniques for the independent producer to identify candidate reservoirs for horizontal drilling in mature oil and gas fields, *in* T. R. Carr, E. P. Mason, and C. T. Feazel, eds., Horizontal wells: Focus on the reservoir: AAPG Methods in Exploration No. 14, p. 205–223.

14

Cost-effective Techniques for the Independent Producer to Identify Candidate Reservoirs for Horizontal Drilling in Mature Oil and Gas Fields

Saibal Bhattacharya

Kansas Geological Survey, University of Kansas
Lawrence, Kansas, U.S.A.

Timothy R. Carr

Kansas Geological Survey, University of Kansas
Lawrence, Kansas, U.S.A.

Paul M. Gerlach[1]

Kansas Geological Survey, University of Kansas
Lawrence, Kansas, U.S.A.

ABSTRACT

Horizontal wells have exploited successfully the remaining oil potential in mature reservoirs around the world. Because a typical horizontal well costs 1.3 to 4 times that of a vertical well, it must produce significantly greater volumes of oil to be considered an economic success. Previous studies have concluded that poor selection of target reservoirs has been the principal cause of failure of horizontal wells. Many mature fields in the Midcontinent of the United States have significant volumes of residual reserves, and vertical wells have proved to be uneconomic for producing these unswept assets. Small independent producers with limited financial and technological resources operate most of these fields. In Kansas, few horizontal infill wells have been drilled, and results have been mixed. Operator concerns for an appropriate economic return and the difficulty in cost-effectively identifying candidate reservoirs have restricted application of horizontal-drilling technology in many mature production areas of the Midcontinent.

Recent declines in cost factors have brought horizontal-drilling technology within the economic reach of small independent producers, but they have been constrained by the lack of low-cost tools and methodology to screen, evaluate, and target horizontal wells to produce incremental reserves in mature areas.

We present several low-cost approaches that can be used to evaluate candidate reservoirs for potential horizontal-well applications. These cost-effective, efficient screening techniques apply at the field scale, lease level, and well level, and enable the small, indepen-

[1]Present affiliation: Charter Development Inc., Wichita, Kansas, U.S.A.

dent producer to identify candidate reservoirs and predict the performance of horizontal-well applications. Field examples have been used to demonstrate the application of each technique. The demonstrated tools use easily available, standard spreadsheet and mapping packages to analyze production data, map geologic data, integrate and compare geologic and production data, conduct detailed petrophysical analyses, carry out field- and lease-level volumetric analyses, and conduct material-balance calculations. The methodology that the independent operator might follow to identify prospective areas in a production region or field can include any combination of these tools. This paper describes the use of PC-based freeware simulators to history-match well and field production, map residual reserves on a field scale, and predict performance of targeted horizontal infill wells. This process of identifying candidate reservoirs or leases and evaluating their productive potential for horizontal infill drilling will enable independent producers to study the viability of horizontal applications before drilling, and thereby help them select targets appropriately.

INTRODUCTION

Horizontal-well technology has been applied successfully to exploit different types of reservoirs throughout the world. The 1980s were the developing years for horizontal drilling, with the technology maturing and gaining acceptance through the 1990s. Many mature fields with significant volumes of remaining reserves are present in the continental United States. Mature fields, as a result of their history of development and production, may have a bank of geophysical and petrophysical log and core data, and production history and test results. These data are invaluable in developing reservoir models, which are used to simulate the viability of horizontal-well application. Independent producers with limited financial and technical resources operate many of these older oil and gas fields.

Although advances in drilling technology have made horizontal-well applications economical for independent producers, one of the principal causes of failure of horizontal wells is poor evaluation and selection of targets (Coffin, 1993; Joshi and Ding, 1996). Typically, a horizontal well costs 1.4 to 3 times more than a vertical well (Joshi, 1991). Lacy et al. (1992) suggest, as a rule of thumb, that to be an economic success, a horizontal well should recover volumes from two to three times that of a vertical well. This makes identification of reservoirs that are viable candidates for horizontal drilling of crucial importance, especially for independent producers with limited resources.

This paper highlights cost-effective tools that can be employed to select prospects for horizontal drilling. Prospective areas are selected by applying one or a combination of different tools, such as (1) production data analysis, (2) geologic mapping, (3) integration of geologic and production data, (4) field-level and lease-level volumetric calculations, (5) detailed lease-level petrophysical analyses, and (6) reservoir simulation. Each method is described and illustrated with a field example from a mature Kansas oil and gas province, where most operators are independent producers.

BACKGROUND

Proper application of horizontal drilling can revive the productive potential of mature fields by mobilizing reserves that cannot be drained economically by vertical wells. The volume and scale of residual hydrocarbons left in mature basins is illustrated by the Welch-Bornholdt-Wherry fields (Figure 1) in Rice County, Kansas. These fields, discovered in 1964, produce from a stratigraphic trap (Osagian, Mississippian). About 1200 vertical wells have been drilled in this area, and the cumulative production as of 1997 was about 60 million bbl of oil. Average values of porosity, thickness, and fluid saturation for the producing zone were obtained from the petrophysical logs of type wells selected as representative in each quarter section of the fields. These values were used to map initial oil in place for each quarter section. Lease-production data were used to calculate the total volume of oil produced per quarter section and thereby map the remaining volume of oil in these fields. Figure 1 shows areas with significant volumes (about 7 million bbl per quarter-section) of residual oil. The limited drainage potential of vertical wells, coupled with reservoir heterogeneity, results in mature basins with substantial unswept hydrocarbons.

The Oppy South field, located in Hodgeman County, Kansas, was discovered in 1962 and developed with 12 vertical wells. Cumulative production as of 1997 was 800

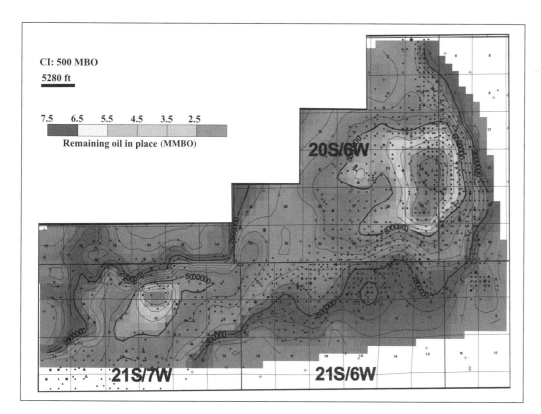

MBO, and daily rates had declined to 300 BOPD. A horizontal infill well was drilled in 1997 to rejuvenate the field (Figure 2). The economic impact of this well was dramatic, spiking the field production rate to 5000 BOPD (Figure 3). The advantage of a horizontal well is that it can drain a large reservoir contact area if optimally directed, thus minimizing the number of infill wells required to effectively redevelop a mature area. In addition, the ability of horizontal wells to produce large fluid volumes under controlled drawdown accelerates production. These wells not only produce at a higher rate than vertical wells but also enhance the producible reserve volume by their extended reach.

Horizontal wells have been applied successfully to exploit thin-bedded and compartmentalized reservoirs, to produce attic oil in fractured reservoir systems, to produce from reservoirs with water or gas coning problems and in low-permeability gas reservoirs. They also are used for enhanced oil-recovery (EOR) operations as injectors and producers and in steam-assisted gravity drainage (SAGD) thermal-recovery operations. Figure 4 highlights different producing zones in Kansas where horizontal technology could be applied; Figure 5 shows the status of horizontal-drilling activity in the state as of 1999. As of 2002, only 22 horizontal wells had been drilled in Kansas.

Small, independent oil and gas producers operate the majority of producing wells in Kansas. Almost 90% of these producers employ fewer than 20 employees (Carr et al., 1998) and generally have no access to advanced commercial technologies that would allow them to screen production areas, identify candidate locations, and study the viability of infill horizontal wells. This and the higher drilling costs of horizontal wells are two major reasons

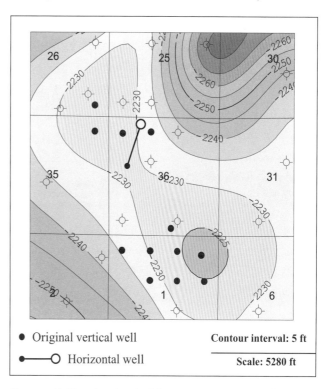

FIGURE 2. Map showing the Mississippian structure in Oppy South field, Hodgeman County, Kansas, and the location of a horizontal well drilled in the field.

FIGURE 3. Impact of a horizontal well on the production of the Oppy South field, Hodgeman County, Kansas.

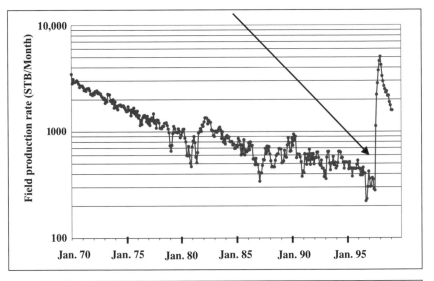

FIGURE 4. Map showing locations of fields in Kansas that produce from zones that have potential for horizontal-well applications.

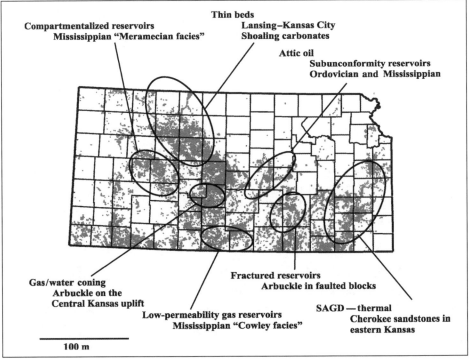

FIGURE 5. Horizontal-drilling activity in Kansas as of 1999.

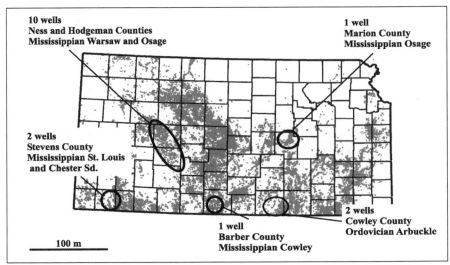

why the horizontal-drilling potential of Kansas has not been exploited. It is the intent of this paper to present a list of cost-effective techniques that independent producers can use to identify candidates for horizontal-well applications.

COST-EFFECTIVE TOOLS TO IDENTIFY CANDIDATE RESERVOIRS FOR HORIZONTAL DRILLING

PRODUCTION DATA ANALYSIS

Fields in a mature production area, such as Kansas, have a long history of production, and production volume data is commonly available. High vertical permeability in aquifer-driven reservoirs may cause a rapid decline in oil cut, which results in poor sweep efficiency. A plot of the production data of a well such as Ritchie No. 1B Moore, Schaben field, Ness County, Kansas (Figure 6), clearly shows a steep decline in oil cut. The operator attempted to contain the water cut by squeeze-cementing the bottom of the perforations, but after a temporary surge in oil rate, the oil cut declined again. Well-production profiles such as these can be used to identify areas with inefficient horizontal sweeps that may have significant unswept reserves and can be considered for horizontal-well application.

Production data can also be used to compare the distribution of initial production (IP) rates in a field with a cumulative production map. Figure 7 maps the cumulative lease production, from initial production in 1924 to 1996, of an area in the Welch-Bornholdt-Wherry fields. The well locations shown on the cumulative lease-production map are those of the first wells drilled in each lease. High IP rates are the result of pay thicknesses of as much as 24 m (80 ft). A comparison of the two maps reveals that the IP peaks coincide with that of cumulative production highs only in certain pockets of the field. The cumulative production in areas such as that circled "A" in Figure 7 is not proportionate to the high IP rate. Such comparisons may highlight areas where vertical wells have been ineffective in draining reserves, and are therefore indicative of significant remaining reserves.

The ability of a horizontal well to drain a significantly larger reservoir volume makes it an ideal application to recover reserves left as a result of excessive well spacing in a field. Figure 8 is a structure map of the producing Mississippian zone in the Aldrich field, Ness County, Kansas. This field was discovered in 1929, and by 1973, it had produced 1.044 MMBO from 15 wells. Field production, however, had declined to less than 400 bbl of oil per month. Clearly, the original well spacing was insufficient to drain the field, and eight vertical wells were drilled as part of the infill-drilling program. Figure 9 shows the effect of the infill wells on monthly field production. By mid-1997, field production was about 900 bbl of oil per month, and an additional 553 MBO had been recovered since the onset of infill drilling.

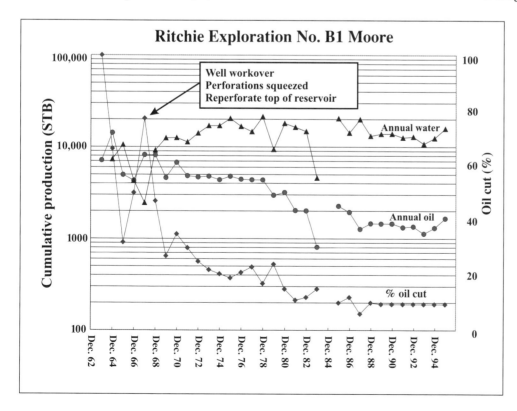

FIGURE 6. Production profile of Ritchie Exploration No. B1 Moore well in Schaben field, Ness County, Kansas.

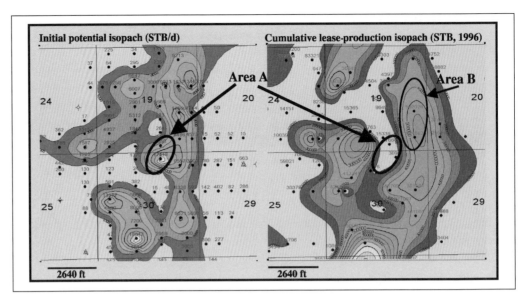

FIGURE 7. Isopach maps of initial production and cumulative lease production as of 1996 of an area in the Welch-Bornholdt-Wherry fields, Rice County, Kansas.

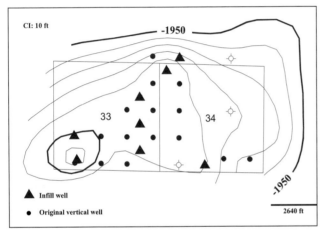

FIGURE 8. Structure map of Mississippian top and well locations in the Aldrich field, Ness County, Kansas.

The Aldrich field is a typical example of a mature field left with significant recoverable reserves because of inadequate drainage. Such fields can be drained efficiently with far fewer horizontal wells than vertical wells, which is of economic benefit for independent producers. An effective method of identifying candidate fields for horizontal infill drilling is to compare field production before and after infill drilling in analogous reservoirs with similar well spacing.

MAPPING GEOLOGIC DATA

The structure map on the Mississippian zone in an area of the Welch-Bornholdt-Wherry fields (Figure 10) does not emphasize subdued features. The first derivative of the structure map highlights the change in the local dip of the formation, which helps to determine the location of folds and faults in the subsurface. The circled area B delineates the axis of the fold. Structural folding may cre-

ate an associated fracture network, which can enhance flow permeability. Such phenomena may be the cause of high cumulative production volumes (as shown in Figure 7) in area B. The first-derivative map also shows that the axis of the structural folding stops short of area A. Significantly lower cumulative-production volumes in area A may have resulted from absence of fold-induced fractures. Areas such as this can be spotted on first-derivative structure maps and analyzed along with the production history, particularly for mature production areas, to evaluate potential of reserves left unswept by vertical wells. Such areas may be well suited for a horizontal well, with its extended drainage.

An important application of horizontal drilling is the recovery of attic reserves. Figure 11 compares the structure map of the Simpson sand in Hollow-Nikkel field, Harvey County, Kansas, with the first-derivative map of the same structure. The attic axis is clearly defined on the first-derivative map and serves as the target location for an exploratory horizontal well to confirm the presence of the attic. Depending on the results of the exploratory well, the same wellbore can be used to drill a second lateral in a southwesterly direction to delineate the extent of the axis.

Murray (1965) postulated that maximum fracture intensity could be predicted by mapping the rate of change of dip or the radius of the curvature of a structure. Clay models and analytical studies demonstrate that fracture porosity and permeability are highest where the curvature of the fold is highest. Because of structural complexity, however, first-derivative maps are unable to delineate fold axis in formations such as the Niobrara Shale in northwestern Colorado (Stright and Robertson, 1993). Stright and Robertson suggest using the maximum second-derivative values to determine the curvature of structural surfaces in complex situations such as the Niobrara structures. Cumulative production can be overlaid on the second-derivative structure map to establish the minimum amount of curvature needed to generate reservoir-quality fractures. These workers determined that fractured reservoirs of the Niobrara have a recoverable potential ranging from 1200 to 1400 bbl of oil per acre if the value of the second derivative of the structural surface exceeds $10^{-4.5}$ ft^{-1}. Structural discontinuities such as faults also will show on second-derivative maps, and will help one to understand the fault-generated fracture system. Detailed mapping procedures to evaluate fracture-enhanced production, along with field applications, are discussed in Stright and Robertson (1993). Candidate reservoirs with a productive fracture system can be exploited with horizontal drilling because of a higher probability of intersecting a fracture network than with vertical wells.

MAPPING PRODUCTION AND GEOLOGIC DATA

Horizontal wells are effective tools to drain reserves trapped in thin pays that may be difficult to exploit with vertical wells because of limited sweep efficiency. The Welch-Bornholdt-Wherry fields produced from an updip stratigraphic trap. On its northwestern side, the reservoir layer subcrops against the overlying Pennsylvanian sediments, whereas the oil-water contact truncates the trap on the southern side. Figure 12 is an overlay of the cumulative oil production per quarter section as of 1997 on the pay isopach of the field. The thick black contour lines show the boundaries of 100 MBO of cumulative production. The red contour lines mark the border of pay greater than 9 m (30 ft). Vertical wells in this field have pay cut-

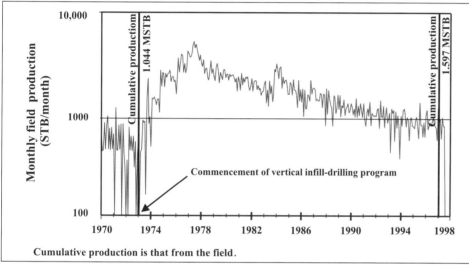

FIGURE 9. Effect of infill drilling on monthly field production in Aldrich field.

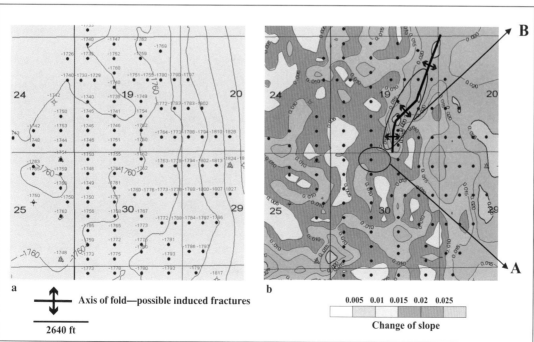

FIGURE 10. Comparison of the structure map (a) of the Mississippian zone in an area in the Welch-Bornholdt-Wherry fields with the first derivative of the structure (b), which identifies the fold axis.

FIGURE 11.
Structure map (a) and first derivative (b) of the structure of Simpson sand identifying the attic axis, in the Hollow-Nickel field, Harvey County, Kansas.

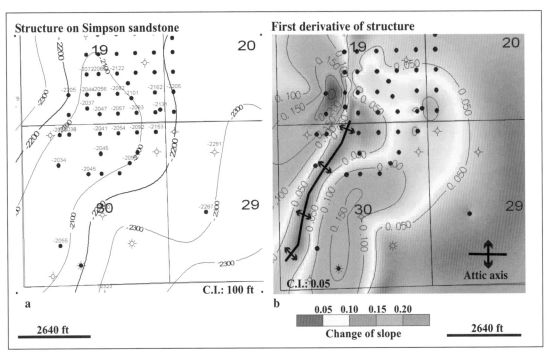

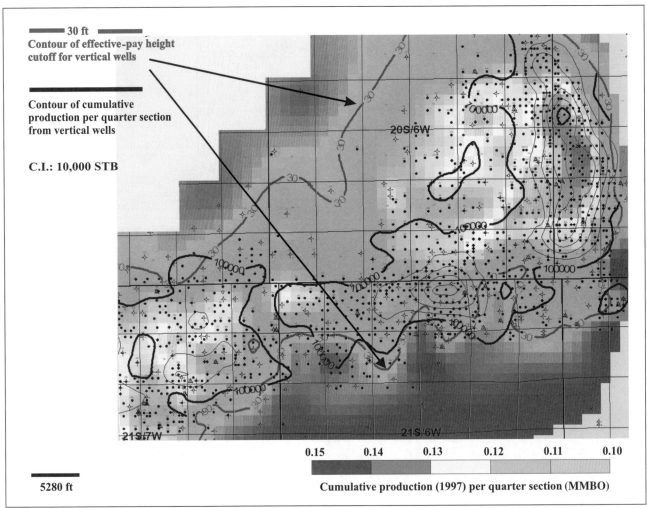

FIGURE 12. Map comparing the cumulative production of oil, as of 1997, per quarter section, with distribution of effective-pay-height cutoff in the Welch-Bornholdt-Wherry fields.

off of 9 m (30 ft), and they are considered economic if they produce more than 100 MBO. In the southern part of the field, the 9-m (30-ft) pay contour closely follows that of the economic production limit, whereas in the northwestern part of the field, the cutoff contours of pay and production are significantly farther apart. The separation between pay and production contours indicates that vertical wells in certain areas have proved uneconomic, even when pay thickness is greater than 9 m (30 ft). Reservoir compartmentalization, together with solution-gas drive, has contributed to poor sweep in the northwestern part of the field. Horizontal wells have been known to drain compartmentalized reservoirs, and they produce at economic rates under lower drawdown than vertical wells. Areas such as the northwestern part of the Welch-Bornholdt-Wherry fields are therefore candidates for horizontal-well application.

MAPPING FIELD-LEVEL VOLUMETRICS

In Canada, horizontal drilling has been applied in conventional reservoirs to accelerate recovery and increase reserves (Faquharson et al., 1992). In the Williston Basin of southeastern Saskatchewan, horizontal drilling has been successful in naturally fractured and stratified Mississippian-age carbonate reservoirs. Horizontal permeabilities in the limestone reservoirs of the Mississippian Midale and Frobisher are low, whereas vertical permeabilities are enhanced by vertical fractures related to dissolution

and collapse of deeper Devonian Prairie salt beds. Although vertical wells are uneconomic in these reservoirs, horizontal wellbores have reached the vertical fractures in proportion to the length of the well.

The Mississippian carbonate (Meramecian facies) reservoirs in south-central Kansas also are naturally fractured and stratified, and application of proper screening methods will help to locate prospective candidates for horizontal drilling. Field-level volumetric calculations are a quick, effective way to start the screening process. The first step is to calculate the original volume of reserves in place. Parameters such as pay thickness, average porosity, and initial saturations are obtained from petrophysical logs that were run in the well during drilling. The Welch-Bornholdt-Wherry fields produce from the Mississippian Osage reservoir, extend over 30 sections, and average about 40 wells per section. Although the extent of the fields and the number of wells may make this task of log interpretation seem daunting, a huge area can be screened effectively by selecting a type well in each quarter-section and obtaining the necessary volumetric parameters from the petrophysical logs of that wells. Figure 13 shows the original oil in place (OOIP) per quarter section (using a formation volume factor of 1.04 reservoir barrels per stock-tank barrel) in the Welch-Bornholdt-Wherry fields. Figure 14 shows the distribution of recovery efficiency in the field; it is useful to focus attention on areas with low recoveries because these areas should be followed up with

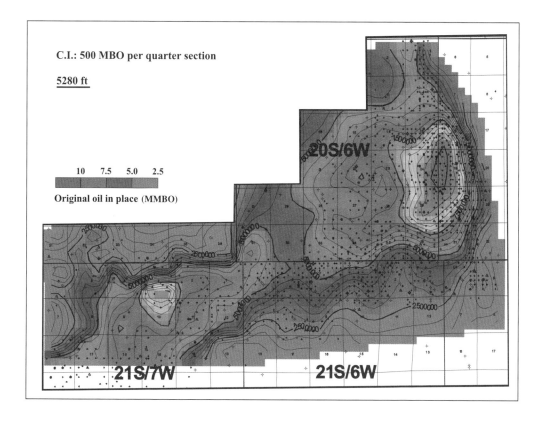

FIGURE 13. Map showing distribution of original oil in place in the Welch-Bornholdt-Wherry fields. Volumetric calculations were carried out using a type well to represent a quarter section.

a more detailed analysis to study the viability of horizontal-well application.

DETAILED PETROPHYSICAL ANALYSES FOR LEASE-LEVEL VOLUMETRICS

Although application of one or a combination of tools described in this paper will result in identification of areas with significant remaining potential, these prospective target areas may extend across leases and may need a more detailed analysis to evaluate the prospects of reserve recovery by horizontal drilling. Cost-effective tools that can be used in lease-level analysis include the Super-Pickett crossplot, the use of pore-size distribution (from capillary pressure data), and nuclear magnetic resonance (NMR) data on core plugs.

Doveton (1994) and Doveton et al. (1996) have described the fundamentals of the Super-Pickett crossplot in detail. Applications of Super-Pickett analysis for integrated petrophysical analysis have been studied by several authors (Watney et al., 1999; Guy et al., 1996; Guy et al., 1997; Bhattacharya et al., 1999). The Super-Pickett crossplot uses Archie equations to plot data from resistivity and porosity logs on log-log axes, with all plotting and calculations carried out in a spreadsheet. Pattern-recognition techniques are used to integrate information about pore character and lithology with porosity, saturation, BVW (bulk-volume water), and BVW_i (bound-volume water). Permeability contours based on empirical correla-

tions between porosity and permeability, such as Timur's equation, are drawn on the Super-Pickett plot to analyze sandstone reservoirs. Various cutoff criteria, such as shaliness, fluid saturation, porosity, and permeability, are included in the analysis to identify "pay." The depth of the free-water level (FWL) is used to generate synthetic capillary-pressure curves for different porosity intervals. These synthetic capillary curves are then compared with laboratory-measured capillary-pressure data obtained from rock samples with comparable porosity and taken from the same interval, to identify petrofacies. NMR log data also can be integrated into the Super-Pickett analysis to highlight zones that probably will produce water-free, i.e., zones where the porosity values are significantly higher than the BVW value, and the BVW and BVW_i values are approximately equal.

Figure 15 is a Super-Pickett crossplot of log data from Ritchie No 2P Lyle Schaben well located in Schaben field, Ness County, Kansas. This well produces from a Mississippian carbonate reservoir, and the crossplot shows a clear transition to 100% water saturation. A detailed petrophysical analysis was carried out on the Schaben field wells (Carr et al., 1998; Bhattacharya et al., 2000) to build a reservoir model for the producing horizon. Results from the Super-Pickett analysis revealed that the BVW_i value ranged from 0.09 to 0.11. High BVW_i values, such as those observed in Schaben field, are caused by the presence of microporosity. Initial water-saturation values in

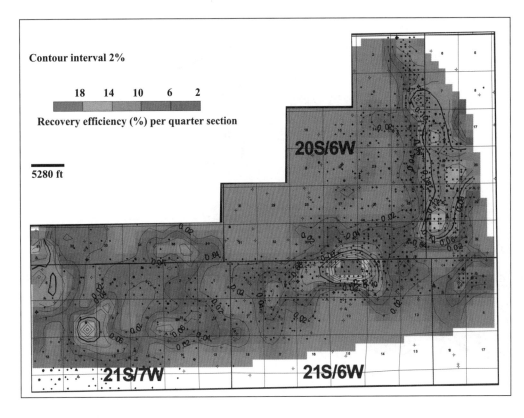

FIGURE 14. Map of recovery efficiency, as of 1997, per quarter section in the Welch-Bornholdt-Wherry fields.

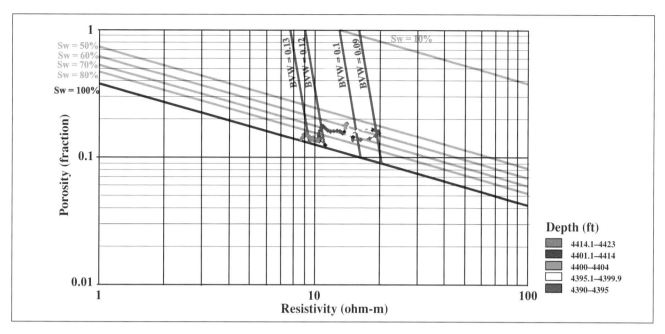

FIGURE 15. Super-Pickett crossplot of Mississippian horizon in Ritchie Exploration No. 2P Lyle-Schaben well in Schaben field, Ness County, Kansas. The Archie constants used are a = 1, m = 2.1, n = 2, and R_w = 0.13. The interval perforated is 4400 to 4404 ft and it produced 85 BOPD and 132 BWPD.

most wells in this field are about 60%, yet most of them have produced water-free for the initial year or two. One possible explanation for this dichotomy is the significant portion of water in the reservoir rock held immobile in the micropores.

A frequency distribution of pore-throat sizes can be generated from capillary-pressure measurements. A qualitative estimate of microporosity and thus its effect on fluid flow can be obtained from this pore-size frequency distribution. Three new wells were drilled and cored as part of the reservoir characterization study of Schaben field. Capillary-pressure measurements were obtained from core plugs taken from the producing horizon of these wells. Figure 16 shows the distribution of micropores (pore throat size ≤5 µm) and macropores (>5 µm) in core plugs taken from the Ritchie Exploration No. 2P Lyle-Schaben well. The pore-size distribution indicates a significant presence of microporosity in the reservoir rock.

Simulation results did not match the initial water-free production phase at many Schaben wells when distributions of log-derived total water saturation were used. Petrophysical logs measure the total volume of water present in the reservoir rock. However, in a reservoir such as the Schaben field, a significant portion of the water is held immobile in the capillary pore spaces (microporosity). NMR measurements provide a quantitative estimate of microporosity and effective macroporosity. A plot of effective macroporosity against total (core) porosity (Figure 17) was generated from NMR measurements on core plugs taken from three wells in the Schaben field.

In this study, it was assumed that the micropores are totally saturated with immobile water, whereas the macropores contained mobile water and hydrocarbons. The effective macroporosity represents the pore volume (i.e., macropore volume) in which fluid flow occurs under reservoir conditions. In the absence of free gas in the reservoir, the oil volume obtained from petrophysical logs was used to calculate the mobile (free) water saturation (Rezaee and Lemon, 1996; Hedges and Moothart, 1996) in the effective macroporosity. Input of effective macroporosity and mobile fluid-saturation distributions

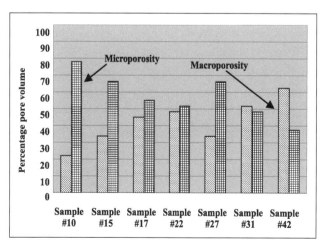

FIGURE 16. Plot comparing the percentage of pore volume occupied by micropore (pore throat ≤5 µm) and macropores (pore throat >5 µm) in core plugs from Ritchie Exploration No. 2P Lyle-Schaben well. The pore print was generated from capillary-pressure data.

(Bhattacharya et al., 2000) resulted in the simulation output to match the initial water-free production period in Schaben field wells. NMR data provide a cost-effective measure of effective pay, macroporosity, and mobile fluid saturations. They relate total water saturation derived from petrophysical logs to the productive potential of a horizon. This detail about pore character and its effects on reservoir fluid flow is essential to characterization studies of reservoirs that are candidates for horizontal drilling.

The petrophysical tools described in this section were used to make a detailed lease-level volumetric analysis on the Wieland and Wieland West fields in Hodgeman County, Kansas. These fields (Figure 18), discovered in

1956, produce from a structural Mississippian Osagian trap. Lease-level volumetric calculations, when compared with cumulative lease production to 1997, revealed that adjacent leases displayed low and often varying recovery efficiencies (Figure 19). The operator of the Wieland West field drilled a commercial horizontal well, No. 1 Antrim-Cossman, between the Cossman and Antrim leases. This well gave a significant boost to the monthly field production of this mature area (Figure 20) and is a good illustration of how detailed lease-level petrophysical studies can result in proper selection of horizontal-well candidates. The success of the horizontal infill well in the Wieland West field, where primary recovery efficiencies ranged from 13% to 15%, indicates that the Wieland field leases are potential targets for horizontal infill applications because of their lower primary-recovery efficiencies (6% to 14%).

COST-EFFECTIVE RESERVOIR SIMULATION

Marginal fields may be exploited efficiently when field-management plans are based on reservoir characterization and simulation studies. In the past, simulation studies required expensive hardware and software, but with the advent of powerful PCs and PC-based software, full-field integrated studies have come within the resource reach of independent producers. In addition, PC-based simulation tools such as BOAST4 and BOAST-VHS, developed by the U.S. Department of Energy (DOE) (U.S. Department of Energy, 1995), have been released as freeware. These simulators have proved to be versatile for full-field studies (Carr et al., 1998; Bhattacharya et al., 2000; Montgomery et al., 2000). They are cost-effective tools

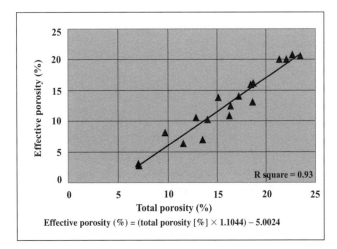

$$\text{Effective porosity (\%)} = (\text{total porosity [\%]}) \times 1.1044 - 5.0024$$

FIGURE 17. Plot of effective versus core porosity constructed from NMR measurements on core plugs obtained from wells in the Schaben field.

FIGURE 18. Pay-zone isopach maps of Wieland and Wieland West fields, Hodgeman County, Kansas.

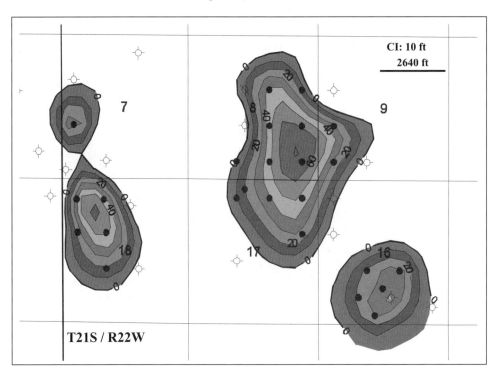

for mapping residual reserves in mature fields and for predicting the performance of horizontal infill wells in candidate reservoirs. Refer to the section titled "PC-based Reservoir Simulation" later in this paper for further discussion of these tools.

The techniques described thus far help to identify leases or areas with geology and recoverable reserve potential suitable for horizontal infill application. Geologic mapping and petrophysical analyses result in the construction of a geomodel, which forms the base for any simulation study. It therefore is prudent to use material-balance calculations to cross-check the robustness of the geomodel before starting reservoir simulation. This makes the exercise of simulation a two-step process—material-balance calculations—followed by field simulation.

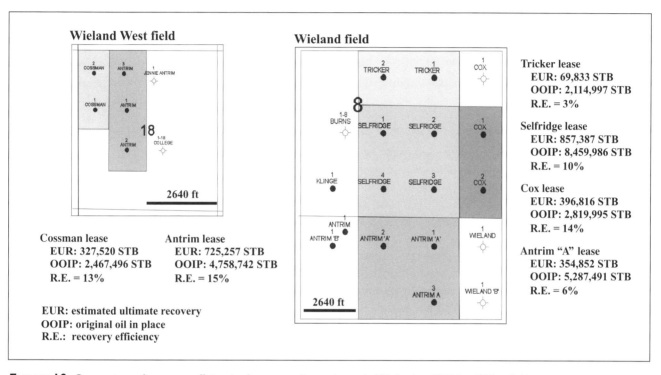

FIGURE 19. Comparison of recovery efficiencies between adjacent leases in Wieland and Wieland West fields.

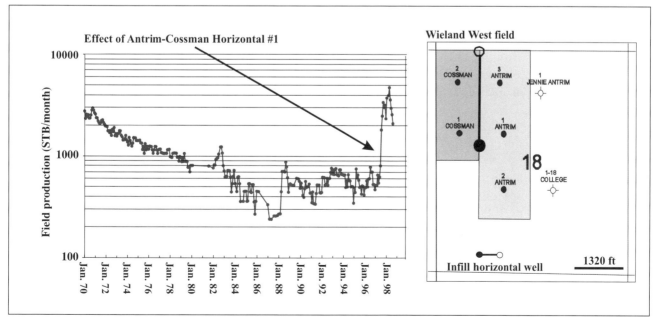

FIGURE 20. Location of infill horizontal well in Wieland West field, and its effect on monthly field production.

SPREADSHEET-BASED MATERIAL-BALANCE CALCULATIONS

Material-balance calculations corroborate the original hydrocarbons in place (OHIP) calculated volumetrically from the geomodel with a geology-independent technique. The inputs required for these calculations include production and pressure histories and PVT (pressure-volume-temperature) parameters of hydrocarbons and water. Material balance calculates the effective OHIP; the difference between volumetric OHIP and effective OHIP is the measure of heterogeneity affecting production performance in the reservoir.

In addition, the material-balance calculation helps to identify the reservoir drive mechanism. For water-driven reservoirs, it helps to define the average aquifer properties and to estimate the volume of influxed water. For reservoirs with gas caps, this technique helps to quantify the volume of the initial gas cap. When pressure and production data are recorded through the life of the field, advanced material-balance calculations can be employed in versatile ways. Such endeavors result in generation of full-field, pseudorelative permeability curves and, for gas-cap driven reservoirs, determination of critical gas saturation and recovery efficiencies at specified abandonment pressures. Dake (1994) has noted that reservoir simulation cannot provide additional clarity when the material-balance calculations show a mismatch with volumetrics. In case of a mismatch, it is prudent to revise the geomodel and its associated petrophysics rather than proceed with the simulation study. The principles of material balance are versatile enough to model different reservoir scenarios, and calculations can be accomplished effectively in a spreadsheet (Dake, 1994).

Material-balance calculations were carried out on the Schaben field. Regular recording of reservoir pressure at each well is necessary for these calculations. However, pressure data through the production life of wells may be unavailable in fields such as Schaben that are operated by independent producers. Fortunately, an estimate of initial reservoir pressure and current operating fluid levels were available for most of the producing wells in the Schaben field. Without a field-pressure history, material-balance calculations cannot be used to confirm volumetric OHIP, but if one assumes that the volumetric OHIP is valid, the technique can be used to confirm the current average reservoir pressure and the reservoir drive mechanism. Identification of reservoir drive mechanism is important because it helps to define the aquifer and to estimate the size of the initial gas cap. Well-production patterns and current fluid levels in the producing wells of Schaben field indicate that the reservoir is supported by strong bottom-water drive. As in most fields, direct measured data of different aquifer parameters such as porosity, permeability, thickness, and rock and fluid compressibilities were not available for Schaben field; they were inferred initially from reservoir properties.

Spreadsheet-based material-balance calculations are suitable for aquifer fitting through trial and error, and they were used for the Schaben field (Carr et al., 1999). Based on the reservoir geomodel, the volumetric estimate of OOIP (original oil in place) for the Schaben field was 37.8 MMSTB. The field has been in production since 1963. DST (drill-stem test) analysis of available data approximated the initial reservoir pressure (P_i) at about 1370 psi. Standard correlations were used to generate PVT profiles for the reservoir fluids, and bubble-point pressure was calculated as 225 psi. The Carter-Tracy formulation was used to calculate the water influx (W_e) from an infinite aquifer. In a spreadsheet, it is easy to change aquifer properties (within geologic and engineering limits) for different average reservoir-pressure distributions until the plot of F/E versus W_e/E appears as a straight line with unit slope (Figure 21). F denotes underground withdrawal of fluids from the reservoir, W_e is reservoir volume of water that influxed from the aquifer, and E represents the sum of the change in volume of oil and dissolved gas and change in volume caused by expansion of connate water and reduction in pore volume. The aquifer properties and reservoir pressure distribution over time were adjusted until the straight line correlation between F/E and W_e/E produced an intercept that translated to an OOIP value lower than but close to that calculated from volumetrics (within 10%). The average reservoir pressure distribution through the production life of the field, obtained under the stated conditions, is shown in Figure 22; it indicates a current average field pressure nearing 800 psi. Available fluid-level data indicate that the majority of

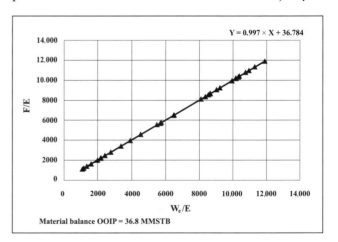

FIGURE 21. Results of material-balance calculations for Schaben field.

wells in the field currently produce against bottom-hole pressures that range from 400 to 1100 psi.

The material-balance study in Schaben field confirmed that the volumetric description of the reservoir-aquifer system, along with the natural bottom-water drive mechanism, can support the reported fluid-production history of the field. This example also demonstrates that the calculated average reservoir pressure lies within the range of current operating fluid-level data. The process of "aquifer-fitting" fine-tuned aquifer parameters, such as height, porosity, permeability, and effective compressibility, as well as reservoir radius. These parameters, along with the reservoir drive mechanism, form essential components of the input file for any reservoir simulation study.

PC-BASED RESERVOIR SIMULATION

Reservoir simulation is the final step in the selection process of a candidate reservoir. This exercise models fluid flow through the reservoir during its producing life and predicts the performance of horizontal wells under different operating scenarios. The results of a simulation study, however, are only as good as the geomodel on which it is based. A combination of cost-effective reservoir simulators, such as the U.S. DOE's BOAST4 and BOAST-VHS, has been applied successfully to carry out full-field studies (Carr et al., 1998; Carr et al., 1999; Bhattacharya et al., 2000). These studies have mapped residual reserves and predicted horizontal infill performance in a mature area such as the Schaben field. They also have demonstrated that the limitations of the pre- and postprocessing tools of these simulators are overcome easily by using

commercially available, inexpensive, user-friendly spreadsheet programs; relational databases; and gridding and mapping packages.

The Schaben field produces from Meramecian and Osagian age (Mississippian) cherty dolostones that lie below the sub-Pennsylvanian unconformity on the western flank of the Central Kansas uplift. The field has been developed on a 40-acre spacing and has been producing since 1963. Data such as core, log, and production history and test results were integrated to develop a 3-D reservoir geomodel. Material-balance calculations then were used to confirm the drive mechanism and to refine aquifer description and properties. Initial simulation of the field was completed on BOAST4, a 3-D, three-phase, isothermal black-oil simulator that uses IMPES (implicit pressure and explicit saturation) solutions to simulate the performance of vertical wells. (This tool can simulate a wide variety of field applications, such as primary depletion [with option to include an aquifer in the model], pressure maintenance by water/gas injection, and performance of waterfloods.) Well-production history of the first 11 years was entered into the simulator, and the permeability distribution was fine-tuned to match the production profile for the next 23 years. After 34 years of production, the oil-saturation map (Figure 23) shows areas of remaining reserves. A saturation-feet map (Figure 24) of the remaining reserves, with a pay-thickness cutoff of 6 m (20 ft) and an oil-saturation cutoff of 40%, was used to design the infill drilling strategy. The boxed area on the saturation-feet map (area A) was selected to compare the potentials of horizontal and vertical infill wells.

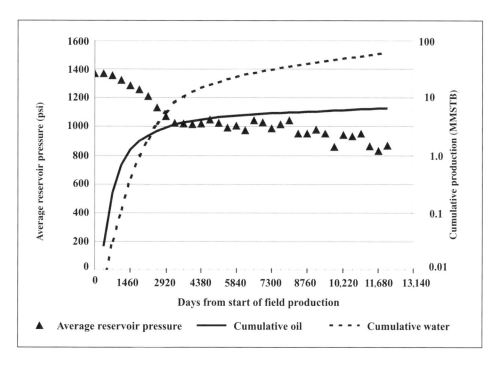

FIGURE 22. Average reservoir pressure profile obtained from material-balance calculations over the producing life of Schaben field.

BOAST-VHS (Chang et al., 1992) is a PC-based, 3-D, three-phase, finite-difference, black-oil simulator that uses IMPES solution techniques to simulate the performance of vertical, slant, or horizontal wells. This tool is recommended for studying problems such as primary depletion, pressure maintenance, and waterflooding. However, because of memory limitations, BOAST-VHS cannot handle more than 810 grid blocks. Thus, in the Schaben simulation project, the entire field was simulated until 1996 in BOAST4. Area A in Figure 25 highlights the portion of the Schaben field that was simulated in BOAST-VHS. Locations of the wells drilled in this area before the onset of the field study are shown by black dots. Red dots denote the locations of the three infill wells drilled by the operator, whose locations were determined on the basis of Figure 24. Of the six original wells, American Warrior No. 2 Witman and Ritchie No. 2D Moore had ceased production by 1996. Remaining fluid saturation and reservoir-pressure distribution data as of January 1997 were obtained for area A from the BOAST4 output and used as input to the BOAST-VHS study. The VHS tool does not have the option to model an aquifer below the reservoir. A bottom-water drive was simulated in the study area by modeling 12 horizontal water-injection wells, spaced uniformly in a layer below the reservoir. Each of these wells was required to inject volumes suffi-

cient to maintain constant pressure in the injection layer, thereby incorporating the effects of a strong bottom-water drive in the reservoir model. Area A was simulated for five years (starting from 1997) in BOAST-VHS to obtain cumulative production from its operating vertical wells. Cumulative production during the same interval of time for area A also was obtained from BOAST4, and the results compared closely with the output from BOAST-VHS. This similarity in production profiles demonstrated that the use of 12 horizontal injectors operating in the layer below the production zone mimicked the effects of a bottom aquifer.

BOAST-VHS simulated two infill development scenarios: (1) area A with three vertical infill wells, along with existing producing wells, and (2) area A with one horizontal infill well, along with existing producing wells. The results of the simulation study are summarized in Figure 26. It shows that at the end of the first year, the three vertical infill wells recovered an additional 53 MSTB, while one horizontal infill well recovered 137 MSTB. Thus, in the first year, the horizontal infill well is expected to produce 250% more than three vertical infill wells, whereas moving an additional 395 MSTB of water. This study also predicts that fluid production from one horizontal infill well during a five-year period will exceed the cumulative production from three vertical infill wells

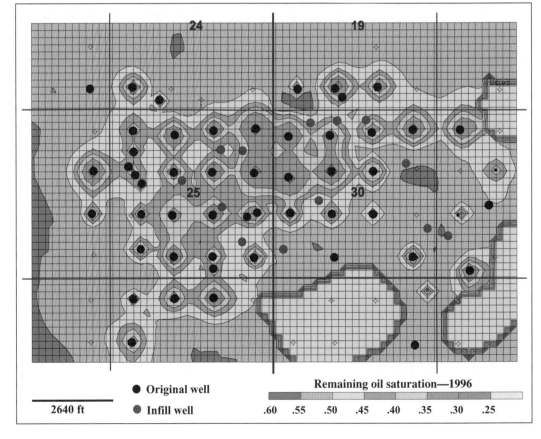

FIGURE 23. Map showing remaining oil saturation in Schaben field, December 1996, developed from results of BOAST4 simulation.

● Original well
● Infill well

2640 ft

Remaining oil saturation—1996

.60 .55 .50 .45 .40 .35 .30 .25

by 103 MSTB of oil and by approximately 2.7 MMSTB of water. In the first year of production, the horizontal infill well appears to deliver half the volume that it will recover ultimately during the five-year period. Such a per-

formance suggests that the operator may be better off to produce the horizontal well for just one year, thus restricting the volume of produced water to 0.6 MMSTB.

In the Schaben field study, various scenarios were

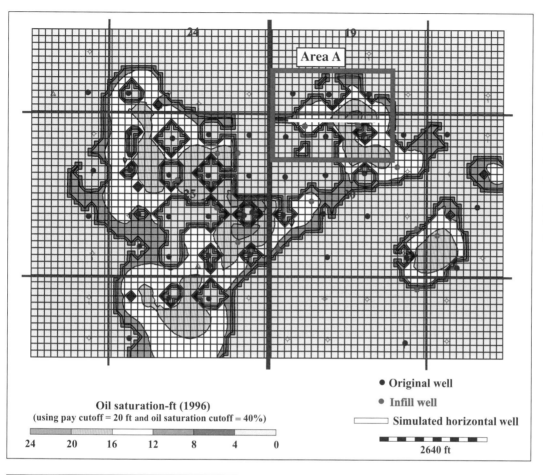

Oil saturation-ft (1996)
(using pay cutoff = 20 ft and oil saturation cutoff = 40%)

24 20 16 12 8 4 0

● Original well
● Infill well
▭ Simulated horizontal well

2640 ft

FIGURE 24. Saturation-feet map of Schaben field, December 1996, developed from results of BOAST4 simulation and using pay cutoff of 6 m (20 ft) and an oil-saturation cutoff of 40%. The boxed area was studied further in BOAST-VHS to compare recoveries from vertical and horizontal infill wells.

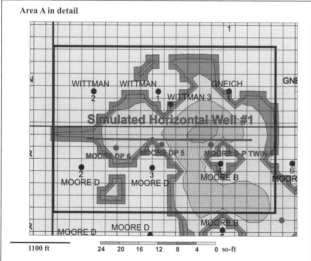

FIGURE 25. Detail map of well locations in area A of Schaben field. The red line shows the location of the simulated horizontal infill well, red dots show the locations of the vertical infill wells, and black dots show the locations of the original wells in this study area.

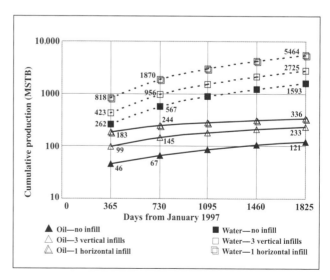

FIGURE 26. Plot comparing the cumulative performance of three vertical infill wells with one horizontal infill well in area A of Schaben field. The infill wells were simulated to start production in January 1997 and produce for five years.

simulated in BOAST-VHS to study (1) the effect of well orientation, (2) producing bottom-hole pressures, and (3) skin factors on the infill production potential. Cost-effective simulation studies reveal information that is critical to evaluating the performance of horizontal drilling at a selected site and to designing efficient production strategies. Production estimates from a horizontal well in a candidate reservoir are important information that allows management personnel to outline a strategy to exploit remaining reserves in mature areas.

CONCLUSIONS

Horizontal drilling has been applied effectively all over the world to exploit reserves under different geologic settings. The technology of horizontal drilling has come of age, but many horizontal wells, although technical successes, are not economical because of poor evaluation and selection of target reservoirs. Mature areas such as the midcontinental United States have significant volumes of unswept hydrocarbons. Independent producers with limited technical and economic resources operate a significant portion of the oil and gas fields in these mature areas. Cost-effective techniques can be used to screen reservoirs in a producing area to identify potential candidates for horizontal drilling.

Mature areas, because of their long production history, may carry an extensive database of geologic and geophysical data, production history, and pressure-test results. Production data such as water-cut, IP, and cumulative production volumes can be mapped over a field to identify areas with inefficient sweep. Analyses of productivity from vertical infill wells reveal whether excessive well spacing is the cause of inefficient sweep. Comparison of a structure map, with its first derivative, may result in highlighting the attic axis or the axis of the fold. Horizontal wells are effective at draining attic oil. Therefore, identified attic axes may be taken up for detailed lease-level analyses to consider their potential for horizontal drilling. Folds may be associated with regional fracture networks, which can be used advantageously to drain hydrocarbons from the reservoir to the horizontal wellbore. Poor sweep by vertical wells in the thin pays of updip stratigraphic traps can be identified by overlaying cumulative production distribution on the pay-zone isopach map. Cumulative production volumes, when mapped over second-derivative maps of structure surfaces, also have proved successful at delineating "fracture fairways." By using a type well per quarter section, regional volumetric calculations can be completed and mapped to highlight areas with low-recovery efficiency. A combination of these techniques quickly shifts the focus of investigation from a large producing basin or region to prospective leases (or small fields).

Application of a cost-effective, state-of-the-art tool, the Super-Pickett crossplot, for petrophysical analyses at well and lease level provides a common platform to integrate petrophysical-log data, lithologic information, pore-size distribution, capillary pressure, and NMR data to identify petrofacies and to predict the productive potential of pay zones. Frequency distribution of pore sizes, derived from capillary-pressure data, have been used to obtain a qualitative measure of microporosity, whereas NMR data from core plugs quantify effective macroporosity and mobile (free) water saturation. Constituent well models, which are constructed from detailed petrophysical analyses, are integrated to build a reservoir geomodel. Spreadsheet-based material-balance calculations verify the geomodel against production, PVT, and pressure history of the reservoir. Finally, field-scale application of cost-effective simulators, such as BOAST4 and BOAST-VHS, can be used to map residual reserves and to compare recovery efficiencies of vertical and horizontal infill wells. An independent producer can use any combination of these tools to select candidate reservoirs and to predict the performance of horizontal infill wells.

REFERENCES CITED

Bhattacharya, S., P. M. Gerlach, T. R. Carr, W. J. Guy, S. Beaty, and E. K. Franseen, 2000, PC-based reservoir characterization and simulation of Schaben field, Ness County, Kansas, in K. S. Johnson, ed., Platform carbonates in the southern Mid-continent: 1996 symposium, Oklahoma Geological Survey Circular 101, p. 171–182.

Bhattacharya, S., W. L. Watney, J. H. Doveton, W. J. Guy, and G. C. Bohling, 1999, From geomodels to engineering models—Opportunities for spreadsheet computing, in T. F. Hentz, ed., Proceedings of 19th Annual Research Conference—Advanced Reservoir Characterization for the 21st Century: Gulf Coast Section, Society for Sedimentary Geology (SEPM) Foundation, p. 179–191.

Carr, T. R., D. W. Green, and G. P. Willhite, 1999, Improved oil recovery in Mississippian carbonate reservoirs of Kansas—near term—class 2: DOE Contract no. DE-FC22-93BC14987, Annual Report, January 1, 1998 to December 31, 1998, 57 p.

Carr, T. R., D. W. Green, and G. P. Willhite, 1998, Improved oil recovery in Mississippian carbonate reservoirs of Kansas—near term—class 2: DOE Contract no. DE-FC22-93BC14987, Annual Report, September 18, 1994 to March 15, 1997, 168 p.

Chang, M., P. Sarathi, R. J. Heemstra, A. M. Cheng, and J. F. Pautz, 1992, User's guide and documentation manual for "BOAST-VHS for the PC": Agreement no. DE-FC22-83FE60149, National Institute for Petroleum and Energy Research, Bartlesville, Oklahoma, 83 p.

Coffin, P., 1993, Horizontal well evaluation after 12 years: Society of Petroleum Engineers 68th Annual Technical Conference and Exhibition, Houston, Texas, SPE Paper 26618, p. 97–102.

Dake, L. P., 1994, The practice of reservoir engineering: Amsterdam, Elsevier Science B.V., 534 p.

Doveton, J. H., 1994, Geologic log analysis using computer methods: AAPG Computer Applications in Geology No. 2, 169 p.

Doveton, J. H., W. J. Guy, W. L. Watney, G. C. Bohling, S. Ullah, and D. Adkins-Heljeson, 1996, Log analysis of petrofacies and flow-units with microcomputer spreadsheet software: 1995 AAPG Midcontinent Meeting Transactions, Tulsa, Oklahoma, p. 224–233.

Faquharson, R. G., L. M. Spratt, and P. C. M. Wang, 1992, Perspectives on horizontal drilling in Canada, *in* J. W. Schmoker, E. B. Coalson, and C. A. Brown, eds., Geological studies relevant to horizontal drilling—Examples from western North America: Rocky Mountain Association of Geologists, Denver, Colorado, p. 15–24.

Guy, W. R., T. R. Carr, E. K. Franseen, S. Bhattacharya, and S. Beaty, 1997, Combination of magnetic resonance and classic petrophysical techniques to determine pore geometry and characterization of a complex heterogeneous carbonate reservoir: AAPG Annual Meeting Abstracts, p. A45.

Guy, W. R., J. H. Doveton, W. L. Watney, T. R. Carr, and S. Bhattacharya, 1996, Reservoir characterization utilizing a low cost resistivity-porosity crossplot and an interactive spreadsheet: AAPG Annual Meeting Abstracts, p. A58.

Hedges, P. L., and S. Moothart, 1996, Infill well water-cut estimates based on open hole log data in a mineralogically complex reservoir—Kuparuk River field, Alaska: Society of Petroleum Engineers Western Regional Meeting, Anchorage, Alaska, SPE Paper 35684, p. 343–350.

Joshi, S. D., and W. Z. Ding, 1996, Horizontal well application—Reservoir management: Society of Petroleum Engineers International Conference on Horizontal Well Technology, Calgary, Alberta, Canada, SPE Paper 37036, p. 105–113.

Joshi, S. D., 1991, Horizontal well technology: Tulsa, Oklahoma, PennWell Publishing Company, 535 p.

Lacy, S. L., W. Z. Ding, and S. D. Joshi, 1992, Perspectives on horizontal wells in the Rocky Mountain region, *in* J. W. Schmoker, E. B. Coalson, and C. A. Brown, eds., Geological studies relevant to horizontal drilling—Examples from western North America: Rocky Mountain Association of Geologists, Denver, Colorado, p. 25–32.

Montgomery, S. L., E. K. Franseen, S. Bhattacharya, P. M. Gerlach, A. Brynes, and W. L. Guy, 2000, Schaben field, Kansas—Improving performance in a Mississippian shallow-shelf carbonate: AAPG Bulletin, v. 84, p. 1069–1086.

Murray, G. H., 1965, Quantitative fracture study—Spanish Pool, McKenzie County, North Dakota: AAPG Bulletin, v. 52, p. 57–65.

Rezaee, M. R., and N. M. Lemon, 1996, Petrophysical evaluation of kaolinite-bearing sandstones—Water saturation (S_w), an example of the Tirrawarra sandstone reservoir, Cooper Basin, Australia: Society of Petroleum Engineers Asia Pacific Oil and Gas Conference, Adelaide, Australia, SPE Paper 37023, p. 539–549.

Stright, D. H., and R. D. Robertson, 1993, An integrated approach to evaluation of horizontal well prospects in Niobrara Shale: Society of Petroleum Engineers Rocky Mountain Regional/Low Permeability Symposium, Denver, Colorado, SPE Paper 25923, p. 755–766.

U.S. Department of Energy, 1995, User's guide and documentation manual for "BOAST 3", Version 1.5: Contract no. DE-AC22-91BC14831, U.S. Department of Energy, Bartlesville, Oklahoma, 64 p.

Watney, W. L., W. J. Guy, J. H. Doveton, S. Bhattacharya, P. M. Gerlach, G. C. Bohling, and T. R. Carr, 1999, Petrofacies analysis—A petrophysical tool for geologic/engineering reservoir characterization, *in* R. Schatzinger and J. Jordan, eds., Reservoir characterization—Recent advances: AAPG Memoir 71, p. 73–90.

Techniques

Stockhausen, E. J., G. E. Smith, J. A. Peters, and E. T. Bornemann, 2003, Flexible well-path planning for horizontal and extended-reach wells, *in* T. R. Carr, E. P. Mason, and C. T. Feazel, eds., Horizontal wells: Focus on the reservoir: AAPG Methods in Exploration No. 14, p. 227–248.

15

Flexible Well-path Planning for Horizontal and Extended-reach Wells

E. J. Stockhausen

ChevronTexaco Exploration and Production
Technology Company, Houston, Texas, U.S.A.

G. E. Smith

ChevronTexaco Exploration and Production
Technology Company, Houston, Texas, U.S.A.

J. A. Peters

ChevronTexaco Exploration and Production
Technology Company, Houston, Texas, U.S.A.

E. T. Bornemann

Schlumberger Oilfield Services
Houston, Texas, U.S.A.

ABSTRACT

The value and success of horizontal and extended-reach wells depend on overall well-path design and the effective wellbore placement in the reservoir. Significant incremental value can be obtained by increasing the precision of well placements within or across stratigraphic lobes and by positioning the wells relative to fluid contacts. Four main factors that significantly impact the success of these types of wells are (1) geologic uncertainty, (2) borehole-position uncertainty, (3) unanticipated buildup rates and doglegs, and (4) communication and understanding between the geologist and the directional driller.

Previous articles on geosteering have discussed uncertainty factors that impact a drilling project and have recommended the development of contingency plans. This paper reviews the impact these factors have on the success of horizontal-drilling projects and describes a methodical approach to well planning and drilling that includes specific contingency-plan recommendations. This approach integrates local geologic information with a flexible well-path design.

The flexible well-path design method anticipates making path adjustments while drilling in order to improve the precision of well placement. The steps of this method include:

- defining the project's uncertainties
- evaluating the need for drilling a pilot hole and for running specialized steering tools
- designing a site-specific, flexible well-path plan using adjustable tangents to address uncertainties while drilling
- designing a "key marker-bed tangent" to locate one or more key marker beds above the target objective, using real-time geologic and directional data

- designing a "soft-landing tangent" to land the well at the desired well orientation in the target objective
- applying advanced steering techniques in the lateral section of a horizontal well to maintain the desired position and orientation in the target objective

Use of this well-planning approach gives the well-construction team flexibility to react to real-time geologic and directional complexities without causing alarm. By understanding the complexities of the project and deploying these techniques, operators can realize a 20% or more increase in net present value (NPV) in their horizontal and extended-reach drilling projects.

INTRODUCTION

Horizontal-drilling projects are key components in the business plans of many operating companies. Their anticipated high production rates and sweep efficiencies can significantly impact the realization of a company's production goals; therefore, there is little or no room for failure.

Lack of flexibility in the well-path design and/or lack of contingency planning tend to result in hasty decisions to adjust the well when geologic targets do not occur as anticipated or when the well path drifts off course. This situation can happen as a result of three uncertainty factors that are often overlooked by well-construction teams.

UNCERTAINTY ISSUES AND THEIR IMPORTANCE

Three uncertainty factors are (1) geologic uncertainty, (2) borehole-position uncertainty related to directional survey accuracy, and (3) unanticipated buildup rates and doglegs.

What happens when these uncertainty factors are overlooked?

- The well may be poorly placed in the reservoir or it may entirely miss the reservoir.
- Well footage in the reservoir may be reduced greatly or even forfeited entirely.
- Operators may be unable to place or operate completion equipment.
- Operators may never fully understand what happened and therefore will not be able to prevent a recurrence of these problems.

The overall result is lost reserves and value.

Lesso and Kashikar maintain that "40% of geosteering projects sustain a major structural interpretation change, which means that the well has to be sidetracked, or that option is actively considered" (Lesso and Kashikar, 1996, p. 137). This statement implies that many horizontal and extended-reach drilling projects have significant errors in their geologic model.

This paper presents the argument that both directional survey and geologic uncertainties contribute to unexpected geologic results. In addition, because of directional survey uncertainties, many wells that are thought to be placed ideally in their reservoirs are actually misplaced. These arguments support a significant need for improved methodology.

Placement of highly deviated wells is further complicated by the uncertainty associated with the directional driller's ability to predict and control build rates while drilling.

FLEXIBLE WELL-PATH PLANS

To maximize the potential of a horizontal or extended-reach well project, operators need a systematic and site-specific method for planning and executing each well that addresses the real-time complexities and uncertainties of the drilling prospect. This paper describes a methodology for dealing with these complex issues by designing a flexible well-path plan.

This method anticipates and plans for the need to make adjustments while drilling, taking into account the combined effect of the three uncertainty factors. A backward well-planning design is used, starting with the completion. The geologic model is incorporated into the well design and is used to refine it, incorporating adjustable tangent sections based on key geologic markers. If a key marker bed is encountered at depths shallower or deeper than its anticipated position on the predrill plan, the tangent section is shortened or lengthened, effectively realigning the well path with the target position. In many projects, use of this design method can eliminate the need for a pilot hole.

FIVE KEY STEPS IN PLANNING HORIZONTAL AND EXTENDED-REACH WELLS

To increase a project's success margin, the well-construction team should use the following planning techniques:

1) Define the lateral section and completion design requirements.
2) Define the range of each of the project's major uncertainties and evaluate the need to drill a pilot hole.
3) Define the value and risk for the well placement in terms of barrels lost per foot off depth or lobe not penetrated.
4) Design a flexible well-path plan, incorporating key geosteering decision points around two or more tangent sections designed to locate key marker beds, and softly land the well in the target reservoir at the desired angle and orientation.
5) Execute the geosteering plan by determining the current position of the well relative to the objective target position and redesigning and realigning the well plan to hit or maintain the borehole's position in the target.

The value and success of horizontal and/or extended-reach drilling projects can be equated with the effort expended on the well-path design plan and the execution of the plan by the well-construction team to place the well precisely. The reservoir engineer generally will recommend the ideal positioning of the well to maximize drainage of the reservoir. For horizontal and/or extended-reach projects, this may consist of

- placing the well at a particular depth below or above a fluid contact
- staying in a particular structural and/or stratigraphic position relative to the top of the zone for a defined lateral-section length
- slanting the well up, down, or sideways across the objective zone
- snaking the well back and forth across the objective zone (either up and down or sideways)

The well-construction team should determine the value of the placement or misplacement of the well in terms of barrels lost per foot off depth or lobe not penetrated together with the associated risk and cost for various placement methods and tools. Questions to consider are (1) whether to drill a pilot hole, (2) which logging tools to run while drilling, and (3) whether to employ a geosteering service. With this information, an asset team can develop the economic parameters needed to run a decision analysis.

The placement of the well to recover maximum reserves can be a risky action from a drilling and steering point of view. For example, a well placed at the crest of a steeply dipping fault block carries a high risk of drilling across the fault into nonreservoir rock or a water zone. A well placed too close to a fluid contact risks drilling across the contact. In these cases, the additional risks associated with directional survey uncertainty are often overlooked, and the well not being exactly where the directional survey indicates often is not considered. This oversight can lead to the well being accidentally and unknowingly drilled into unwanted fluid zones, creating a state of total confusion for the well-construction team when the well is brought on production.

DIRECTIONAL SURVEY UNCERTAINTY ISSUES

Borehole-position uncertainty is related to systematic errors inherent in directional survey measurements (Wolff and de Wardt, 1981). For highly deviated horizontal and extended-reach wells, these inherent errors create a large amount of uncertainty about the well's target size. Wolff and de Wardt (1981) found that the inherent uncertainty of a survey measurement could be represented by an ellipsoid of uncertainty, as shown in Figure 1. They also found

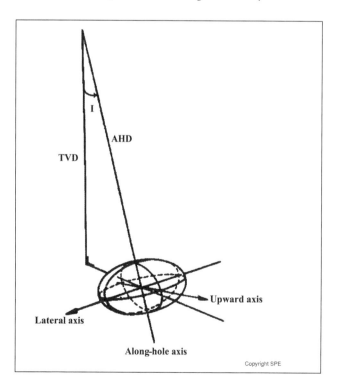

FIGURE 1. Definition of position-uncertainty ellipsoid (after Wolff and de Wardt, 1981, p. 2343).

that the amount of uncertainty for successive survey points increases with depth and can be represented by a cone of uncertainty around and along the wellbore. However, care must be taken to understand what any particular ellipsoid provided by the directional-drilling service industry represents. Different companies use different confidence levels for defining ellipsoids.

Directional uncertainty is not equal in all directions because of the different measurement tools used. The lateral axis (left to right) is associated mainly with uncertainty in the direction of magnetic north and with magnetic interference effects, which typically range from ± 1° of azimuth for the commonly used measurement-while-drilling (MWD) tools. This lateral-axis error tends to be the largest. The smallest uncertainty is in the along-hole-axis direction (AHD), which is associated with the accuracy of drill-pipe depth measurements, which range from about ± 0.6 m/304 m (± 2 ft/1000 ft). The upward axis of uncertainty, also referred to as the high-side axis of uncertainty, is associated mainly with the inclination measurement accuracy of about ± 0.25°.

For high-angle and horizontal wells, the upward axis defines the range of the possible true-vertical-depth (TVD) error. The radial (along-hole-direction) axis usually is small compared to the lateral axis, except for wells drilled with a significant amount of turning. In this case, a portion of the lateral axis uncertainty is translated to or added to the along-hole direction as the well path changes azimuth.

More recently, the directional drilling industry has developed sophisticated error models to describe the range of uncertainties for downhole measurement tools (Williamson, 1999). Borehole-position uncertainty equations contain many variables. Directional-drilling service companies should provide an uncertainty report as part of the planning, drilling, and postwell analysis process.

The probability-of-error models used by the industry generally assume a normal distribution (Figure 2). In addition, the industry typically uses a two-standard deviation range, also called the 2σ (2-sigma) deviation range, when describing borehole-position uncertainty ranges for a given survey point. This equates to a 95% confidence interval that the wellbore is within ± 2 standard deviations for a given axis direction. When reported, the 2σ values are stated separately for each of three ellipsoid axes.

Figure 2 shows the various confidence levels for a typical normal-probability distribution. If the calculated 2σ uncertainty range is ± 3.7 m (12 ft) for the high-side axis of a given survey point, then there is a 95% probability that the wellbore is within 3.7 m in the high-side axis direction at that point. A 1σ range represents a 68% confidence interval; i.e., there is a 68% probability that the wellbore will be within ± 1.8 m (6 ft) in the high-side axis direction for this particular survey point. Likewise, for the 2σ values, there is a 5% probability that the wellbore is located more than ± 3.7 m from the survey point in the high-side axis direction. However, no standard error model at present is accepted uniformly by the industry, and stated values vary from one error model to another.

Another issue associated with these error models is that they do not account for human error. Serious problems can occur from human errors, such as using the wrong kelly-bushing elevation or the wrong magnetic-declination correction factor.

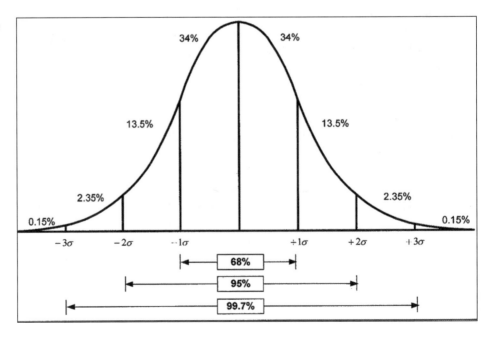

FIGURE 2. Normal distribution of the probability of borehole uncertainty for a given axis.

TYPICAL DIRECTIONAL UNCERTAINTY RANGES

A typical shallow horizontal well landed at 1524 m (5000 ft) TVD (the point where the well reaches 90°) drilled using a 5°/30-m (5°/100-ft) buildup rate with standard MWD directional tools may have a 2σ TVD uncertainty range of ± 3.7 m (12 ft) or more at the landing point. This TVD uncertainty grows to approximately ± 7.6 m (25 ft) TVD at the end of a 610-m (2000-ft) lateral section. These ranges are used in the section "TVD Uncertainty Example: Pinnacle-reef Oil Target" below.

This same well may have a 2σ lateral-axis uncertainty (left and right of the wellbore) of ± 15 m (50 ft) at the landing point that grows to approximately ± 30 m (100 ft) by the end of the lateral section. A deep 9144-m (30,000-ft) extended-reach well can have a 2σ borehole-position uncertainty in the range of approximately ± 30 m (100 ft) TVD and ± 152 m (500 ft) laterally at its target point.

EFFECTS OF IGNORING DIRECTIONAL SURVEY UNCERTAINTY

In general, the geologic and reservoir engineering communities have ignored borehole-position uncertainty, as has, for the most part, the geosteering community. S. W. Poston (1985) discusses several examples of how inaccurate wellbore surveys could result in lost reserves and value.

Ignoring borehole-position uncertainties can cause many problems, which are amplified even more when applied to horizontal and extended-reach wells. It can result in poor placement of the well in the reservoir, shorter than desired lateral sections, or the well entirely missing the reservoir.

We will examine and discuss the effects of TVD depth or high-side axis uncertainty on drilling horizontal wells using a pinnacle-reef oil target example. Following this discussion, we will examine and discuss the affects of lateral axis uncertainty using an example of drilling a horizontal well in a steeply dipping fault block.

TVD UNCERTAINTY EXAMPLE: PINNACLE-REEF OIL TARGET

This example illustrates the issue of directional survey uncertainty with respect to placement of a horizontal well parallel to a fluid contact. Figure 3 displays the case for drilling a horizontal well in an oil leg of a pinnacle-reef-type trap.

In this example, the following information is given:

- The reservoir has a large gas cap and a 12-m (40-ft) oil column, with a gas-oil contact (GOC) at −1513 m

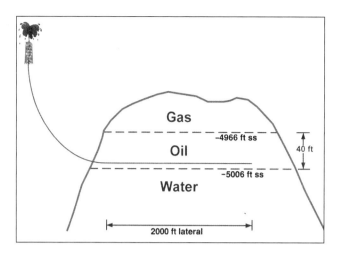

FIGURE 3. Example of pinnacle-reef oil drainage.

(−4966 ft) subsea (ss) and an oil-water contact (OWC) at −1526 m (−5006 ft) ss.
- The reservoir-drive mechanism is gas-cap expansion with no aquifer support.
- The reservoir-simulation model shows that maximum recovery will occur if the horizontal drainhole is placed at −1524 m (−5000 ft) ss, which is 1.8 m (6 ft) above the OWC.

The well is drilled using a 5°/30 m (5°/100 ft) buildup rate, and the directional driller places the drainhole within 0.3 m (1 ft) of the path using only directional MWD tools. No logs are run. The well comes on production making 100% water. What happened?

In this example, the 2σ borehole-position uncertainties of the high-side axis are ± 3.7 m (12 ft) TVD at the landing point and ± 7.6 m (25 ft) TVD at the end of a 610-m (2000-ft) lateral section. The 1σ vertical uncertainties for this well would be ± 1.8 m (6 ft) TVD at the landing and ± 3.8 m (12.5 ft) TVD at total depth. Figure 4 shows that the probability errors for this well indicate a 16% chance that the well was unknowingly landed in the water zone, and approximately a 33% chance that the well was drilled into the water leg by the total depth of the well.

During the field management team's two-month investigation into this problem, borehole-position uncertainty was never taken into account. Because there are little or no data to determine possible explanations, the team assigns a high-risk factor to the offset development project and the offset prospect is canceled.

This not-so-fictitious example demonstrates the problem of drilling only with directional tools. It also demonstrates the need to include borehole-position uncertainty analysis as an additional risk factor. In cases such as this, when drilling near fluid contacts, serious consideration should be given to drilling a pilot hole.

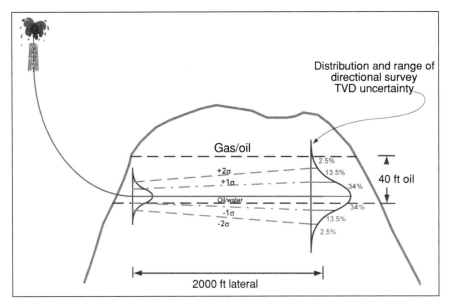

FIGURE 4. Growth and distribution of TVD uncertainty for the pinnacle-reef oil-drainage example shown in Figure 3.

PILOT-HOLE CONSIDERATIONS

A pilot hole is an expendable wellbore designed to acquire geologic information, such as structural control for the top and base of the target reservoir and faults and fluid contacts. It adds stratigraphic control and sometimes can be used to identify future targets. A pilot hole should be designed concurrently with the drainhole and designed to minimize the amount of plug-back and sidetrack footage, thus minimizing the relative position error between the pilot hole and main wellbore. When attempting to place a horizontal well parallel to a fluid contact, a pilot hole can minimize TVD error relative to the contact by zeroing the TVD depth uncertainty based on the fluid contact encountered in the pilot hole, then restarting TVD uncertainty accumulation backward along the pilot hole, and then starting the accumulation forward from the sidetrack point.

For additional examples and ideas on handling TVD uncertainty relative to fluid contacts, refer to Barry et al. (1998). The authors state in their field example, "tying the MWD survey data to the known gas-oil contact during the landing operation reduces the geometric error prior to drilling the horizontal section, while control of the geometric error during horizontal drilling is achieved using real-time resistivity and neutron-density data" (Barry et al., 1998, p. 221).

HOW DIRECTIONAL TVD UNCERTAINTY CAN AFFECT STEERING DECISIONS

Vertical TVD uncertainty can lead to overshooting or undershooting of targets and to poor steering decisions.

Generally, the first course of action when a key marker bed and/or the top of the objective horizon come in higher or lower than anticipated is to revise the structure maps. The second course of action is to make a course correction. Both of these decisions could be wrong.

All options should be considered before making the decision to alter the well design or alter the structure maps, particularly if the apparent depth discrepancy is in the range of directional-survey uncertainty. The drilling team should use multiple working models while landing a well. One model should flex the surface to match the apparent top based on the directional-survey data. This results in altering the structure map, which in turn alters the apparent dip in the plane of the well path and may suggest a need to change the well trajectory to a higher or lower inclination and attack angle for the landing. A second model would assume that the depth discrepancy encountered is associated with directional survey error. In this case, the geologic model should be bulk-shifted up or down to match the pick encountered in the well but still maintain the apparent geologic dip of the predrilled model. This may suggest that the landing depth needs to be adjusted up or down.

Over the next portion of the well, the two models must be tested by predicting the anticipated measured-depth thickness needed to encounter the next marker bed. Testing these models using marker beds at known thicknesses above the objective allows the team to make better decisions about adjusting the well trajectory in order to align the final approach to the target bed. This concept is discussed further in the following section.

LATERAL-AXIS (AZIMUTH) UNCERTAINTY ISSUES

As stated earlier, the lateral-axis (azimuth) uncertainty has the largest error range. For a shallow horizontal well, the lateral-axis uncertainty range (left and right of the wellbore) may be ± 15 m (50 ft) at the landing point of a well and expand to ± 30 m (100 ft) by the end of the lateral section. On a deep (9144-m [30,000-ft]) extended-reach well, the lateral-axis uncertainty can be in the range of ± 152 m (500 ft) laterally at the target depth.

The effect of lateral-axis uncertainty on the value of the project can be negligible when drilling shallow horizontal wells on broad, flat-lying strata. However, when drilling into moderate to steeply dipping beds along the geologic strike direction, lateral uncertainty error can have a significant negative impact on the results of horizontal/extended-reach drilling projects.

LATERAL-AXIS (AZIMUTH) UNCERTAINTY EXAMPLE: DRILLING HORIZONTAL WELLS ALONG STRIKE IN STEEPLY DIPPING BEDS

Figure 5 shows a northwest-dipping reservoir that is trapped updip by a down-to-the-southeast fault. The reservoir has an OWC at −1622 m (−5320 ft) ss and a GOC at −1591 m (−5220 ft) ss. The proposed well plan calls for drilling a horizontal well along the strike direction at −1606 m (−5270 ft) ss in a stratigraphic position that is 9 m (30 ft) vertically below the top of an 18 m- (60 ft-) thick objective interval. The probability distribution for the lateral uncertainty range is shown for the horizontal section of the well. As illustrated, the lateral uncertainty range grows wider from the landing point toward the proposed total depth of the lateral section. The 1σ and 2σ ranges of lateral uncertainty are shown by the inner and outer dashed lines, respectively.

In this example, if the actual well path is to the left or right of the proposed path, the well may be located accidentally in the wrong lobe. Any well path accidentally drilled at the proposed horizontal depth and outside the 1σ range would not encounter even the reservoir. The well would be located in a stratigraphic position, either above or below the objective depending on whether the well is to the left or right of the 1σ range (see X-section view, Figure 5). In this example, approximately a 32% chance exists for the proposed well to miss the reservoir entirely if it is drilled using directional MWD tools only.

Three techniques can be used to reduce the risk of misplacing the well shown in Figure 5.

1) Design a pilot hole to intersect one or both of the fluid contacts, at the same time attempting to minimize the length of the plug-back section.
2) Design the well to approach the reservoir in an apparent dip direction rather than a strike direction, and land the well at 90° inclination at the desired TVD prior to entering the reservoir. Continue drilling in the apparent dip direction until the well intersects the objective, then turn the well, changing the hole azimuth direction to be parallel to the strike of the bed.
3) Use magnetic monitoring (sometimes referred to as "in-field referencing") to reduce the azimuthal magnetic errors in high-cost areas, particularly in high latitudes closer to the earth's magnetic poles (Williamson et al., 1998).

In this example, the well is designed to take advantage of contoured structure changes, landing at a point where the well is drilling in an apparent updip direction. The contours then become parallel to the well path in the lateral section. In this design, there still may be a need to further adjust the path using azimuth changes in a real-steering application. As discussed above for TVD uncertainty, use multiple working models as the well approaches the objective. Structural changes in depth will cause not only a change in apparent dip but also a change in the strike direction.

In situations such as those seen in Figure 5, some operators have designed their wells so as to eliminate the need for a pilot hole. These wells are designed to hit the GOC in the build section, then exit out the base of the zone, and finally reenter the zone at a specific desired distance below the GOC.

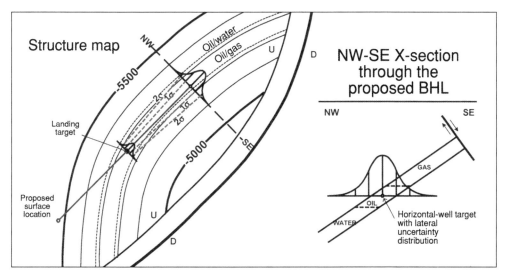

FIGURE 5. Example of lateral axis of uncertainty growth and distribution for a horizontal-well prospect to drill along the geologic strike in a steeply dipping fault block. BHL = bottom-hole location.

GEOLOGIC UNCERTAINTY ISSUES

Just as geologists and reservoir engineers often ignore borehole-position uncertainty, drilling engineers tend to ignore the issue of geologic uncertainty when planning wells. Once a target is established, often it becomes a fixed object in space to be hit at any cost. Defining and communicating the total geologic uncertainty for each target is an important step in the successful implementation of the flexible well-path planning method.

Figure 6 illustrates three components of the total geologic uncertainty that should be defined for each horizontal and extended-reach drilling target. These include (1) structural uncertainty, (2) lateral-dip uncertainty, and (3) lateral-stratigraphic uncertainty.

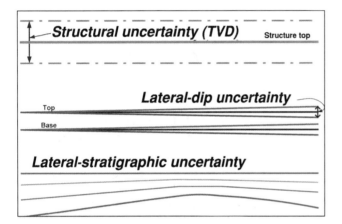

FIGURE 6. Three components of geologic uncertainty that can significantly impact horizontal and extended-reach drilling projects.

Structural Uncertainty

Structural uncertainty, as defined in this paper, is the range of depth uncertainty for encountering a given geologic horizon at a given depth at a given location. Horizontal and extended-reach drilling prospects tend to be field-development or delineation-type projects but still can have a structural uncertainty in the range of ± 3 to 30 m (10 to 100 ft) TVD.

Structural geologic uncertainty is dependent on many factors, including quality and amount of offset well-control and seismic data. Building and calibrating a geologic model is largely dependent on combining many sources and vintages of data. Checks are needed to determine if the regional 3-D seismic survey is corrected to the same Universal Transverse Mercator (UTM) coordinate system as the well and tied to any of the existing surface locations in the field, and if the rotary table (RT) or kelly-bushing (KB) measurements are accurate relative to the reference-height datum measurement (e.g., mean sea level).

Often, errors and discrepancies exist in the data used,

which, when combined with other sources of data, lead to the building of an inaccurate geologic model. For example, a faulty directional survey, a KB elevation error, and a wrong log correlation might be incorporated into the building of the seismic velocity model used to predict the target location.

Borehole-position uncertainty of the offset wells also contributes to structural uncertainty, particularly in older fields where few if any directional surveys exist for wells drilled as straight holes. Modern survey tools run in these straight holes often reveal that these wells had drifted 76 m (250 ft) or more from their surface locations. Older holes, which were surveyed using older, less accurate instruments, have been found to have similar inaccuracies when they are resurveyed with modern, inertial-rate gyro-surveying tools. It is therefore recommended that surveys of these older wells be conducted with modern tools, if practical.

Another source of error that can be incorporated into the building of the geologic model is error in land-surveying measurements. When locating individual wells and seismic shotpoints, contractors and operating companies may use various sources and vintages of base maps, perhaps with differing grid coordinate systems.

Seismic control, particularly high-quality 3-D seismic data, can be very beneficial in reducing structural uncertainty for a given prospect. Even with high-quality 3-D data, however, it is not unusual to see geologic markers that do not tie to the seismic data because of lateral velocity variations or single check-shot surveys used for 3-D project depth conversions. The more accurate the velocity model, the less structural uncertainty will exist.

Lateral-dip Uncertainty: Importance of Knowing Geologic Dip

The objective of many horizontal wells is to stay in a particular stratigraphic lobe for hundreds or thousands of feet. To accomplish this, it is very important to know the geologic dip and the apparent geologic dip in the plane of the proposed well. Lateral-dip uncertainty can significantly impact the actual measured-depth bed thickness (MT) encountered in the lateral section.

For example, when attempting to penetrate and remain in a 3-m (10-ft) true-vertical-thickness (TVT) bed, a 1° change in apparent dip can reduce the MT in the lateral section from infinity to 175 m (573 ft) (see Table 1). A change of 2° shortens the MT to 87 m (286 ft), and a 3° change in apparent dip will reduce the MT to 58 m (191 ft) encountered in the zone.

Table 1 shows the MT that will be encountered in a horizontal well (drilled at 90° inclination) while crossing a 3-m (10-ft) TVT bed at various apparent dip angles. In

TABLE 1. MEASURED-DEPTH BED THICKNESS (MT) AS A FUNCTION OF APPARENT DIP.

	Bed thickness (TVT) (ft)	Apparent dip	Hole angle	Measured-depth thickness (MT) (ft)
Updip down-section drilled wells	10	10°	90°	57
	10	5°	90°	114
	10	4°	90°	143
	10	3°	90°	191
	10	2°	90°	286
	10	1°	90°	573
Flat	10	0°	90°	Infinite
Downdip up-section drilled wells	−10	1°	90°	573
	−10	2°	90°	286
	−10	3°	90°	191
	−10	4°	90°	143
	−10	5°	90°	114
	−10	10°	90°	58

this table, the updip-direction horizontal wells encounter the bed in a normal down-section direction, that is, the well drills from the top of the bed toward the base of the bed. When a horizontal well drilled at 90° is drilled in a downdip direction, it will encounter the bed in a reverse geologic order; that is, the well will drill from the base of the zone back up-section toward the top of the bed. Therefore, for conventional purposes, the thicknesses displayed in Table 1 for downdip and up-section drilled wells are shown as a negative number.

Table 1 also can be used in reverse to solve for the dip of the beds based on the MT encountered when crossing the zone. This assumes that the well stays exactly horizontal and the bed is exactly 3 m (10 ft) thick. For example, if the well encounters 91 m (300 ft) of MT crossing the zone, then the apparent dip in the plane of the well would be slightly less than 2°.

Solving the lateral-dip uncertainty problem is key to the success of many horizontal wells. Once solved, the well path can be redesigned with minor changes in inclination and/or azimuth to stay in the bedding plane. This is accomplished in real-time analyses through correlation and rapid model updating, as well as working directly with the directional drillers.

Lateral-stratigraphic Uncertainty: Define Ranges in Bed Thickness

From the previous discussion on the importance of knowing geologic dip, it was demonstrated that to predict the MT required to drill through a section at a constant hole-inclination angle or to solve for apparent dip rate, the vertical thickness of the bed must be known. Lateral changes in vertical bed thickness can cause problems with

these predictions. For example, if an 87-m (286-ft) MT lobe is encountered while drilling at 90° and the zone is 3 m (10 ft) thick, then the bed would have a 2° apparent dip in the plane of the well. However, if the bed is only 1.5 m (5 ft) thick, the apparent dip in the plane of the well would be 1°. On the other hand, if the bed is 6 m (20 ft) thick, the apparent dip in the plane of the well would be 3°. Each of these solutions could lead to a different steering decision, because a different attack angle would be chosen.

The next step in the process of designing a flexible well-path plan is to quantify the uncertainties associated with lateral thickness changes. This is accomplished by conducting a study of offset-well data to define the thickness range between marker beds and the top and base of the reservoir. The thickness range of same stratigraphic intervals in different wells provides a measure of that uncertainty. The most useful and practical way of measuring bed thickness is by using TVT.

As defined by Tearpock and Bischke (1990, p. 513), "True vertical thickness is the thickness of a bed when measured in a vertical direction." For directional wells drilled in areas with dipping beds, it is important to note that TVT is not the same as the true-vertical-depth thickness (TVDT) derived from a TVD log, nor is it identical with true-bed thickness (TBT), which is also known as true-stratigraphic thickness (TST).

Figure 7 illustrates the relationships among MT, TVDT, and TVT for three wells drilled through a dipping bed. In this example, the bed has a constant thickness and dip; therefore, each of the wells has the same TVT. However, the MT and TVDT thickness of the interval varies in each well. For the straight hole, the MT, TVDT, and TVT are equal. For the updip-drilled well, the TVDT is significantly less than the TVT of the interval. In the downdip-drilled well, the TVDT of the well is significantly thicker than the TVT of the interval. Using TVDT data in this case could lead to the wrong conclusion, i.e., that the geologic model is thinning in the updip direction.

The relationships shown in Figure 7 among MT, TVDT, and TVT tend to become more exaggerated and complicated at higher inclination angles (Figure 8).

In some cases of highly deviated wellbores, the combination of hole angle and bed-dip angle is such that a well path goes beyond the point of drilling down through the beds (down-section drilling) to a point where the well begins to drill back up through the beds. In this case, the well will encounter the beds in a reverse-geologic-section direction, referred to in this paper as up-section drilling. This concept is illustrated by the directional well (Well #2) shown in Figure 8. This cross-section view shows two wells drilled through a series of parallel beds that have an apparent dip of 15°. A reverse-geologic section would be encountered by any well drilled in a downdip direction in the plane of the model well with a hole angle greater than 75°.

Figure 8 further illustrates the problems associated with using true-vertical-depth and measured-depth logs to derive bed thicknesses, or for use in log correlations when drilling highly deviated wells through dipping beds. As observed in the model, the directional wellbore, Well #2, instantaneously changes hole angle by 10° at each bed boundary. Each bed is exactly 30 m (100 ft) TVT, as measured vertically in the straight hole, Well #1. The length of the vertical side of each of the gray triangles along Well #2 shows the TVDT that would be seen on a TVD log of Well #2. The hypotenuse of each of the gray triangles

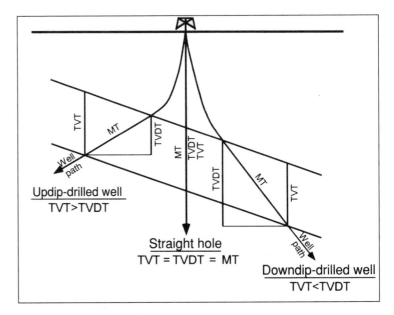

FIGURE 7. Comparison of measured-depth thickness (MT), true-vertical-depth thickness (TVDT), and true-vertical thickness (TVT) as a function of well trajectory and bedding dip.

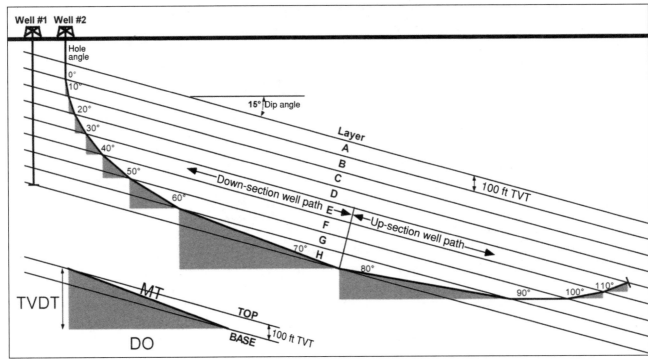

FIGURE 8. Downdip-drilled well-path layer-cake model. DO = distance out.

along Well #2 shows the measured-depth thickness (MT) that would be seen on a measured-depth log of Well #2. Correlations between Well #1 and Well #2 using MD or TVD logs would be impracticable to impossible after the directional hole reaches about 70° inclination.

Additional problems associated with using TVD logs to measure bed thickness or for use in well correlation can be seen in the model in Figure 8. The model shows how the TVDT for each bed penetrated in the directional well steadily increases as the well drills down through the geologic section to the base of layer H. Then, along the 80° leg of the directional well, the TVDT is positive, although the well now is drilling up-section. The 90° leg of the directional well demonstrates the greatest problem with TVD logs when drilling horizontal wells through dipping beds: at 90°, there is zero change in true-vertical-depth. Therefore, no TVD section is recorded on the log even though the well penetrates the entire layer G in a reverse-section direction. As the directional well in the model continues and exceeds 90°, the true-vertical-depth thicknesses begin to reverse. However, the reverse thicknesses recorded are still significantly less than the 30-m (100-ft) TVT that would be encountered in a straight hole at that location.

To solve the issues of correlation and apparent thickness associated with drilling directional wells through dipping beds, the use of TVT logs is recommended. The TVT technique adjusts and corrects log data for the effects of bed dips so that meaningful and accurate correlations can be made, provided that the correct geologic dip has been input. The recommended technique is to integrate changes in borehole azimuth, inclination, and bed dips simultaneously at the sample rate of the measured-depth log (indexed at 0.15 m [0.5 ft]) to create a continuous TVT log.

Currently, very few TVT log-adjustment programs are available for real-time application at the rig site. Even when available, most TVT log programs are limited and incapable of providing a TVT correlation log for portions of wells drilled in a reverse up-section direction.

Some geologists like to use TST, which is the thickness of a bed measured perpendicular to the bedding planes. However, unless the wells represented in Figures 7 and 8 are drilled exactly in the true dip plane, TST cannot be accurately measured on the cross-section view. TVT, however, can be measured on any vertical-section plane.

COMBINING GEOLOGIC AND DIRECTIONAL UNCERTAINTY

Combining the effects of geologic uncertainty and borehole-position uncertainty, as illustrated in Figure 9,

leads to a significantly larger area and range to the target than when considering only one uncertainty factor. Another issue illustrated in Figure 9 is the effect that the directional-well survey data has on measuring the depth of a geologic marker. The actual well path is not necessarily where the directional survey places it, and this can cause problems.

For example, the actual well path could intersect the target horizon at point A, the upper location of the points labeled as possible intersection points between the well and the mapped horizon. However, the directional survey could indicate that the well is directly on the proposed path and that the marker is at the center of the upper ellipse, which is higher than the well actually is placed. This could cause anxiety in a real-time drilling situation. The geologist may lose credibility, because the top of the zone is apparently outside the estimated range of geologic uncertainty. The reservoir engineer, on the other hand, may experience false optimism, believing additional reserves exist to be added to the prospect, because the zone is significantly higher than originally anticipated.

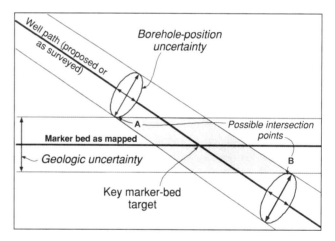

FIGURE 9. Combining geologic uncertainty and borehole-position uncertainty.

In the reverse situation, if the actual well intersected the target horizon at point B but the directional survey said the well is at the center of the lower uncertainty ellipse, the geologist could still loose credibility and the reservoir engineer could experience false pessimism, believing that significantly less reserves exist than originally anticipated, and might wrongly consider killing the project.

DIRECTIONAL DRILLING AND DIRECTIONAL CONTROL

The directional driller's ability to predict and control build rates while drilling is vital to the optimal placement of horizontal and extended-reach wells. The role of the directional driller is no longer to drill a line as defined on a piece of grid paper, but to work as a member of a well-construction team that has responsibility to react to the real-time complexities and uncertainties of the drilling prospect.

Communication, especially between the directional driller(s) and the geosteering engineer/geologist, is of particular importance. Both must be prepared for changes, and each must be aware of the other's capability to react to a given situation. For instance, based on real-time geologic information, the geologist may decide that the target should be moved; the directional driller, therefore, will need to figure out if it is possible to do so without jeopardizing the well. On the other hand, if the directional driller for some reason cannot hold the planned trajectory to the target, the geologist should consider whether it is reasonable to move or enlarge the target. Flexibility in each of these situations may save the team from having to sidetrack a well.

Many factors contribute to difficulties in staying on a proposed well-plan trajectory, including the experience of the directional driller, settings and configuration of the directional tools, the types and positions of directional MWD sensors, and geologic phenomena such as tight streaks and the dips of bedding planes. For wells planned at relatively low buildup rates, for instance 6°/30 m (6°/100 ft), it is not unusual to have unplanned doglegs in the actual well at rates more than 2°/30 m higher than planned. These unplanned doglegs often are created by the bit reacting to tight streaks, as well as other geologic phenomena encountered while drilling.

SCRUTINIZE DIRECTIONAL SURVEYS AND SURVEYING PLAN

When drilling with a standard steerable motor, the directional driller must choose a bend-angle setting in order to accomplish the design build-rate curve. From fear of getting behind on the build-curve design, it is common practice for the directional driller to choose a bend-angle setting that produces a buildup rate higher than the planned trajectory rate. To accomplish building at the lower planned buildup rate, the directional driller will employ a combination of slide and rotary drilling. In the slide mode of drilling, the bottom-hole assembly is oriented with a set tool-face angle, and drilling is accomplished with the aid of a downhole mud motor to turn the bit. In rotary mode, the drill string typically is designed to hold and maintain a constant inclination and direction, although in practice this does not always occur.

With the advent of top-drive rigs, it has become common practice to drill 27-m (90-ft) stands. Over long sections of the well, the directional driller will often set a pattern of slide drilling/rotary drilling for a portion of each stand. Unfortunately, it has also become common practice to take only one directional survey per stand (27 m [90 ft]), in which case the calculated directional survey reflects only the average build/drop rate (dogleg severity for 3-D well paths). The calculated directional survey is based on the assumption that the well path is drilled on a smooth, circular arc between each consecutive set of survey points. In actuality, the well path follows a more tortuous path consisting of a series of shorter-radii arcs, with higher dogleg severities, divided by a series of relatively straight tangent sections.

The practice of taking only 27-m (90-ft) surveys can cover up significant doglegs that could negatively impact the project. In addition, the calculated TVD and hole position will be affected. Therefore, the surveying plan should ensure proper data spacing between surveys. Taking and recording a survey at the point that coincides with the beginning and ending of each slide section is therefore recommended.

Another part of the surveying plan is to record and communicate the maximum dogleg-severity limit that each piece of equipment that is run into the hole can withstand, including casing and completion equipment. Because the well can be drilled does not mean that it can be completed or operated successfully. It is therefore important that these limits be known by each member of the well-construction team. The survey data should be closely scrutinized while drilling to ensure that excessive doglegs do not exist.

CONTINUOUS SURVEYS

Some of the newer MWD directional tools have near-bit inclination sensors that provide continuous inclination readings to the directional driller, even while the tools are rotating. Continuous, real-time directional data, which includes hole azimuth as well as inclination, are also available but are generated farther behind the bit (approximately 18 m [60 ft] back). These tools can be very useful to the directional driller for predicting buildup-rate tendencies, as described in Lesso et al. (2001). These tools also can be extremely useful geosteering aids.

Continuous surveys are taken while drilling in a near-continuous fashion, using MWD systems. Stationary (standard) surveys, on the other hand, are usually taken during connections while the drill pipe is suspended just off bottom when, according to Lesso et al., "MWD tools can measure the wellbore inclination and azimuth every 90 seconds, even while rotating. This means that a survey can be taken every 2–3 ft (or less) while drilling (under normal drilling conditions), instead of 30 or 90 foot intervals" (Lesso et al., 2001, p.1).

Data from continuous directional and inclination measurements clearly show how directional-well paths are more tortuous than calculated from static surveys, particularly when the static surveys are taken every 27 m (90 ft) (Figure 10).

Figure 10 shows 152 m (500 ft) of a build section from a directional well drilled using a standard steerable-system motor (Lesso et al., 2001, p. 10). Lesso et al. describe the drilling situation as follows:

> This well is a slant well trajectory where the plan was to build from vertical to 60° at 4.0°/100ft. Here, the well is building from 44° to 56° while azimuth is turning . . . slightly to the right. There are five large slide sections ranging from 35 to 45 feet in duration. The tool face angle stays close to zero with some variation up to 10° right. This is a common build section. The standard surveys yield no more information. However, the continuous inclination shows that something abnormal is going on. A sharp build of 10.1°/100ft is the result of the slide sections, while a drop rate ranging from −4.2°/100ft to −5.8 °/100ft occurs in the rotary sections. Meanwhile, the standard survey data indicates [sic] that the well is building at 2.4 deg/100ft (Lesso et al., 2001, p. 473).

This example is a relatively severe case, but it clearly shows how 27-m (90-ft) survey intervals can completely mask what is actually happening with the steerable-system curve rate and the rotary build/drop rate. A review of the continuous direction and inclination data is therefore recommended while drilling. Unfortunately, no standard set of procedures exists for incorporating these data into directional survey calculations to arrive at dogleg severity or hole positioning. Additionally, these data often are not saved or included with the data tape provided by the directional-drilling service company at the end of the job and ultimately are lost. A best practice for the future is to save these data so that they are available when directional-drilling technology evolves enough to use the sophistication of the new measurements.

TVD Accuracy of Static Versus Continuous Surveys

Modeling of various sequences of slide-rotary drilling using steerable motors indicates that the well-survey position calculated using a static survey could result in significant differences. Test cases using 27-m (90-ft) survey intervals show TVD differences in the range of ± 3 to 9 m (± 10 to 30 ft) for some typical long-radius, build-curve designs and bend-angle settings. Setting the bend angle higher than that needed to drill the planned trajectory results in a larger range in the calculated well-position points.

The implications of these results when applied to horizontal-well applications are far-reaching. Inaccurate surveys lead to poor steering decisions and errors in creat-

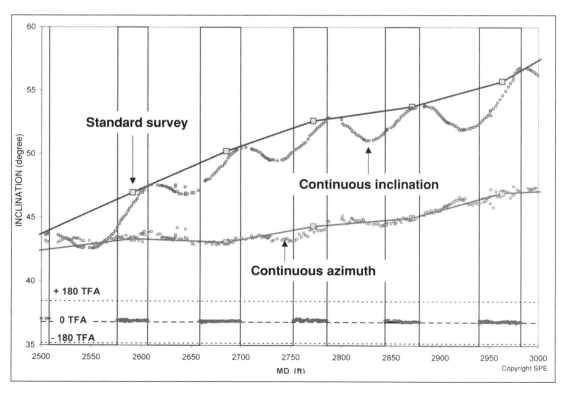

FIGURE 10. Continuous direction and inclination versus standard survey trajectory in a conventional build section— 500 ft of data (from Lesso et al., 2001, p. 10). TFA = tool-face angle.

ing structure and reservoir isopach maps. These errors cause further mistakes in reserve calculations, recovery efficiencies, and field management decisions. The magnitude of error for some fields has been calculated to be as much as 100,000 barrels per foot of TVD error.

DIRECTIONAL COURSE CORRECTIONS AND STEERING

A few large doglegs may not impair the well, but a large number of small doglegs (associated with trying to get back to the proposed drilling line or too many geosteering decisions) may impact drilling productivity and may jeopardize the planned completion of the well.

Another problem, which leads to serious torque and drag problems, is oversteering by the geosteering engineer or the well-site geologist. Steering up and down multiple times when you are not sure where the bed boundaries are can cause serious problems. The directional driller and the geosteering engineer must work together, keeping in mind that adjustments made while drilling must not impair the ability to drill and complete the well or to operate production equipment.

BIT-DEVIATION TENDENCIES VERSUS GEOLOGIC DIP

Figure 11 illustrates the bit-deviation tendency versus bedding as a function of hole inclination and angle of bed dip. It has been known for many years that the bit tends to deviate or walk in the updip direction at low hole-inclination angles. However, in areas with high bed dip or in high-angle wells drilled into low-dipping strata, the bit-deviation tendency is toward the downdip direction. This tendency to walk in the downdip direction generally increases as the hole inclination becomes nearly parallel to the bedding planes. This often leads to higher than expected doglegs just above the landing point of a horizontal well, which may impair the landing of the well in the target zone.

A hard or tight streak at the top of a horizontal-well objective can be particularly troublesome and practically impossible to drill through once the hole angle becomes parallel to the tight streak. It is therefore critical for the geologist to communicate to the directional well planner, early in the planning phase of the well, the probability of encountering tight streaks above the target. The directional plan can be modified to a slightly higher attack (or incident) angle to get the bit through these zones. After a successful landing in the objective, a tight streak above the well can be a great advantage when drilling, because it will be difficult to exit from the zone through the streak.

The point at which the bit becomes parallel (tangent) to the bedding plane along a curved section of wellbore is referred to as the bit-deviation inflection point. This is, coincidentally, the same point where the well encounters a reversal in direction of geologic section being cut. (Refer to the discussion of Figure 8, up-section versus down-section drilling). When building an angle past a deviation inflection point (see Figure 11), the deviation (walking) tendencies reverse themselves, which is often perplexing to an unaware directional driller. This can be particularly bad when the directional sensors are far behind the bit. By the time the directional drillers realize what is happening, the well may have drilled out of the zone in a direction opposite to that anticipated.

The need to incorporate a detailed geologic model (structure maps and cross sections) with the well-path design plan cannot be overemphasized. These maps and cross sections communicate geologic dip, the presence of tight streaks, and other geologic phenomena to the geologist, directional driller, and other well-construction team

FIGURE 11. Bit-deviation tendency versus hole-inclination angle and geologic dip. Adapted from and used with permission of Baker Hughes INTEQ.

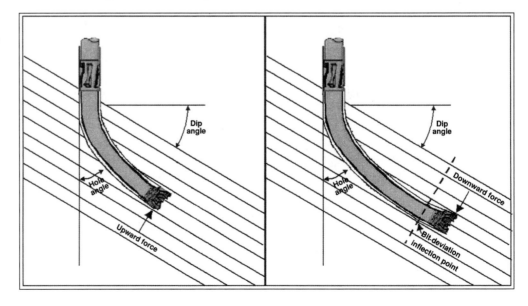

members. With this information, a good directional driller can often develop a better feel than the geologist for the geologic dip.

BIT-DEVIATION TENDENCIES WHEN DRILLING ALONG STRIKE

Horizontal and extended-reach wells often are drilled in or along the geologic strike direction, perpendicular to the geologic dip azimuth. In higher-dipping beds, this leads to a tendency for the well to walk toward the down-dip direction and off the proposed vertical-section plane. To compensate for this tendency, the directional driller may have to turn the tool-face orientation toward the updip direction during the build section. When the well is horizontal, the borehole will be parallel to the bedding planes while drilling along strike. However, if the hole angle increases beyond 90°, the well will begin to cut up-section, the bit-walking tendency will reverse, and the well will try to walk in the updip direction off the proposed vertical-section plane.

DIRECTIONAL-CONTROL EFFECTS ON BED THICKNESS

The objective of the horizontal or extended-reach project is to stay in the target zone for as long as possible, or to cut across multiple beds. To accomplish either of these objectives requires an understanding of the relationship between hole direction and geologic-dip direction.

Table 2 shows the MT that will be encountered while crossing a 3 m- (10 ft-) thick, flat-lying bed at high hole-inclination angles. Note the similarity to Table 1.

Not all horizontal wells are truly horizontal. When dealing with multilobed reservoirs, many horizontal-well

designs require the well to slant across individual lobes with a desired measured-depth length per lobe. From Table 2, we see that the hole angle required to drill across a 3 m- (10 ft-) thick flat-lying bed, if the desired measured-depth length was 75 m (250 ft), would be approximately 92.5° if the well is designed to slant up, or 87.5° if the well is designed to slant down.

The typical horizontal or extended-reach well encounters a combination of varying bed thicknesses and geologic dips, along with all the uncertainties previously discussed. The driller and geosteering engineer must determine and answer questions such as:

- What are the precise hole angle and direction to use?
- How far do we drill?
- How many turns do we make?
- Can we get there from here?
- How will we know we have reached the target?
- Where are we, relative to the bed boundaries?
- Now that we are at a certain objective location, is this where we want to be?

SOLUTIONS TO MAXIMIZE THE POTENTIAL OF A HORIZONTAL OR EXTENDED-REACH WELL

This section describes a methodology for dealing with the complex issues described above in designing a flexible well-path plan. Many of the following concepts were introduced by Frank Schuh (1989) in his paper "Horizontal Well Planning—Build Curve Design." This section builds on his foundation. As stated earlier, the flexible well-path planning method anticipates and plans for the need to make adjustments while drilling. It incorporates the key geosteering decision points that will be made during the execution of the project. At each of the key decision points, the position of the well relative to the objective target position is determined by using real-time geologic and directional data. The well-path plan then can be realigned if necessary to hit or maintain position relative to the target.

HANDLING UNCERTAINTIES WITH FLEXIBLE WELL-PATH DESIGNS

A flexible well plan differs from the traditional, static, single-line plan in that it is dynamic and uses one or

TABLE 2. MEASURED-DEPTH BED THICKNESS (MT) AS A FUNCTION OF HOLE ANGLE.

	Bed thickness (TVT) (ft)	Apparent dip	Hole angle	Measured-depth thickness (MT) (ft)
Down-section drilled wells	10	0°	80°	57
	10	0°	85°	114
	10	0°	86°	143
	10	0°	87°	191
	10	0°	88°	286
	10	0°	89°	573
Flat	10	0°	90°	Infinite
Up-section drilled wells	−10	0°	91°	573
	−10	0°	92°	286
	−10	0°	93°	191
	−10	0°	94°	143
	−10	0°	95°	114
	−10	0°	100°	58

more flexible (adjustable) tangent sections to accommodate both drilling and geologic uncertainties encountered while drilling. It is a site-specific plan that incorporates the local geology and drilling requirements into the design. Using real-time geologic and directional data, tangent sections are used to locate one or more marker beds above the target and to softly land the well in the target bed at precisely the desired well orientation. Incorporating these tangent sections allows the drilling (steering) team to react to variations from the predrill geologic model and direction plan in real time and without alarm.

The first paradigm shift that operators must overcome if they want to maximize the potential of their horizontal or extended-reach project is to embrace a flexible well plan that anticipates and plans for drilling changes. The well-planning team should consist of a cross-functional group chartered to develop well-path contingency plans and to preplan decision and steering points. The team should include drilling, completion, production, reservoir, and geosteering engineers, and earth scientists from both the operating company and the vendor service company.

DESIGN OF A FLEXIBLE WELL PLAN: BACKWARD FROM THE COMPLETION

Designing a horizontal or extended-reach well is an iterative and interactive process. The first step in the flexible well-plan design should be designing the completion. Highly deviated horizontal and extended-reach wells should be designed backward, starting with the completion process and completion equipment, which dictate many drilling parameters such as hole size and maximum allowable dogleg severity. Some completion equipment, such as electrical submersible pumps, may need to be placed in a straight tangent section in the well to operate efficiently. In these cases, this tangent section must be designed first, because it dictates whether a completion will be successful. (This tangent should not be modified during drilling without concurrence of the well-planning team.)

In the flexible well-path planning method, adjustable tangent sections are designed around key marker-bed horizons to handle the current total range of combined uncertainty factors for that portion of the well. Figure 12 illustrates this concept, using a single tangent section. In this figure, the plan calls for placing a horizontal well at a precise depth below the marker bed. Although the bed dips are flat, a relatively large uncertainty range exists for the depth at which the marker bed will be encountered. To handle this uncertainty, a tangent section is designed to start at the shallowest depth at which the bed could be encountered. The tangent angle is set at a borehole inclination arrived at by designing the well backward and

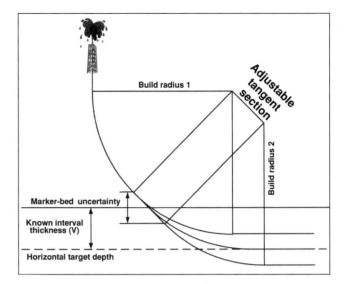

FIGURE 12. Flexible well-path plan.

knowing the vertical height (V) between the marker bed and the target and the desired final build rate. During the drilling operation, the directional driller holds the tangent angle until the marker bed is encountered. The final target TVD is then readjusted, and the lower build section can begin at the depth appropriate to accommodate a perfect landing.

The following equations from Frank Schuh's 1989 paper (p. 48–49) can be used to calculate build-curve designs for 2-D well paths drilling in flat-lying beds. These equations are based on calculating circular arcs for the build sections and straight lines for the tangent sections (see Figures 13 and 14).

The key equations for calculating vertical height (V), displacement (D), and measured-depth length (L) of a vertical circular arc are:

$$R = \frac{5730}{B} \tag{1}$$

$$V = R \times (\sin I_2 - \sin I_1) \tag{2}$$
$$D = R \times (\cos I_1 - \cos I_2) \tag{3}$$
$$L = \frac{100\,(I_2 - I_1)}{B} \tag{4}$$

where B = build rate (deg/100 ft), R = build radius (ft), I_1 = initial inclination (degrees), I_2 = final inclination (degrees), V = vertical height (ft), D = displacement (ft), and L = measured-depth length (ft).

The appropriate equations for the straight adjustable tangent section intervals are:

$$L^2 = V^2 + D^2 \tag{5}$$
$$V = L \cos I \tag{6}$$
$$D = L \sin I \tag{7}$$

where I = inclination (degrees), V = vertical height (ft), D = displacement (ft), and L = measured-depth length (ft).

Figure 15 illustrates an ideal, flexible well-path plan having both "key marker-bed" and "soft-landing tangent" sections. The first step in planning the backward well path is to design the objective target section. This is followed by designing the final build section, the soft-landing tangent section, the intermediate build section, and the key marker-bed tangent section(s) and then back-calculating the well to the surface. A key marker bed is a good, easy-to-recognize, geologic correlation marker, ideally within 61 vertical m (200 vertical ft) of the top of the target objective zone for long-radius well-path designs. Of course, this is not always possible.

A spreadsheet similar to that shown in Table 3 can be created based on equations (2) and (3) to begin the backward well-path design process for a horizontal well in flat-lying beds. This table shows the vertical-height difference (V) and displacement lengths (D) for landing a well at 90° starting at various initial inclination angles and using three different build rates. If the drilling engineer suggests using a build rate of 5°/30 m (5°/100 ft) and the reservoir engineer recommends placing the lateral section 4.6 m (15 ft) below the top of the objective, then the soft-landing tangent angle should be designed at an inclination of approximately 81°.

Similarly, the key marker-bed tangent-section design would follow. During drilling, the well path is redesigned immediately after locating the marker bed along the tangent section and is realigned to hit the next key marker bed or the top of the objective along a soft-landing tangent. A similar process occurs again when the well encounters the top of the objective zone in the soft-landing tangent section. The well is realigned to softly land at the desired well orientation in the objective. While drilling the lateral section of a horizontal well, the geosteering team continuously monitors the current position of the well relative to the objective target position and makes adjustments as necessary to maintain the position in the target.

Before beginning the site-specific, flexible well-path planning process, the team must evaluate the total range of uncertainties. In particular, it is important to know the uncertainty in geologic dip rate and the thickness between the key marker beds and the top of the objective. It may take several iterations to complete the plan. For instance, the team will have to use an approximate directional-survey uncertainty factor until the final directional survey is planned. It is often better and quicker to follow a conservative approach with longer tangent sections than overly precise, short-tangent sections.

DESIGNING THE OBJECTIVE TARGET SECTION

Selecting a Line or Curved-line Target Well-path Type

Horizontal and extended-reach well targets differ from conventional well targets. Instead of a point or series of points with a given tolerance radius or box, they are a continuous line or curved-line target with little or no tolerance. The uncertainty issues related to horizontal and extended-reach well drilling mean that these lines are not

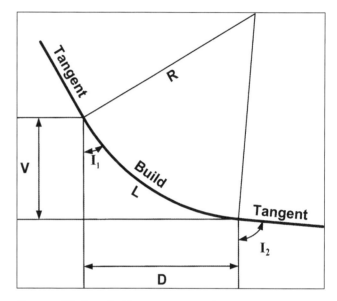

FIGURE 13. Basic build-curve geometry (vertical section view).

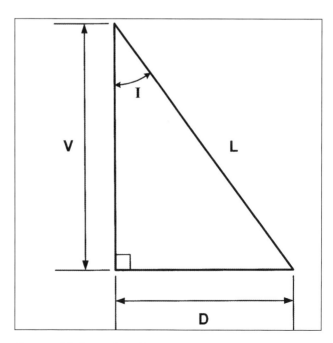

FIGURE 14. Basic adjustable tangent-section geometry (vertical section view).

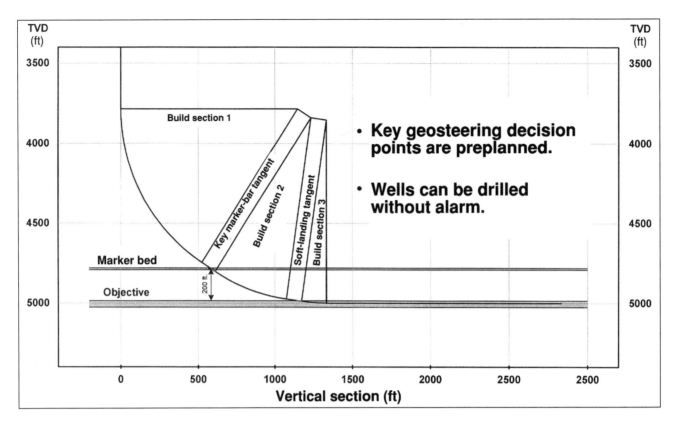

FIGURE 15. Flexible well-path design with key marker-bed and soft-landing tangent sections.

TABLE 3. VERTICAL-HEIGHT (V) AND DISPLACEMENT (D) DIFFERENCE BETWEEN TWO POINTS ALONG A DEFINED BUILD-RATE CURVE.

Bld. rate (deg/100 ft) Radius (ft)		4 1432	4 1432	5 1146	5 1146	6 955	6 955
Inclination 1	Inclination 2	V (ft)	D (ft)	V (ft)	D (ft)	V (ft)	D (ft)
89	90	0.2	25.0	0.2	20.0	0.1	16.7
88	90	0.9	50.0	0.7	40.0	0.6	33.3
87	90	2.0	75.0	1.6	60.0	1.3	50.0
86	90	3.5	99.9	2.8	79.9	2.3	66.6
85	90	5.5	124.8	4.4	99.9	3.6	83.2
84	90	7.8	149.7	6.3	119.8	5.2	99.8
83	90	10.7	174.6	8.5	139.7	7.1	116.4
82	90	13.9	199.4	11.2	159.5	9.3	132.9
81	90	17.6	224.1	14.1	179.3	11.8	149.4
80	90	21.8	248.7	17.4	199.0	14.5	165.8
75	90	48.8	370.7	39.0	296.6	32.5	247.2
70	90	86.4	489.9	69.1	391.9	57.6	326.6
65	90	134.2	605.4	107.4	484.3	89.5	403.6
60	90	191.9	716.2	153.5	573.0	127.9	477.5
55	90	259.0	821.6	207.2	657.3	172.7	547.7

necessarily fixed in space; they tend to be a moving line target until the well is landed in the zone, whereupon the lines becomes more restricted. Lesso and Kashikar (1996) referred to targets in these high-angle or horizontal wells as "point moving targets," and the geosteering process and precise placement of wellbores as "the management of these point moving targets."

Five common line-target well-path types are used for the completion of horizontal and extended-reach wells (four of these are shown in Figure 16). These well-path types include:

Defined vertical-depth wells

- This well type maintains a constant TVD distance from oil-water, gas-oil, or gas-water contact(s).
- Pilot holes designed to find the critical contact(s) might be highly desirable because of directional-survey TVD uncertainties.

Defined structural-position wells (for single-lobed targets)

- This well type maintains a constant position relative to the top or base of the reservoir. In general, the fluid contacts for this type of well are distant and not a major concern.
- Using a flexible well-path planning approach may eliminate the need for pilot holes.
- Note: When drilling a long lateral section in a zone that has a similar log character at its top and base, it is difficult to tell whether the well is exiting from the top

or base of the objective. In this case, it may be prudent to run special azimuthal geosteering tools that can take readings from the high side or low side of the borehole and thus determine if the wellbore is exiting the top or base of the zone.

- If geosteering tools are not used, remember the rule "When lost—turn up." If you make the wrong decision and the well exits from the top of the zone, generally it is easy to make a low-side sidetrack without having to plug back the well.

Zigzag wells (for single-lobed targets)

- This well type is designed to slowly cut back and forth from the top to the base of thin objective zones at incident angles such that changes in MWD log responses, which indicate an approaching bed boundary, can be attributed to the correct boundary. The well is then redesigned at a new attack angle to head back across the target zone. In a dipping formation, zigzag wells can be drilled at a defined vertical depth by snaking back and forth sideways using azimuth changes (instead of inclination changes).

Slant wells (for single or multiple-lobed targets)

- Slant wells are drilled at a low-incident angle from the top to the base or from the base back to the top of the objective interval, generally to intersect multiple lobes.
- A slant well is often chosen because of concerns about vertical connectivity of a multiple-lobed objective.

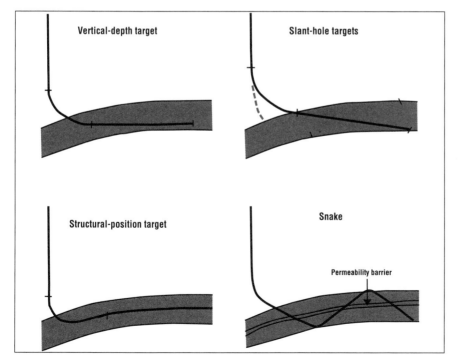

FIGURE 16. Common horizontal-well target types.

- These wells can be affected significantly by slight changes in hole dip angle.
- A slant well also can be drilled as a defined vertical-depth well by drilling horizontally in an updip or downdip direction.

Snake wells (for multiple-lobed targets)

- Snake wells are designed to drill at low-incident angles back and forth from the top to the base and back to the top of the objective, or from the base to the top and back to the base of the objective one or more times.
- The objective of a snake well is to have multiple intersections of the various lobes in a multiple-lobed sandstone. This type of well is often chosen because of concerns about vertical connectivity of a multiple-lobed objective and/or the desire to have multiple take-points per lobe.

DETERMINATION OF COMPLETION REQUIREMENTS AND EQUIPMENT LIMITATIONS

Because a well can be drilled does not mean that it always can be completed or operated efficiently. The well-planning team must communicate to the drilling team the limits on build-curve rates or doglegs that the drilling tools and completion hardware can withstand. In a drilled well, it is not uncommon to have doglegs that are more than 1.5 times those found in the planned design. Therefore, the directional well plan should be conservative to avoid pushing the operational limits of completion tools.

Certain completion tools, such as electric submersible pumps, need to be placed in a smooth, straight, tangent section of the wellbore and at a particular angle to obtain maximum operating efficiency. Tangent sections for these types of tools should be planned separately from the tangent sections used for steering decisions. The drilling team should avoid making steering or well-path adjustments over these tangent sections.

The drilling team also must determine the bending limits of the drilling equipment and bottom-hole assemblies and include this information in the well plan. While drilling, however, the added factor of rotation of the drill string in the curve can lead to tool failure; therefore, the team also must know the bending limits of these tools while they are rotating.

Other equipment-related factors that often need to be addressed concern surface facilities. Many older fields were not designed to handle a 2000 to 10,000 BOPD well, particularly when water cuts increase. The effective choke on the well may not be at the wellhead but downstream at the separator. Therefore, it is prudent to include a facilities engineer and field operations personnel in early design meetings and when developing well-productivity requirements to avoid having a well that, although capable of producing 20,000 BOPD, is producing at severely restricted rates because of facilities limitations.

DESIGNING FOR A SOFT LANDING

After determining the line-target type, selecting the drilling and completion equipment to run, and establishing the proposed build-rate range, the team should design a soft landing as part of the backward well-path design process. The objective of a soft landing is to land the well precisely in the objective at the correct borehole orientation and at the correct depth below the top of the objective. The desired depth below the top of the objective varies, but with horizontal wells in general, it is about 3 m (10 ft).

The soft-landing design approach presented here is based on use of a key marker-bed tangent in conjunction with a soft-landing tangent. The relative current geologic and directional uncertainties of the soft-landing target will be reduced significantly once the well encounters the key marker bed. For example, the relative combined uncertainty between the key marker bed and the soft-landing target would be reduced to ± 4.5 m (15 ft) vertically in a typical well. In addition, by that time, the directional driller will have a good understanding of directional tool performance.

Procedure Summary

1) The well-planning team begins to design the well path by using the planned build rate from the final landing point and calculating backward toward the top of the objective.
2) A tangent angle is set at a point below the top of the objective. This design provides the reaction time needed for the geosteering team to evaluate data while drilling. For a flat-bed situation, this angle is often in the 80° to 85° range for a horizontal well designed at 90°.
   - The reaction time needed is related to the measured-depth drill footage required to decide where the top of the objective was penetrated, usually at 3 to 18 m (10 to 60 ft). The reaction time generally is related to the distance of the formation-evaluation sensors from the bit, how quickly the information reaches the surface, and the experience of the geosteering team.
   - The determination of exactly where the top of the objective is penetrated can be complicated when using resistivity measurements that are affected by

anisotropy and/or the presence of polarization horns.

- If the directional driller has a feel for how the bit reacts to different lithologies, he often knows where the top of the objective is encountered before it is seen on the MWD log data.

3) The tangent angle is held constant in a backward direction (uphole) to the shallowest point where the top of the objective could be encountered (based on the shallowest value in the uncertainty range developed during the preplanning uncertainty analysis).

Note: A set of 3-D equations for handling both inclination and azimuth changes in the approach to the soft landing is not presented in this paper. The general recommendation is to line up the well azimuth on the objective target at the position of the key marker-bed tangent.

DESIGNING A KEY MARKER-BED TANGENT SECTION

The objective of the key marker-bed tangent is to reduce the well-placement uncertainty, thus enabling the wellbore to be refined and in line for the soft landing. The range of total combined uncertainty factors for planning may be ± 9 m (30 ft) TVD for shallow horizontal wells to more than ± 30 m (100 ft) TVD for deep horizontal or extended-reach wells.

Procedure Summary

1) The team designs the key marker-bed tangent by continuing the build rate backward (uphole) from the top of the soft-landing tangent. The well path is designed using the planned build rate.

2) The tangent angle is set at a point below the top of the key marker-bed objective to allow for the amount of reaction time needed for the geosteering team to evaluate data.

3) The tangent angle is held backward (uphole) to the shallowest point where the top of the key marker-bed objective could be encountered (based on the marker-bed uncertainty range).

4) From the top of the key marker-bed tangent section, the well can be back-calculated to the surface location using a combination of build, turn, and tangents as needed, while simultaneously considering the dogleg-severity limitations of the drilling and completion equipment.

If the key marker bed or the objective reservoir top comes in deeper than anticipated but still within the range of uncertainty, it is assumed that the well will continue to be drilled at the tangent angle until the bed is encountered, and the well angle should be maintained. If the key marker-bed depth or top of the objective is not encountered within the defined uncertainty range, then it may be a good time to pick up the bit off bottom and make a conditioning trip to allow time for evaluation and team comment.

HELPFUL HINTS FOR DESIGNING FLEXIBLE WELL-PATH PLANS

- Designing horizontal and extended-reach wells is often an iterative and interactive process. The team must be prepared to spend considerably more time in planning these types of wells than in planning a vertical-hole project.

- The team should use a conservative build rate to avoid pushing the bending limits of the drilling and operating tools.

- The team should design the flexible-tangent sections too long rather than too short.

- For some horizontal or extended-reach wells, it is desirable to drill in a continuous sliding mode during each of the build sections. In this case, the team should design a drilling envelope based on the predicted range of build for a given bottom-hole assembly starting from the target and working backward. An example of a build range for a particular well may be between 5°/30 m (5°/100 ft) to 7.5°/30 m (7.5°/100 ft). The objective is to arrive at the same kickoff point in the backward design.

- If possible, design the well to intersect a known shallower fluid contact on the way to a deeper objective to reduce TVD depth uncertainty during drilling.

- For some wells drilled in areas with dipping beds, the lateral-position uncertainty issue can be very significant when drilling along the strike of the beds. Plan an approach to the objective in an apparent dip direction, landing the well at the desired TVD depth. Plan to hold a tangent until the well penetrates the top of the zone, and then change azimuth to the strike direction.

- The well-planning team should include information about the completion equipment in their plan. Such considerations add additional but important complications to the well-path design. In some instances, a special tangent section may be needed for the placement of completion equipment. The drilling team should avoid making steering or well-path adjustments over these tangent sections.

CONCLUSIONS

Use of the flexible well-path planning method in drilling horizontal and extended-reach wells increases the accuracy of well placements, thereby increasing the net present value (NPV) by an estimated 20% or more. This method plans for the need to make trajectory adjustments while drilling, recognizing the uncertainties and inaccuracies that exist in geologic models, directional surveys, and ability to predict precisely and control buildup rates and doglegs for drilling projects.

A flexible well-path plan differs from the traditional, static "drill-the-line" plan in that it is dynamic. Key geosteering decision points are preplanned based on the identification of one or more key marker horizons with known interval thicknesses to the proposed target. The flexible well-path plan incorporates one or more adjustable tangent sections in the design. These tangents are designed across key marker beds and handle the current total range of combined uncertainty factors for that portion of the well. Each tangent section is designed at a specific borehole inclination angle, which is determined from the desired buildup rate and required borehole angle needed to align the well with the next marker bed or to softly land at the precise well orientation in the target. Advanced steering techniques are then applied as necessary in the lateral section to maintain the desired position and orientation in the target bed.

The flexible well-path planning approach gives the well-construction team flexibility to react without alarm to real-time geologic and directional complexities. The increased precision of well placements and understanding of the well's position allow operators to realize higher recoveries and sweep efficiencies. In addition, operators will realize lower costs because of fewer sidetracks and better overall understanding of the position of the well in the reservoir.

REFERENCES CITED

Barry, A., P. Burnett, and C. Meakin, 1998, Geosteering horizontal wells in a thin oil column: Presented at the 1998 Society of Petroleum Engineers Asia Pacific Oil and Gas Conference and Exhibition, October 12–14, 1998, Perth, Australia, paper SPE-50072, p. 221–233.

Lesso Jr., W. G., and S. V. Kashikar, 1996, The principles and procedures of geosteering: Presented at the 1996 International Association of Drilling Contractors/Society of Petroleum Engineers Drilling Conference, March 12–15, New Orleans, Louisiana, paper SPE-35051, p. 133–147.

Lesso Jr., W. G., I. M. Rezmer-Cooper, and M. Chau, 2001, Continuous direction and inclination measurements revolutionize real-time directional drilling decision-making: Presented at the 2001 International Association of Drilling Contractors/Society of Petroleum Engineers Drilling Conference, February 27–March 1, 2001, Amsterdam, Netherlands, paper IADC/SPE-67752, p. 470–484.

Poston, S. W., 1985, Inaccurate wellbore surveys can result in lost reserves, part 1: World Oil, v. 200, no. 5, p. 71–72, 74.

Poston, S. W., 1985, Inaccurate wellbore surveys can result in lost reserves, part 2: World Oil, v. 200, no. 6, p. 71–73, 75.

Schuh, F. J., 1989, Horizontal well planning—Build curve design: Presented at the 1989 Centennial Symposium, Petroleum Technology into the Second Century, October 18–19, New Mexico Institute of Mining and Technology, Socorro, New Mexico, paper SPE-20150, p. 47–61.

Tearpock, D. J., and R. E. Bischke, 1990, Applied subsurface geological mapping: Englewood Cliffs, New Jersey, Prentice-Hall, Inc., 672 p.

Williamson, H. S., P. A. Gurden, D. J. Kerridge, and G. M. Shiells, 1998, Application of interpolation in-field referencing to remote offshore locations: Presented at the 1998 Society of Petroleum Engineers Annual Technical Conference and Exhibition, September 27–30, New Orleans, Louisiana, paper SPE-49061, p. 387–398.

Williamson, H. S., 1999, Accuracy prediction for directional MWD: Presented at the 1999 Society of Petroleum Engineers Annual Technical Conference and Exhibition, October 3–6, Houston, Texas, U.S.A., paper SPE-56702, 16 p.

Wolff, C. J. M., and J. P. de Wardt, 1981, Borehole position uncertainty—Analysis of measuring methods and derivation of systematic error model: Journal of Petroleum Technology, v. 33, no. 12, p. 2339–2350.

Morton, A. C., P. J. Spicer, and D. Ewen, 2003, Geosteering of high-angle wells using heavy-mineral analysis: The Clair field, west of Shetland, U.K., *in* T. R. Carr, E. P. Mason, and C. T. Feazel, eds., Horizontal wells: Focus on the reservoir: AAPG Methods in Exploration No. 14, p. 249–260.

16

Geosteering of High-angle Wells Using Heavy-mineral Analysis: The Clair Field, West of Shetland, U.K.

Andrew C. Morton[1]

HM Research Associates
Woodhouse Eaves
Leicestershire, U.K.

Patrick J. Spicer

Ryton Exploration
GeoScience Ltd.
Barnet, Hertfordshire, U.K.

David Ewen

BP Exploration
Aberdeen, U.K.

ABSTRACT

Conventionally, a combination of biostratigraphy, logging-while-drilling (LWD) data, and routine cuttings description is used to monitor continually the geology encountered during drilling of high-angle wells. However, the biostratigraphic component of the geosteering tool kit cannot be used if there is insufficient diversity of microfossils in the sequence to be drilled. In such circumstances, alternative, less conventional geosteering methods can be used. This paper presents the application of one such approach, heavy-mineral analysis (HMA), in monitoring high-angle wells in the Clair field, west of Shetland, U.K. The reservoir sequence in the Clair field comprises Devonian-Carboniferous nonmarine fluvial and eolian sandstones that lack a continuous, diverse suite of palynomorphs. Consequently, there is no high-resolution biostratigraphic framework for reservoir correlation. By contrast, heavy minerals occur throughout the sequence and therefore offer a potential for correlation and discrimination of different sandstone units.

During appraisal of the Clair field in 1996 and 1997, high-angle wells targeted Clair Unit V, a unit approximately 50 m thick with the best overall reservoir quality. The overlying Unit VI has poor reservoir quality, and therefore it was important that the wells avoided drilling through significant lengths of this unit. The reservoir quality of the underlying Unit IV is poorer than in Unit V but is better than in Unit VI. Therefore, drilling of Unit IV sandstones could be tolerated but ideally would be avoided.

Prior to the application of real-time HMA at well site, it was necessary to establish whether the method would provide adequate, repeatable distinction among units IV, V, and VI. This was assessed first using core material from the "reference" well 206/8-8. On the basis of variations in a number of key parameters, two major events were observed. One of these corresponds closely to the V–VI boundary, and the other lies within the top part of

[1]Also with the Department of Geology and Petroleum Geology, University of Aberdeen, Aberdeen, U.K.

Unit IV, approximately 20 m below the IV–V boundary. These promising results were tested further by analysing two uncored sequences, which established that these events occur elsewhere in the field and that they can be identified by analysis of cuttings samples. It was therefore considered that HMA provided sufficient resolution for the method to be applied at the well site as part of the geosteering tool kit. The stratigraphic implications of the HMA data are illustrated by reference to the two wells (206/8-10Z and 206/8-11Z) where the method has been used to date.

INTRODUCTION

When drilling high-angle wells, frequently it is crucial that the wellbore remains within a particular stratigraphic unit. Conventionally, geosteering of high-angle wells is achieved using a combination of cuttings description by the well-site geologist, logging while drilling (LWD), and high-resolution biostratigraphy. However, in some reservoir sequences, the biostratigraphic component of the geosteering suite may not be available. This occurs if biostratigraphic events lack sufficient resolution or if the sequence is entirely barren biostratigraphically. Under such circumstances, an alternative stratigraphic approach may be required.

One such technique is heavy-mineral analysis (HMA), which characterizes sandstones by determining the miner-alogical composition of the high-density (>2.8 gm cm^{-3}) grain component. HMA provides a basis for correlation and subdivision of sandstone sequences by identifying stratigraphic changes in mineralogical characteristics. If the target formation has a distinctive mineralogical composition, it can be identified by analysis of cuttings material, enabling continuous real-time monitoring of the wellbore during drilling.

This paper describes the application and impact of well site HMA during drilling of two high-angle wells in the Clair field, west of Shetland (Figure 1). HMA was used because of the scarcity of biostratigraphic markers in the reservoir sequence and because provenance studies of the sandstones had revealed the existence of potentially useful stratigraphic changes in mineralogy (Allen and Mange-Rajetzky, 1992).

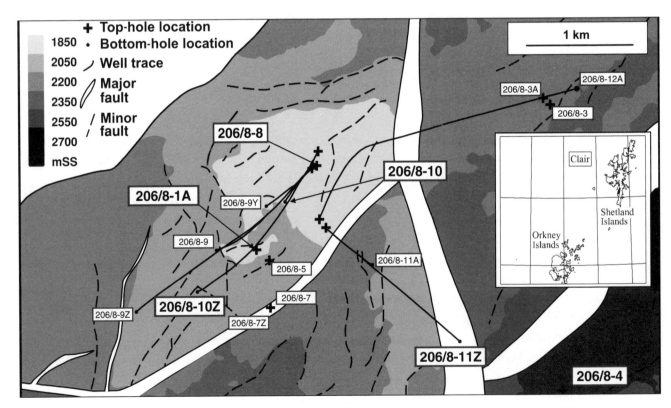

FIGURE 1. Map showing location of wells in the Clair field, including 206/8-1a, 206/8-4, 206/8-8, 206/8-10Z, and 206/8-11Z. Inset shows location of Clair field, west of Shetland.

CLAIR FIELD

The Clair field was discovered in 1977 by well 206/8-1A. The reservoir sequence (Figure 2) comprises Devonian to possibly Early Carboniferous nonmarine clastics, overlain by Cretaceous mudstones and Tertiary sandstones and mudstones, and underlain by Archean (Lewisian) gneisses (Ridd, 1981; Blackbourn, 1987; Coney et al., 1993). The sequence has been subdivided into two major lithostratigraphic units, the Upper Clair Group and the Lower Clair Group (Allen and Mange-Rajetzky, 1992), and 10 informal units (Units I–VI in the Lower Clair Group and VII–X in the Upper Clair Group). The Lower Clair Group, which was deposited in an areally restricted intermontane basin with limited external drainage (Allen and Mange-Rajetzky, 1992), represents a single first-order cycle of fluvial advance followed by fluvial retreat (McKie and Garden, 1996). Within the Lower Clair Group, three second-order cycles have been recognized (McKie and Garden, 1996). The first of these cycles comprises fluvial conglomerates (Unit I), fluvial sandstones (Unit II), and eolian sand sheets (Unit III). The next comprises Unit IV, which is dominated by fluvial sandstones with minor eolian reworking, and Unit V, which contains eolian sand sheets with minor fluvial deposits. The final second-order cycle comprises the lacustrine transgressive-regressive Unit VI. The Upper Clair Group also comprises a single major cycle (Coney et al., 1993). It was deposited by a larger fluvial system draining a wider hinterland (Allen and Mange-Rajetzky, 1992), and conse-quently a major change in provenance is recognized at the Lower/Upper Clair Group boundary. Units VII, VIII, and IX comprise mainly fluvial sandstones, with marginal marine and distributary bay deposits becoming prevalent in Unit X.

The results of the discovery well 206/8-1A and the adjacent well 206/7-1, which flowed 1500 and 960 BOPD of 25° API oil, respectively, were considered promising. However, subsequent appraisal drilling in the later 1970s and the 1980s failed to demonstrate a commercially viable project (Coney et al., 1993). It was not until after 3-D seismic data had significantly enhanced understanding of the field and the subsequent drilling of two wells in 1991 that significant progress was made. A carefully sited vertical well, 206/8-8, was drilled in the same field segment as the discovery well ("core area"), and a horizontal well (206/7a-2) was drilled through Upper Clair Group sediments and into the basement. Well 206/8-8 flowed more than 3000 BOPD from each of two intervals, one of which was Unit V, and well 206/7a-2 flowed 2100 BOPD after acid stimulation.

The most important finding of the 1991 drilling was identification of an open-fracture system concentrated in the interval from upper Unit IV to lower Unit VII (but also present in the basement). Probably equally important was the realization that significant fluid flow from the main part of the reservoir depends on the presence of both good-quality matrix and open fractures (Coney et al., 1993). These requirements are met best by Unit V in the core area of the Clair field. Thus, Unit V was the ob-

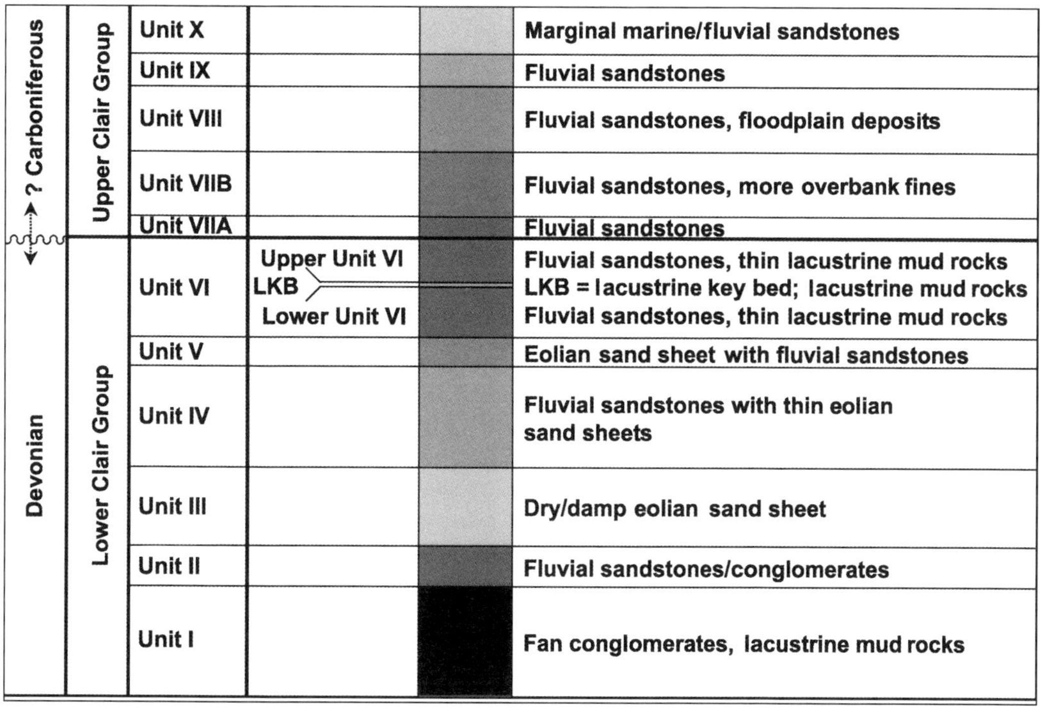

? Carboniferous	Upper Clair Group	Unit X			Marginal marine/fluvial sandstones
		Unit IX			Fluvial sandstones
		Unit VIII			Fluvial sandstones, floodplain deposits
		Unit VIIB			Fluvial sandstones, more overbank fines
		Unit VIIA			Fluvial sandstones
Devonian	Lower Clair Group	Unit VI	Upper Unit VI LKB Lower Unit VI		Fluvial sandstones, thin lacustrine mud rocks / LKB = lacustrine key bed; lacustrine mud rocks / Fluvial sandstones, thin lacustrine mud rocks
		Unit V			Eolian sand sheet with fluvial sandstones
		Unit IV			Fluvial sandstones with thin eolian sand sheets
		Unit III			Dry/damp eolian sand sheet
		Unit II			Fluvial sandstones/conglomerates
		Unit I			Fan conglomerates, lacustrine mud rocks

FIGURE 2. Stratigraphy of the Clair field.

jective for penetration and testing by the first high-angle well targeted at the Lower Clair Group, which was well 206/8-9, drilled and subsequently sidetracked as 206/8-9Z in 1992. The sole reason for the sidetrack was suboptimum penetration of the Unit V target, a sandstone approximately 45 m thick. Too much of the 206/8-9 wellbore was in the poorer-quality, overlying unit, Unit VI. More effective geosteering tools were therefore high on the agenda when it came to planning the subsequent high-angle wells 206/8-10Z and 206/8-11Z, which also targeted Unit V.

For HMA to be a useful component of the geosteering suite, Unit V must have a mineralogical signature that distinguishes it from the underlying Unit IV and the overlying Unit VI. Therefore, mineralogical studies were initiated on adjacent offset wells to investigate possible mineralogical variations over the Unit IV–VI interval. Because there is an error of approximately ±30 m in the prediction of the depth of the top Unit V event (because of seismic resolution and velocity uncertainty), analysis was conducted throughout Unit V, through the lowest 40 m of Unit VI, and through the top 40 m of Unit IV. Although the ultimate objective was to identify stratigraphic units from analysis of cuttings samples at well site, the first phase of the analytical program concentrated on core material from well 206/8-8, to provide the best and least equivocal template of stratigraphic variations in mineralogy.

HEAVY-MINERAL STRATIGRAPHY: THE UNIT IV–VI INTERVAL

Heavy-mineral assemblages are inherited primarily from the sediment source area but are also affected by processes that operate during the sedimentation cycle. The most significant alterations to heavy-mineral assemblages result from hydraulic and diagenetic processes (Morton and Hallsworth, 1994, 1999). The heavy-mineral parameters used to provide a high-resolution stratigraphic framework ideally should reflect only changes in sediment provenance.

There are two types of provenance-sensitive criteria available: ratio data and varietal data (Morton and Hallsworth, 1994). Ratio data determine the relative abundance of minerals with similar hydraulic and diagenetic behavior, such as apatite and tourmaline. Varietal data quantify differences shown by individual mineral populations, such as garnet. The most objective type of varietal data is generated by single-grain geochemical data on composition of individual mineral populations, such as garnet (Morton, 1985) or tourmaline (Henry and Dutrow, 1992) using electron microprobe techniques, but

this approach is obviously not applicable at the well site. However, it is possible to undertake varietal studies that differentiate on the basis of optical properties such as color, habit, or internal structure (Mange-Rajetzky, 1995). In this context, it was considered significant that Allen and Mange-Rajetzky (1992) observed variations in apatite roundness in the Clair Group succession related to the extent of eolian influence, especially because Unit V is considered to have significantly greater eolian influences than Unit VI and, to a lesser degree, Unit IV (McKie and Garden, 1996).

In addition to provenance-sensitive criteria, it may also prove useful to employ other criteria for correlation, provided that limitations on the use of such approaches are fully appreciated. For example, it may be possible to correlate on the basis of variations in the abundance of minerals that are unstable in the subsurface, but it would be inadvisable to attempt to use this approach for long-distance correlation or for correlating sequences across a wide burial-depth range.

The pilot study on the basal Unit VI–top Unit IV interval in well 206/8-8 identified variations in several heavy-mineral parameters (Figure 3), enabling the definition of three heavy-mineral units, termed VIm, Vm, and IVm.

Unit VIm (down to 1860 m) has consistently high apatite-tourmaline index (ATi) and garnet-zircon index (GZi) values (see Figure 3 for definition), both >90. The unit has low apatite roundness index (ARI), although roundness increases toward the base. The unstable minerals epidote and titanite are both present, together forming approximately 10% of the total assemblage.

Unit Vm (1860–1932 m) includes sandstones with lower ATi and GZi than in Unit VIm, although some samples with high ATi and/or high GZi also occur. ARI is high throughout. Unstable minerals are generally absent, although occasional samples have appreciable quantities, giving rise to some spikes in the curve.

Unit IVm (below 1932 m) is comparable to Unit VIm in that ATi and GZi are consistently high and ARI is low, although variations in the latter parameter are gradational. Unstable minerals are very scarce or entirely absent.

These variations are interpreted as the interplay of three processes. Differences in ATi and GZi indicate differences in provenance, with the prevalent high ATi–high GZi source being diluted by low ATi–low GZi material in Unit Vm. One possible explanation for this is that the additional low ATi–low GZi detritus was introduced by eolian processes, which had greater influence in this part of the sequence (McKie and Garden, 1996). The increase in apatite roundness in Unit Vm also is attributable to the greater amount of eolian activity, with mechanical abrasion causing this relatively soft mineral to become rapidly abraded and rounded. The gradational decline in ARI in the basal part of Unit VIm indicates either continued eolian activity or reworking of Unit Vm sand. The presence of unstables in Unit VIm results from the lower porosity in this part of the sequence (McKie and Garden, 1996). This has inhibited pore-fluid movement, thus preserving the relatively unstable minerals. The occasional presence of sandstones containing epidote and titanite in Units Vm and IVm is attributed to the same process, reflecting the presence of tight zones in these units.

The differentiation of three heavy-mineral units over the Unit VI–Unit IV interval in well 206/8-8, therefore, depends partly on changes in provenance, partly on depositional facies, and partly on the extent of diagenesis. The mineralogical events associated with the changes in provenance and facies are likely to be recognizable on a semiregional scale, but greater caution is needed when using the variation in the abundance of unstables as a correlation tool. To establish whether all the variations observed in well 206/8-8 also occur elsewhere in the Clair field, two further wells were analyzed: 206/8-1A and 206/8-4 (Figure 1). This phase of the study also was designed to establish whether the features observed in the core from well 206/8-8 also could be identified by analysis of cuttings material, although some core was analyzed from 206/8-1A.

As shown in Figure 4, the 206/8-1A sequence shows virtually identical features to 206/8-8, the only significant difference being that the baseline ATi level from the cuttings in Unit IVm is lower than in 206/8-8 or than in Unit VIm in 206/8-1A. This could be a result of downhole contamination of the Unit IVm section by material from the overlying Unit Vm. Alternatively, it may have resulted from loss of the relatively soft mineral apatite through the mechanical action of the drill bit. The pattern shown by well 206/8-4 (Figure 5) also is closely comparable to well 206/8-8. However, there is little downhole variation in unstables, which are relatively low in abundance throughout, and this results in less contrast in this parameter. The generally lower abundance of unstables is because of the greater burial of the 206/8-4 sequence, with higher temperature pore fluids causing mineral dissolution even in the less permeable Unit VIm.

The results of the pilot study on the three wells, therefore, was considered to be very encouraging, in that it proved possible to identify clear downhole variations in several parameters in three wells from the Clair field area, from both core and cuttings samples. The only significant inconsistency is that the upward appearance of unstable minerals at the Unit VIm–Vm boundary in wells 206/8-8 and 206/8-1A is less pronounced in well 206/8-4 because of the more advanced dissolution in the deeper well. This was not considered a problem during drilling of wells 206/8-10Z and 206/8-11Z, both of which are locat-

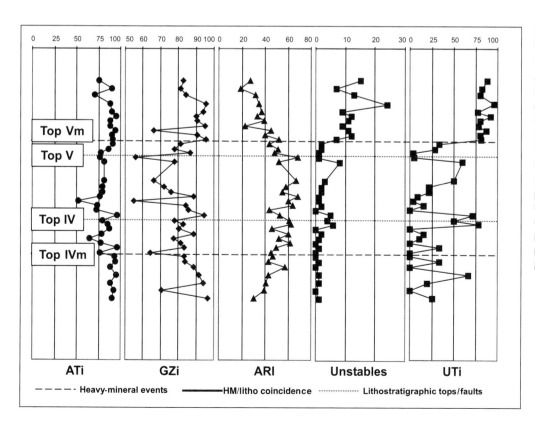

FIGURE 3. Downhole heavy-mineral profile for the cored well 206/8-8. ATi = apatite-tourmaline index (% apatite in total apatite + tourmaline); GZi = garnet-zircon index (% garnet in total garnet + zircon); ARI = apatite roundness index (% rounded apatite in apatite population); Unstables = % epidote + titanite; UTi = unstables-tourmaline index (% epidote + titanite in total epidote + titanite + tourmaline).

ed in the shallower parts of the field and thus within the "unstables" preservation window. The boundary between the heavy-mineral units Vm and VIm coincides closely with the boundary as defined on lithostratigraphic and petrophysical parameters in all three wells, although there is a small discrepancy in the case of well 206/8-8. There-fore, the heavy-mineral data are able to discriminate be-tween the lower net-to-gross reservoir rocks of Unit VI and the high net-to-gross reservoir rocks of Unit V.

By contrast, the base of Unit V is not marked by any clear mineralogical change, because the boundary be-tween heavy-mineral units Vm and IVm occurs approxi-

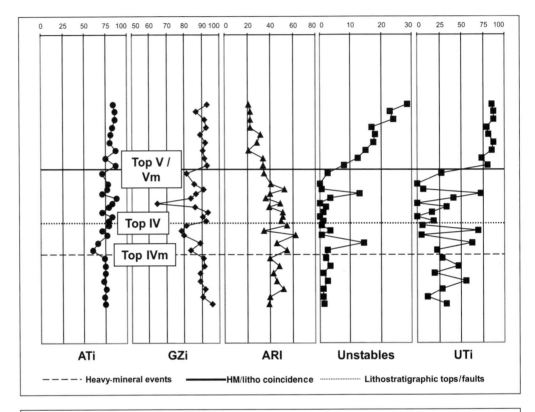

FIGURE 4. Downhole heavy-mineral profile for well 206/8-1A. See Figure 3 for key.

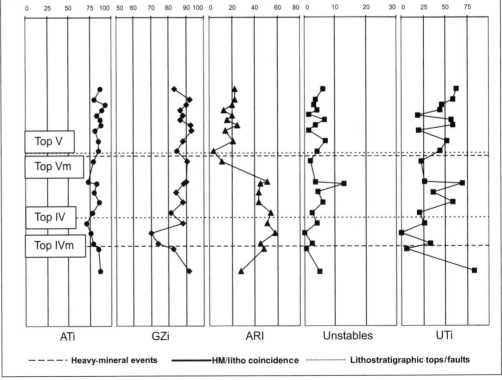

FIGURE 5. Downhole heavy-mineral profile for well 206/8-4. See Figure 3 for key.

mately 20–40 m below base Unit V, as defined lithostratigraphically. Consequently, heavy-mineral data do not provide a basis for discrimination between Units IV and V but are able to diagnose whether the well track is substantially below the base of Unit V. The recognition of these two semiregionally consistent heavy-mineral events, one at the top of Unit V and one in the upper part of Unit IV, was considered sufficiently encouraging for the technique to be applied offshore.

APPLICATION AT THE WELL SITE

To date, HMA has been used to monitor the progress of the track of two wells, 206/8-10Z and 206/8-11Z. In both cases, sample preparation and heavy-mineral separation were carried out in a conventional offshore palynology preparation unit. The cuttings samples, which were drilled using synthetic oil-based mud, were cleaned in an ultrasonic unit using a combination of sodium carbonate and detergent. The 63–125-μm fraction was extracted by sieving, and heavy minerals were separated by gravity-settling in bromoform. Heavy-mineral residues were mounted on glass slides and analyzed using a petrographic microscope. There was, therefore, no difference in preparation and analysis between the offshore and onshore environments. Heavy-mineral preparation took approximately 60–75 minutes, with optical analysis taking approximately 30 minutes, enabling data to become available in less than two hours from receipt of each sample at the surface. Given that key LWD sensors were located more than 20 m behind the bit and that average penetra-

tion rates were 5–10 m/hr, heavy-mineral data were generally acquired before logging data were available. Furthermore, although a single HMA determination could be highly indicative of a formation change, generally at least 10–15 m of log trace would be needed to identify the same change with any confidence.

WELL 206/8-10Z

The well track of 206/8-10Z is shown in Figure 6. This cross section is based on immediate postwell interpretation, integrating the well-site heavy-mineral work with log picks for the lithostratigraphy and the seismic and fracture data (believed indicative of the location of faults), but predated full review of the seismic data and availability of the well seismic calibration. The cross section shows that the well crossed (as expected) a number of faults. When combined with the error bars on the well prognosis caused by seismic and depth conversion uncertainty, mentioned above, this constituted a significant challenge in steering the well. The goal for well placement was to ensure that the well stayed at the right depth to optimize penetration of Unit V. Thus, we needed to interpret stratigraphic data on an ongoing basis to know the location of the bit and its position relative to the well prognosis. This knowledge would allow timely decision making about whether to steer the well up, to steer the well down, or to maintain attitude. Timely decision making was crucial, both in terms of the drilling cost of errors necessitating a plugback and sidetrack, but also in terms of well evaluation. A suboptimum well path would be expected to limit well performance on test. In addition,

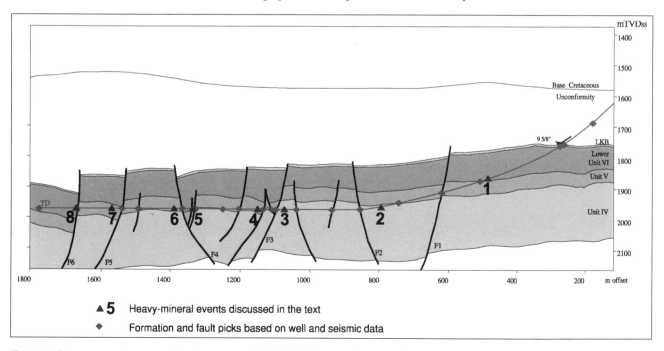

FIGURE 6. Interpreted cross section along trace of 206/8-10Z showing location of heavy-mineral events #1–#8.

delay could have had a serious negative impact on the duration of the planned long-term test, because there were potential weather-window constraints at the end of the program.

The observed variations in heavy-mineral parameters along the well track are shown in Figure 7. The high-angle sidetrack commenced in Unit VI, which was recognized on the basis of high abundances of unstable minerals, high ATi and GZi values, and low ARI. The top of Unit V was heralded by the gradual increase in ARI in the lower part of Unit VI. It was diagnosed by the major decline in abundance of unstable minerals and the fall in ATi values, along with the continued increase in ARI at point #1 in Figures 6 and 7. Further indications of Unit V are given by subsequent reductions in GZi (Figure 7).

The change to high ATi, high GZi, and low apatite roundness at point #2 (Figures 6 and 7) indicated penetration of Unit IVm, suggesting that the wellbore at this point was approximately 20 m below base of lithostratigraphic Unit V. As shown in Figure 6, this may result from the well having gone downsection stratigraphically. Alternatively, it may indicate that the wellbore had intersected either the seismic-scale fault (F2 in Figure 7) bounding an upthrown fault block identified prior to drilling (but incorrectly located on the section) or a related subseismic-scale fault in front of F2. Consistently high ATi and GZi, along with low apatite roundness, indicated continued drilling of this upthrown block over a significant distance.

At point #3, penetration of Unit Vm is indicated by a marked fall in GZi values allied to an ongoing fall in ATi values. This event occurs within an interval where ARI values are typical of Unit Vm. Because the seismic data strongly indicated that the well up to this point should be at its deepest penetration of Unit IV, it is thought that the higher ARI values prior to point #3 arise from an interval

FIGURE 7. Variations in heavy-mineral parameters in 206/8-10Z, showing heavy-mineral events #1–#8. See Figure 3 for key.

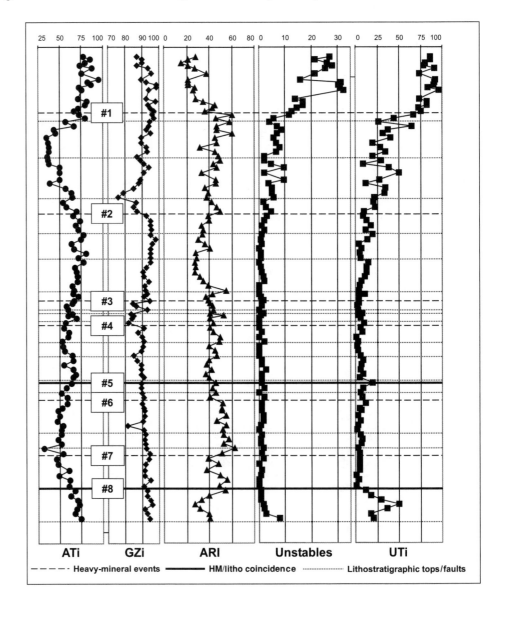

in Unit IVm. Although the data showed some subtle variations, Unit Vm indications continued from point #3 as far as point #8. At this point, the cuttings descriptions by the well-site geologist indicated penetration of Unit VI, suggesting that the well track had passed across the major fault predicted prior to drilling (F6 in Figure 6). The fault was subsequently confirmed by the LWD tool. However, the expected dramatic rise in abundance of unstable minerals, coupled with a decrease in apatite roundness and increases in ATi and GZi, failed to materialize immediately, and penetration of Unit VIm became obvious only after about 50 m of continuous drilling.

The inability to recognize Unit VIm was interpreted as the result of contamination, because the well at this point had been close to horizontal for nearly 1 km, drilling almost exclusively through Units V and IV. This caused considerable dilution of the Unit VIm heavy-mineral signal when it was first penetrated. Because of this problem, we decided to determine a new parameter—the ratio of unstable minerals to tourmaline (UTi). We considered this parameter to be a more sensitive indicator of Unit VIm, because unstables are abundant and tourmaline is scarce in Unit VIm, whereas the converse is true for units IVm and Vm. When this parameter was determined, the Unit Vm–VIm boundary became clear (Figure 7). Having positively identified Unit VIm, it was decided to terminate drilling rather than attempt to drill through the downthrown block back into Unit Vm, as originally intended. Determination of UTi during reexamination of heavy-mineral preparations from higher in the well also showed high UTi at point #5, suggesting that the wellbore may have entered Unit VIm briefly, a possibility predicted prior to drilling.

The parameter UTi was subsequently computed for the reference wells (Figures 3, 4, and 5). As in well 206/8-10Z, this parameter is a more dramatic indicator of the Unit VIm–Unit Vm boundary, particularly in well 206/8-4.

As already noted, Unit Vm indications were continuous from point #3 to point #8. As such, the geosteering tool served its purpose by providing evidence that the well attitude was correct. Subsequent analysis of the heavy-mineral data combined with seismic and log data have suggested, however, that the subtle changes observed probably are meaningful. Thus, the fall in ATi at point #4, coincident with a sharp change in GZi and allied with continued high ARI values, is now thought to represent a fault penetration with a change in level in Unit Vm. At point #5, there is an increase in UTi, suggesting a possible Unit VIm penetration, consistent with evidence from seismic data (Figure 6). At point #6, there is another increase in ARI allied to falling ATi. Taken in conjunction

with the seismic indication of a fault, this suggests fault penetration into Unit Vm. Similarly, at point #7, there is a marked fall in ARI with virtually no change in GZi, UTi, or unstables and only a gradual increase in ATi. This is suggestive of an intra–Unit Vm fault, which can also be inferred from the seismic data. These interpretations lend support to the view that heavy-mineral analysis can be an effective geosteering tool.

WELL 206/8-11Z

Unlike well 206/8-10Z, the high-angle sidetrack 206/8-11Z commenced in the Upper Clair Group. This provided an opportunity to determine the mineralogical characteristics of the Upper Clair Group in the vicinity of the well, so that penetration of Upper Clair Group predicted at the toe of the well (Figure 8) could be readily identified. The HMA profile for 206/8-11Z is shown in Figure 9.

From commencement of the high-angle sidetrack down to point #1, the sequence is characterized by extremely high GZi values, low ATi values, low apatite roundness, relatively high RuZi (rutile-zircon index), moderate unstable abundances, and high UTi values. Staurolite is a conspicuous and common component, forming 2.5% to 7.5% of the assemblages. Chloritoid is present in the lower part of the zone. These assemblages are markedly different from those seen previously in Units IV–VI but are similar to those described from the Upper Clair Group by Allen and Mange-Rajetzky (1992). This section therefore has the heavy-mineral characteristics of the Upper Clair Group.

At point #1, there is a marked change in mineralogy. ATi values are consistently high, whereas GZi and RuZi are markedly lower. Staurolite becomes rare and sporadic. Unstables show a significant downhole increase, although they drop back to lower values approaching point #2. According to well-log data, point #1 does not correspond precisely with the base of the Upper Clair Group but is apparently in Unit VIIA. This suggests that the fundamental change in provenance in the Clair Group occurs slightly above the base of the Upper Clair, in Unit VIIA. At point #2, there is a major downhole decrease in UTi, along with an increase in apatite roundness. ATi shows a downhole decline, but GZi remains similar to the overlying interval. These parameters indicate penetration of Unit Vm.

The section between points #1 and #2, therefore, comprises Unit VIm (Lower Clair Group) as well as Unit VIIA (Upper Clair Group). There are some variations in mineralogy in this part of the well track, suggesting that HMA may be able to provide a higher-resolution stratigraphic breakdown in the Unit VIm-VIIA interval.

At point #3, there is a change to lower ARI, along

with high ATi and high GZi. If the well had been drilling downsection, this would have been consistent with penetrating Unit IVm. However, the well at this point was essentially horizontal, close to bed parallel, and had not penetrated a normal (vertical) thickness of Unit Vm. Accordingly, mineralogical variations beneath point #3 have to be interpreted as indicating either subtle stratigraphic changes because of small differences between bed dip and well attitude or more dramatic changes caused by crossing seismic-scale or subseismic-scale faults.

Between points #3 and #4, the sequence shows an alternation of two mineralogical types. The first type is that described above, immediately beneath point #3. The high ATi and GZi values are not definitive; although they are consistent with both Unit VIm or IVm, some Unit Vm samples have these characteristics. The ARI data also are equivocal, but the change to lower roundness-index values suggests a stratigraphic position in either Unit VIm or basal Unit Vm–upper Unit IVm. Most of the samples with this mineralogy have low unstable-mineral contents and low UTi values, suggesting that Unit VIm can be ruled out. However, there are some samples with higher unstable-mineral abundances. In some cases, the unstable minerals probably are contaminants; several of the samples with high unstable-mineral contents were collected immediately after a change of drill bit. During the subsequent reentry into the hole, cavings or held-up cuttings from higher in the section, rich in unstables, easily could have been pushed in front of the bit along the low side of the hole and mixed with the genuine cuttings. In other samples, however, the high unstable-mineral content appears to be genuine. In such cases, well logs indicate a

high degree of cementation, indicating that the presence of unstable minerals is the result of preservation in low-permeability cemented intervals. Taking all these factors into account, this mineralogical type is considered to indicate basal Unit Vm–upper Unit IVm, heavily carbonate-cemented in places. Along the well track, this mineralogical type alternates with another type having relatively low ATi and slightly higher apatite roundness, clearly of Unit Vm affinity.

The significant change at point #3 was unexpected, and the original well prognosis was consequently reviewed in light of the position of the change on the cross section along the well trace. The most probable explanation was that the wellbore had penetrated a fault that had not been resolved on the seismic data, crossing from a position in the main part of litho–Unit V (HM Unit Vm) into sediments probably belonging to HM Unit IVm (Figure 8). From our offset data, these sediments must lie at least 20 m beneath the top of litho–Unit IV on the upthrown side of the fault. Subsequent penetration of sandstones with Unit Vm characteristics, allied to the bedding attitude deduced from the predrill mapping of the top Unit V surface, suggested that the wellbore was tracking close to the Unit Vm–Unit IVm boundary. Because this position was suboptimum, it was decided to steer the well upsection to try to regain a position nearer the midpoint of litho–Unit V. This demonstrates the value of the HM data, because this conclusion could not have been made on the basis of the cuttings and LWD data alone.

A major change in mineralogy takes place at point #4 (Figures 8 and 9). ATi and apatite roundness drop dramatically, GZi rises to nearly 100, and RuZi also increas-

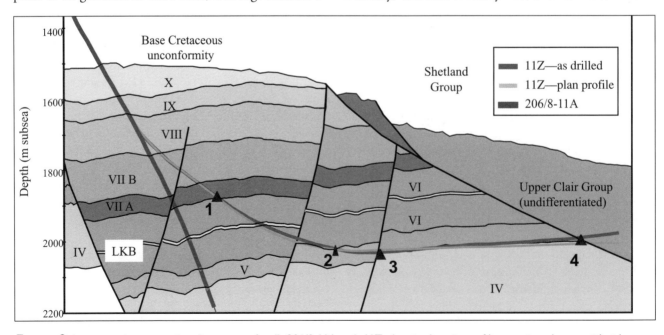

FIGURE 8. Interpreted cross section along trace of wells 206/8-11A and -11Z, showing locations of heavy-mineral events #1–#4.

es significantly. Staurolite reappears as a major component of the assemblages. These features all indicate penetration of Upper Clair Group (Unit VIIB or VIII) strata. Unstable-mineral abundances are slightly lower than previously seen in Unit VIIB or VIII, probably because of dilution by material from the stable-mineral-dominated high-angle section. This may also be why the UTi value levels out at approximately 80, compared with >90 for the equivalent section penetrated prior to point #1. The well was terminated shortly after this event was recognized.

IMPACT

Throughout drilling of well 206/8-10Z, with the exception of the toe section of the well noted above, HMA made a significant contribution to the interpretation of well position and thus facilitated daily decision making on the well profile. Although in practice few decisions to change well course were required, the high confidence that HMA interpretations gave in well position was a sig-

nificant aid in decision making. The high-confidence interpretations also reduced the danger of an erroneous decision to steer off course because of uncertainty in other data sources.

In well 206/8-11Z, the well prognosis was less accurate than for well 206/8-10Z, and HMA made a significant contribution to understanding the deviation from plan. It had a major influence on steering decisions aimed at optimizing the remaining well course without plugback and sidetrack after the well unexpectedly crossed a fault out of Unit V into Unit IV. In the absence of input from HMA, decision making would have been very difficult.

CONCLUSIONS

The first offshore applications of HMA in monitoring high-angle wells were successful, in terms of the ability to acquire data in a short time frame, in the ability to produce high-resolution stratigraphic data, and in contributing to optimizing drilling decisions. After the min-

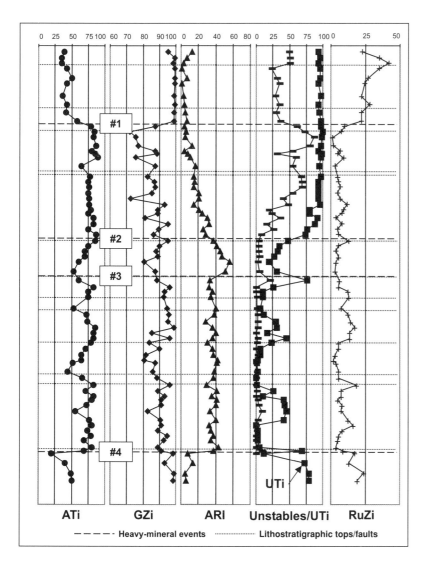

FIGURE 9. Variations in heavy-mineral parameters in 206/8-11Z, showing heavy-mineral events #1–#4. See Figure 3 for key. RuZi = rutile-zircon index (% rutile in total rutile + zircon).

eralogical characterization of different stratigraphic units in adjacent vertical wells, we were able to identify with reasonable certainty which parts of the sequence the wellbore encountered. This enabled building at the correct angle as the wellbore approached the top of Unit V and identified major changes in stratigraphy as the well tracked across fault zones. We learned that with prolonged horizontal drilling, resolution of stratigraphic events becomes more difficult because of contamination from the formation. This was successfully countered by determination of higher-resolution parameters. In well 206/8-11Z, contamination proved to be a particular problem immediately after the drill bit was run back into the hole, which caused a temporary rise in abundance of unstable minerals.

Another important lesson, learned from well 206/8-10Z, was that data quality improved dramatically when use of the turbine was discontinued. Significantly greater heavy-mineral yields were acquired from samples drilled using the positive-displacement mud motor compared with those from the interval drilled using the turbine.

In view of the successful application of this technique on the Clair field, there is no reason to doubt its potential effectiveness elsewhere, particularly for other biostratigraphically barren reservoirs. The only prerequisite is that laterally persistent vertical changes in mineral assemblage exist that can be recognized by analysis of cuttings material. This can be readily ascertained by studies of exploration or appraisal wells prior to drilling high-angle wells.

ACKNOWLEDGMENTS

We are grateful to BP Exploration and the Clair field partners for permission to submit this paper, and to Alan Repper and Mike Halfpenny for their excellent technical support offshore on *Sedco Explorer*.

REFERENCES CITED

Allen, P. A., and M. A. Mange-Rajetzky, 1992, Sedimentary evolution of the Devonian-Carboniferous Clair Field, offshore northwestern UK: Impact of changing provenance: Marine and Petroleum Geology, v. 9, p. 29–52.

Blackbourn, G. A., 1987, Sedimentary environments and stratigraphy of the late Devonian–Carboniferous Clair Basin, west of Shetland, *in* J. Miller, A. E. Adams, and V. P. Wright, eds., European Dinantian environments: London, John Wiley and Sons, p. 75–91.

Coney, D., T. B. Fyfe, P. Retail, and P. J. Smith, 1993, Clair appraisal: The benefits of a co-operative approach., *in* J. R. Parker, ed., Petroleum geology of northwest Europe: Proceedings of the 4th Conference: Geological Society (London), p. 1409–1420.

Henry, D.J., and B. L. Dutrow, 1992, Tourmaline in a low grade clastic metasedimentary rock: An example of the petrogenetic potential of tourmaline: Contributions to Mineralogy and Petrology, v. 112, p. 203–218.

Mange-Rajetzky, M. A., 1995, Subdivision and correlation of monotonous sandstone sequences using high resolution heavy mineral analysis, a case study: The Triassic of the Central Graben, *in* R. E. Dunay and E. A. Hailwood, eds., Dating and correlating biostratigraphically-barren strata: Geological Society (London) Special Publication 89, p. 23–30.

McKie, T., and I. R. Garden, 1996, Hierarchical cycles in the non-marine Clair Group (Devonian), UKCS, *in* J. A. Howell and J. F. Aitken, eds., High resolution sequence stratigraphy: Innovations and applications: Geological Society (London) Special Publication 104, p. 139-157.

Morton, A. C., 1985, A new approach to provenance studies: Electron microprobe analysis of detrital garnets from Middle Jurassic sandstones of the northern North Sea: Sedimentology, v. 32, p. 553–566.

Morton, A. C., and C. R. Hallsworth, 1994, Identifying provenance-specific features of detrital heavy mineral assemblages in sandstones: Sedimentary Geology, v. 90, p. 241–256.

Morton, A. C., and C. R. Hallsworth, 1999, Processes controlling the composition of heavy mineral assemblages in sandstones: Sedimentary Geology, v. 99, p. 3–29.

Ridd, M. F., 1981, Petroleum geology west of the Shetlands, *in* L. V. Illing and G. D. Hobson, eds., Petroleum geology of the continental shelf of north-west Europe: London, Graham & Trotman, p. 414–425.

Index